Lecture Notes in Networks and Systems 751

The series "Lecture Notes in Networks and Systems" publishes the latest developments in Networks and Systems—quickly, informally and with high quality. Original research reported in proceedings and post-proceedings represents the core of LNNS.

Volumes published in LNNS embrace all aspects and subfields of, as well as new challenges in, Networks and Systems.

The series contains proceedings and edited volumes in systems and networks, spanning the areas of Cyber-Physical Systems, Autonomous Systems, Sensor Networks, Control Systems, Energy Systems, Automotive Systems, Biological Systems, Vehicular Networking and Connected Vehicles, Aerospace Systems, Automation, Manufacturing, Smart Grids, Nonlinear Systems, Power Systems, Robotics, Social Systems, Economic Systems and other. Of particular value to both the contributors and the readership are the short publication timeframe and the world-wide distribution and exposure which enable both a wide and rapid dissemination of research output.

The series covers the theory, applications, and perspectives on the state of the art and future developments relevant to systems and networks, decision making, control, complex processes and related areas, as embedded in the fields of interdisciplinary and applied sciences, engineering, computer science, physics, economics, social, and life sciences, as well as the paradigms and methodologies behind them.

Indexed by SCOPUS, INSPEC, WTI Frankfurt eG, zbMATH, SCImago.

All books published in the series are submitted for consideration in Web of Science.

For proposals from Asia please contact Aninda Bose (aninda.bose@springer.com).

Kelly Cohen · Nicholas Ernest · Barnabas Bede ·
Vladik Kreinovich
Editors

Fuzzy Information Processing 2023

Springer

Editors
Kelly Cohen
University of Cincinnati
Cincinnati, OH, USA

Nicholas Ernest
Thales Avionics
Cincinnati, OH, USA

Barnabas Bede
DigiPen Institute of Technology
Redmond, WA, USA

Vladik Kreinovich
Department of Computer Science
University of Texas at El Paso
El Paso, TX, USA

ISSN 2367-3370 ISSN 2367-3389 (electronic)
Lecture Notes in Networks and Systems
ISBN 978-3-031-46777-6 ISBN 978-3-031-46778-3 (eBook)
https://doi.org/10.1007/978-3-031-46778-3

This Springer imprint is published by the registered company Springer Nature Switzerland AG
The registered company address is: Gewerbestrasse 11, 6330 Cham, Switzerland

Paper in this product is recyclable.

Fuzzy Information Processing 2023

Why fuzzy. In many practical situations, we need to make decisions. Many real-life decisions are difficult. For example, in air traffic control, we need to make very fast decisions on which planes should land on which runway. In such situations, we use computers to help us make good decisions.

Very often, we have only partial knowledge about the situation. An important part of our knowledge comes from experts, and experts often formulate their knowledge by using imprecise ("fuzzy") words from natural language like "small." To utilize this knowledge in our computer-assisted decision making, we need to translate imprecise expert knowledge into precise computer-understandable (i.e., numerical) terms. Techniques for such translation—pioneered by Lotfi Zadeh—are known as *fuzzy* techniques.

Contents of this volume. This volume contains papers presented at the 2023 Annual Conference of the North American Fuzzy Information Processing Society NAFIPS'23 (Cincinnati, Ohio, May 31–June 2, 2023) and at the related Workshop on Constraint Programming and Decision Making CoProD'23.

Papers forming this volume deal both with general theoretical and computational aspects of fuzzy techniques—in particular, their relation to deep learning—as well as specific applications to various areas including aerospace engineering, agriculture, cosmology, demographics, finance, geosciences, medicine, and robotics.

Our thanks. We are thankful to the local organizing team led by Kelly Cohen and Nicholas "Nick" Ernest, to all the authors, to all the reviewers, to all the participants, and—last but not the least—to Springer-related helpers, especially to Janusz Kacprzyk, editor of this series of book. Without all this help, this volume would not have been possible.

Kelly Cohen
Nicholas Ernest
Barnabas Bede
Vladik Kreinovich

Contents

Accurate and Explainable Retinal Disease Recognition via DCNFIS

Mojtaba Yeganejou, Mohammad Keshmiri, and Scott Dick(✉)

University of Alberta, Edmonton, Canada
{yeganejo,sdick}@ualberta.ca

Abstract. The accuracy-interpretability tradeoff is a significant challenge for Explainable AI systems; if too much accuracy is lost, an explainable system might be of no actual value. We report on the ongoing development of the Deep Convolutional Neuro-Fuzzy Inference System, an XAI algorithm that has to this point demonstrated accuracy on par with existing convolutional neural networks. Our system is evaluated on the Retinal OCT dataset, in which it achieves state-of-the-art performance. Explanations for the system's classifications based on saliency analysis of medoid elements from the fuzzy rules in the classifier component are analyzed.

Keywords: Neuro-fuzzy systems · Deep learning · XAI

1 Introduction

Deep neural networks are now the preferred solution for problems such as image classification [44], Natural Language Processing and Information Retrieval [7], robotic surgery [40], and many others. Medical diagnostics are another application domain, which AI researchers have studied for decades [30]. Recently, deep learning methods have significantly improved medical image analysis, especially in diagnosing breast cancer [8], skin melanoma [1], retinal diseases [29], etc.

However, the medical community has historically been reluctant to adopt AI-based diagnostic tools, in large measure because the tools cannot explain their reasoning process to the satisfaction of medical professionals [30]. Comprehending medical data and making clinical judgments is a complex and demanding task, as symptom occurrence and presentation can vary greatly between two people with the same disease [11,27,36,56]. The "explainability" problem is exacerbated today because the core knowledge extracted in a deep neural model is encoded in a complex pattern of possibly billions of weighted connections [16]. This format is incomprehensible for human beings, and thus deep models are considered uninterpretable black boxes [10], making them unsuitable for clinical usage [35]. Explainable AI (XAI) approaches are often suggested as a pathway towards increasing user trust and acceptance of AI algorithms, including in the medical field. Clinicians must be able to inspect an AI, and agree that its diagnoses are based on sound evidence and reasoning, before they can accept it into

K. Cohen et al. (Eds.): NAFIPS 2023, LNNS 751, pp. 1–12, 2023.
https://doi.org/10.1007/978-3-031-46778-3_1

their practice. However, XAI algorithms have been found to lose some classification accuracy in making their inferential models comprehensible, a well-known phenomenon usually referred to as the accuracy-interpretability tradeoff. More precisely, XAI algorithms today are evaluated in multiple dimensions; accuracy is one, functionality (speed, modality, content, user interface) is another, and explanation effectiveness (e.g. user understanding, trust) is the third [15]. If AI is to be used in diagnostics, new XAI algorithms are needed that offer effective explanations, in forms readily understood by clinicians, and which still maintain state-of-the-art accuracy.

In this paper, we test a new end-to-end trainable and explainable Deep Convolutional Fuzzy Inference System (DCNFIS) [53] on a medical diagnosis problem. DCNFIS replaces the dense classifier layers of a CNN (which are themselves uninterpretable) with a modified version of ANFIS network [22]. This new architecture was found to be as accurate as the base classifiers it was developed from (LeNet, ResNet, Wide ResNet). However, the fuzzy-rule-based design of ANFIS allows us to treat each rule as a cluster, and select the medoid element of each cluster. Medoids are representative objects of a cluster whose sum of dissimilarities to all the objects in the cluster is minimal [42]. These medoids are our proposed explanations, following [52]. We evaluate DCNFIS on an Optical Coherence Tomography (OCT) dataset. OCT is an important retinal imaging method as it is a non-invasive, high-resolution imaging technique for observing the human retina and pathologies within it [32]. A dataset of OCT images was assembled in [25]. We find that DCNFIS achieves state-of-the-art performance on this dataset, and saliency maps computed from the cluster medoids do highlight the clinically-relevant features of three out of four classes. Note that saliency maps are specific heat map visualizations that highlight features (pixels) in the input to an AI model that were significant for model's predictions [39]. Considering the fact that extracted medoids of DCNFIS are class representatives, important features of a medoid can be considered as a global explanation of how an AI model works. This differs from using saliency maps to explain individual classifications, as those are local rather than global interpretations. As ordinary deep networks do not perform any clustering of their inputs, DCNFIS thus offers a novel means of generating explanations.

The reminder of this paper is organized as follows. In Sect. 2, we review essential background and related work. In Sect. 3 we discuss our methodology, including the OCT dataset, the DCNFIS architecture, and our experimental procedures. We discuss the results of our experiments, comparing the accuracy of DCNFIS against the state-of-the-art, in Sect. 4. In Sect. 5 we discuss the global explanations that we derive for this dataset, and we close with a summary and discussion of future work in Sect. 6.

2 Background Review

2.1 Deep Learning and Fuzzy Systems

The assumption made by fuzzy classifiers [31] is that an object has partial membership in each class within a continuous, overlapping area that separates two adjoining classes. These classifiers use linguistic if-then principles to represent complex models in a straightforward and intelligible manner. Over 25 years have passed since researchers began examining hybrids of fuzzy logic and shallow neural networks [21], but hybrids of fuzzy logic and deep networks have been studied for a much shorter time. Deep fuzzy networks using both extensions of type-1 fuzzy sets and type-1 fuzzy logic are found (e.g., fuzzifying the inputs to a deep MLP network [37]). Zheng et al. used stacked denoising autoencoders with Pythagorean fuzzy sets [55]. A second reservoir was added to echo state networks using fuzzy clustering for feature reinforcement (combatting the vanishing gradient problem) in [54]. In [12], fuzzy clustering is combined with the training of stacked autoencoders. A ResNet model for segmenting lips within human faces incorporates fuzzy logic in [14].

Deep neuro-fuzzy systems include a hybrid ANFIS, recurrent network, and long short-term memory addressing the contact-force problem in remote surgery [5]. For a stacked TSK system, an adversarial training technique is proposed in [13]. A fuzzy Choquet integral-impersonating neural network is proposed in [20]. John et al. used a deep network as a feature extractor before clustering in [23]. This last method is somewhat similar to that in [50] and [51,52], in that an alternate classifier built from a clustering technique is used to replace the densely-connected layers at the end of a deep network. The last two, however, use the Gustafson-Kessel (GK) fuzzy clustering algorithm and Fuzzy C-Means (FCM), respectively.

2.2 Deep Learning in Medical Diagnosis

Diabetic Retinopathy (DR) can be diagnosed via deep learning using digital photographs of the fundus of the eye [34]. Plis et al. [33] used Deep Belief Networks to extract features from functional MRI (fMRI) images and MRI scans of patients with Huntington's disease and schizophrenia. A review of brain MRI segmentation was written by Akkus et al. [2]. Amoroso et al. introduced an XAI framework for breast cancer therapies in [3]. Dindorf et al. suggested an interpretable and pathology-independent classifier for spinal posture in [9]. They use Support Vector Machines (SVM) and Random Forests (RF) as classifiers. The Local Interpretable Model-agnostic Explanations (LIME) algorithm then provides explanations for their predictions. Gu et al. introduced VINet, a computer-aided diagnosis system [13], which offers diagnostic visual interpretations.

3 Methodology

3.1 Dataset

We have chosen a commonly used benchmark dataset for deep learning in medical diagnosis OCT [25]. This dataset consisted of Retinal optical coherence tomography (OCT), an imaging technique used to capture a high-resolution cross section of the retinas of a patient. Three diseases (Choroidal Neovascularization (CNV), multiple DRUSEN, and Diabetic Macular Edema (DME)) are represented, as well as normal OCT images. Normal OCT images of healthy retina have no signs of abnormalities or diseases, CNV is an abnormal growth of blood vessels under the retina, which can cause fluid leakage, bleeding, and scarring. OCT images of CNV show the presence of fluid or blood in the retina or under the retina. DME is a common complication of diabetes that affects the macula, the central part of the retina responsible for sharp, detailed vision. In DME, fluid accumulates in the macula, causing it to swell and leading to vision loss. OCT images of DME show the presence of fluid in the macula. DRUSEN are yellow or white deposits that accumulate under the retina, typically associated with aging and age-related macular degeneration (AMD). OCT images of DRUSEN show the presence of these deposits, which can affect the thickness and integrity of the retina [24,48].

The class distribution of this dataset is presented in Table 1 given by the dataset oweners, and in order to be able to compare our results with other methods we will keep the same training and testing dataset. The source images are single channel, and their sizes are (384–1536) * (277–512) [48]. As the images are gathered from different hospitals, in some images there is a white area at the top and the images are not originally centralized [24,48]. The images have been resized to 224*224 [18,19].

Table 1. Distribution of OCT images in each class among training and test dataset.

Classes	Training set	Test set
CNV	37205	250
DME	11348	250
DRUSEN	8616	250
Normal	26315	250

3.2 Proposed Architecture

Our fuzzy classifier is based on the Adaptive Neuro-Fuzzy Inference System (ANFIS) architecture [22,43]. ANFIS is known to be a universal approximator [22], and is thus theoretically equal to the fully-connected layer(s) it replaces in the ResNet or LeNet architectures. ANFIS is a layered architecture where the first layer computes the membership of an input in a Gaussian fuzzy set using Eq. 1, for fuzzy rule i and input j. We compute the natural logarithm of the membership values, for numerical stability in the tails.

$$\begin{aligned} M_{ij} &= \exp(-\beta_{ij} \cdot (x_j - \mu_{ij})^2) \\ \log M_{ij} &= -\beta_{ij} \cdot (x_j - \mu_{ij})^2 \end{aligned} \tag{1}$$

Rule firing strengths are computed in layer 2 using the product t-norm; this obviously becomes a sum in the log domain as showed in the following equation:

$$\begin{aligned} \omega_i &= \prod_j M_{ij} \\ \log \omega_i &= \sum_j \log M_{ij} \end{aligned} \tag{2}$$

In layer 3, we sum the log firing strength and a linear combination of the input variables.

$$\begin{aligned} \zeta_i &= \bar{\omega}_i \cdot (\sum_j W_{ij} x_j + b_i) \\ \zeta_i &= \log \bar{\omega}_i + (\sum_j W_{ij} x_j + b_i) \end{aligned} \tag{3}$$

Note that we do not take the logarithm of the linear combination of inputs. This allows us to create fuzzy regions(clusters) along with an end-to-end trainable deep convolutional backend without causing an error gradient explosion.

Finally, layer 4 computes the class probabilities using the softmax activation function:

$$P\left(c_i \middle| \, x\right) = y_i = \frac{\exp\left(\zeta_i\right)}{\sum_o \exp\left(\zeta_o\right)} \tag{4}$$

This architecture mimics the operation of a Fuzzy Inferential System (FIS), which is a rule-based expert system (and thus highly interpretable). The hybrid learning rule for ANFIS (least squares during the forward pass for the Layer 4 parameters, backpropagation for the Layer 1 parameters) allows the fuzzy rule-base to be induced from a dataset. Thus, we can directly translate a trained ANFIS into fuzzy if-then rules relating its inputs (the outputs of the final CNN layer) to output classes. For more information about DCNFIS see [53].

3.3 Experimental Setup

The adaptive parameters of our network are trained using the Adam optimizer [26] for 300 epochs in batches of 32. The initial learning rate is 0.001 and then it will be reduced to 10^{-6} per an annealing schedule. Data augmentation techniques are not applied. We create four rules in DCNFIS, one corresponding to each class.

4 Performance Evaluation

In this section, we compare the performance of DCNFIS-ResNet11V2 with other methods. Table 2 compares the performance of DCNFIS with some state-of-the-art models like Swin-Transformer, ViT, Kermany, CliqueNet, and LLCT applied to OCT2017. As shown in this table, DCNFIS reaches 100% accuracy on the test set. As far as we are aware, this result is now the state-of-the-art.

Table 2. Performance comparison of DCNFIS and other methods

Model	Testing Accuracy
Our Method	**100**
Kermany [24]	96.6
ViT [46]	99.0
CliqueNet [45]	98.6
SwinTransformer [28]	99.7
Swin-Poly Transformer [17]	**99.8**
LLCT [47]	98.8

We present the confusion matrices produced by DCNFIS on the training dataset in Table 3. Note that the training accuracy is **99.72%** which means our classifier has only miss-classified **228** samples out of 83484.

Table 3. Confusion matrix of DCNFIS

	CNV	DME	DRUSEN	NORMAL
CNV	37020	23	162	0
DME	1	11340	0	7
DRUSEN	24	0	8591	1
NORMAL	2	3	5	26305

According to the confusion matrix, the model predicted 37,006 CNV samples as CNV, with 23 misclassified as DME and 162 misclassified as DRUSEN. This latter misclassification covers 71% of all of our misclassifications. In Sect. 5, we discuss the CNV class in more detail.

We showed in [53] that the performance of DCNFIS is comparable to the CNNs it is derived from, and even sometimes performs better. In this paper, we have again shown that DCNFIS is as accurate as existing deep learners. We next examine how explanations may be derived for DCNFIS, and demonstrate that these explanations match the actual rationales for three out of four classes in the medical literature.

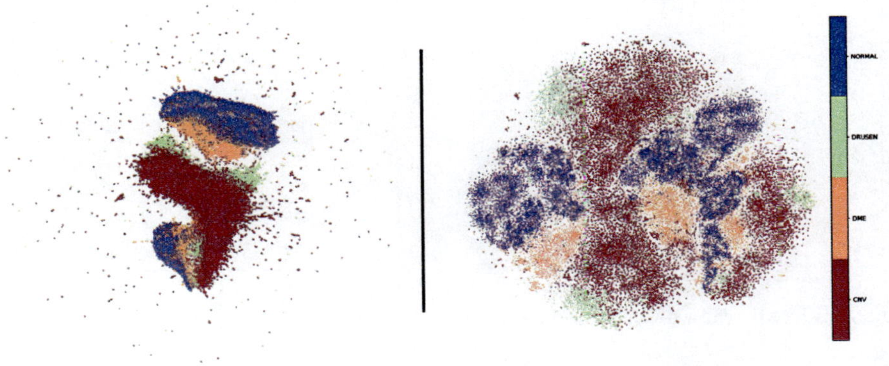

Fig. 1. 2D visualization of extracted feature space on Retinal OCT dataset using UMAP(left) and TSNE(right) with default parameters.

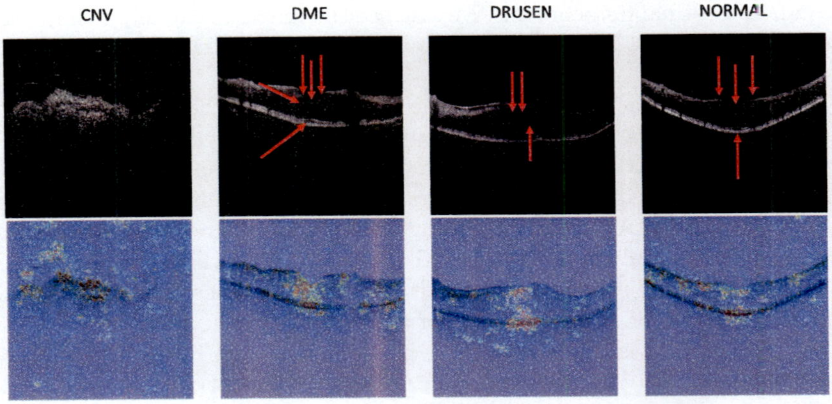

Fig. 2. Medoids of the 4 classes in dataset extracted by DCNFIS (top row) and their saliencies (bottom row)

5 Interpretability

Figure 1 presents 2D visualizations of the Retinal OCT training dataset using UMAP (left) and TSNE (right) with default parameters. Our method for generating explanations from DCNFIS was described in [53]. In brief, the firing degree of each data point for a given rule is treated as the point's membership in a fuzzy cluster corresponding to that rule. We then determine the medoid element of that fuzzy cluster, and compute a saliency map for it. That saliency map is our explanation for the cluster (and thus, the whole class it belongs to). Following our work in [49] we tested various visualization techniques and finally chose the Gradient-Input [4] method for visualization of the extracted medoids and the saliencies.

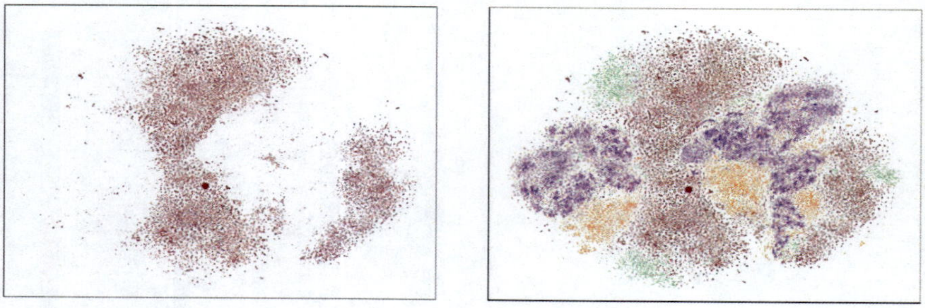

Fig. 3. Visualization of CNV and samples and location the CNV class representative.

Figure 2 shows the medoids extracted from DCNFIS for all four classes. CNV will be discussed later, as this class is more challenging than DME, DURSEN, and the Normal class. Saliency analysis of the DME medoid highlights an optic nerve cup and retinal detachment. We also see the macular edema at the bottom of our saliency image. These are the features that experts use to diagnose DME [45].

As indicated by the red arrows in the input image of the DURSEN medoid, the medically significant elements are outer plexiform[1] layer and outer nuclear layer shape. These parts are abnormally shaped and unusually thick [38]. As we see, these are captured in the saliency map as highly significant regions of the image. In the NORMAL class the Fovea[2] and Choroid[3] are highly salient, and we also see the retinal layers and foveal depression and their specific curve in the saliency map (red arrows in the input images) [38]. These are the medically-significant elements of an OCT image of a normal retina.

The CNV medoid is, in fact, a rather noisy image. Figure 3 plots the location of the CNV medoid (dark red point) in the entire dataset(right image) and in the CNV images (left image) using 2D t-SNE. As shown in this figure, the class is formed of multiple disjunctions that seem to have led to choosing a relatively poor image as the class representative. Now, both Fig. 1 and Fig. 3 indicate that all four classes appear to be composed of multiple disjunctions; the question is why this only appears to negatively impact the CNV class. In fact, medically CNV is divided into four sub-classes [41]: occult CNV, its subtype Polypoidal Choroidal Vasculopathy (PCV), classic CNV, and Retinal Angiomatous Proliferation (RAP). Each produces different features in the retinal OCT, and so the observed medoid is perhaps the least bad choice of a single class representative. We also note that the bulk of our training errors occur on the CNV class. By contrast, the other disease classes, and the Normal class, seem to have less

[1] A layer of neuronal synapses in the retina of the eye [6].

[2] A small depression within the neurosensory retina where visual acuity is the highest [6].

[3] A thin layer of tissue that is part of the middle layer of the wall of the eye, between the sclera (white outer layer of the eye) and the retina [6].

variability; while they appear to be different disjunctions in the visualizations, *conceptually* they seem relatively homogeneous in this dataset. While DCNFIS remained as accurate as other algorithms in both of these datasets, the *explanation effectiveness* of the resulting medoids seems weak for CNV.

6 Conclusion

In this paper we have evaluated a new XAI architecture in a medical imaging task. The DCNFIS architecture replaces the dense classification layers of an arbitrary CNN with an ANCFIS network, modified so that the entire architecture is end-to-end trainable. The explanation mechanism for DCNFIS treats each fuzzy rule as a fuzzy cluster in the feature space created by the convolutional component of the network, and extracts the medoid element of the cluster as a representative image. A saliency analysis of that image then provides the explanation for the whole rule (and thus, the class it describes).

Future work in this topic will focus on the problem of classes formed from multiple disjunctions. As was seen in the current paper, in some cases (although not all), classes formed from multiple disjunctions were not effectively explained by the medoids. We intend to test a modification of DCNFIS that will permit multiple rules to define a class, rather than forcing a 1:1 correspondence between rules and classes. We believe that this will improve both classification accuracy and explanation effectiveness in DCNFIS.

References

1. Abayomi-Alli, O.O., Damasevicius, R., Misra, S., Maskeliunas, R., Abayomi-Alli, A.: Malignant skin melanoma detection using image augmentation by oversamplingin nonlinear lower-dimensional embedding manifold. Turk. J. Electr. Eng. Comput. Sci. **29**(8), 2600–2614 (2021)
2. Akkus, Z., Galimzianova, A., Hoogi, A., Rubin, D.L., Erickson, B.J.: Deep learning for brain MRI segmentation: state of the art and future directions (2017). https://doi.org/10.1007/s10278-017-9983-4
3. Amoroso, N., et al.: A roadmap towards breast cancer therapies supported by explainable artificial intelligence. Appl. Sci. **11**(11), 4881 (2021)
4. Ancona, M., Ceolini, E., Öztireli, C., Gross, M.: Towards better understanding of gradient-based attribution methods for deep neural networks. arXiv preprint arXiv:1711.06104 (2017)
5. Aviles, A.I., Alsaleh, S.M., Montseny, E., Sobrevilla, P., Casals, A.: A deep-neuro-fuzzy approach for estimating the interaction forces in robotic surgery (2016). https://doi.org/10.1109/FUZZ-IEEE.2016.7737812
6. Binder, S.: The Macula: Diagnosis, Treatment and Future Trends. Springer Science & Business Media, Vienna (2004). https://doi.org/10.1007/978-3-7091-7985-7
7. Chahardoli, R., Barbosa, D., Rafiei, D.: Relation extraction with synthetic explanations and neural network. In: 2021 International Symposium on Electrical, Electronics and Information Engineering, pp. 254–262 (2021)

8. Dhungel, N., Carneiro, G., Bradley, A.P.: The automated learning of deep features for breast mass classification from mammograms. In: Ourselin, S., Joskowicz, L., Sabuncu, M.R., Unal, G., Wells, W. (eds.) MICCAI 2016. LNCS, vol. 9901, pp. 106–114. Springer, Cham (2016). https://doi.org/10.1007/978-3-319-46723-8_13
9. Dindorf, C., et al.: Classification and automated interpretation of spinal posture data using a pathology-independent classifier and explainable artificial intelligence (XAI). Sensors **21** (2021). https://doi.org/10.3390/s21186323
10. Dong, Y., Su, H., Zhu, J., Zhang, B.: Improving interpretability of deep neural networks with semantic information, vol. 2017-January (2017). https://doi.org/10.1109/CVPR.2017.110
11. Eigner, I., Bodendorf, F., Wickramasinghe, N.: Predicting high-cost patients by machine learning: a case study in an Australian private hospital group (2019). https://doi.org/10.29007/jw6h
12. Feng, Q., Chen, L., Chen, C.L.P., Guo, L.: Deep fuzzy clustering-a representation learning approach. IEEE Trans. Fuzzy Syst. **28** (2020). https://doi.org/10.1109/TFUZZ.2020.2966173
13. Gu, D., et al.: Vinet: a visually interpretable image diagnosis network. IEEE Trans. Multimed. **22** (2020). https://doi.org/10.1109/TMM.2020.2971170
14. Guan, C., Wang, S., Liew, A.W.C.: Lip image segmentation based on a fuzzy convolutional neural network. IEEE Trans. Fuzzy Syst. **28** (2020). https://doi.org/10.1109/TFUZZ.2019.2957708
15. Gunning, D., Vorm, E., Wang, Y., Turek, M.: Darpa's explainable AI (XAI) program: a retrospective. Authorea Preprints (2021)
16. Haykin, S.: Neural Networks and Learning Machines, vol. 3 (2008). 978–0131471399
17. He, J., Wang, J., Han, Z., Ma, J., Wang, C., Qi, M.: An interpretable transformer network for the retinal disease classification using optical coherence tomography. Sci. Rep. **13**(1), 3637 (2023)
18. He, K., Zhang, X., Ren, S., Sun, J.: Deep residual learning for image recognition, pp. 770–778 (2016). https://doi.org/10.1109/CVPR.2016.90
19. He, K., Zhang, X., Ren, S., Sun, J.: Identity mappings in deep residual networks. In: Leibe, B., Matas, J., Sebe, N., Welling, M. (eds.) ECCV 2016. LNCS, vol. 9908, pp. 630–645. Springer, Cham (2016). https://doi.org/10.1007/978-3-319-46493-0_38
20. Jafari, R., Razvarz, S., Gegov, A.: Neural network approach to solving fuzzy nonlinear equations using z-numbers. IEEE Trans. Fuzzy Syst. **28** (2020). https://doi.org/10.1109/TFUZZ.2019.2940919
21. Jang, J., Sun, C., Mizutani, E.: Neuro-fuzzy and soft computing-a computational approach to learning and machine intelligence [book review]. IEEE Trans. Autom. Control **42** (2005). https://doi.org/10.1109/tac.1997.633847
22. Jang, J.S.R.: ANFIS: adaptive-network-based fuzzy inference system. IEEE Trans. Syst. Man Cybern. **23** (1993). https://doi.org/10.1109/21.256541
23. John, V., Mita, S., Liu, Z., Qi, B.: Pedestrian detection in thermal images using adaptive fuzzy c-means clustering and convolutional neural networks (2015). https://doi.org/10.1109/MVA.2015.7153177
24. Kermany, D.S., et al.: Identifying medical diagnoses and treatable diseases by image-based deep learning. Cell **172**, 1122–1131.e9 (2018). https://doi.org/10.1016/J.CELL.2018.02.010
25. Kermany, D.S., et al.: Large dataset of labeled optical coherence tomography (oct) and chest x-ray images. Mendeley Data **2** (2018)
26. Kingma, D.P., Ba, J.: Adam: a method for stochastic optimization. arXiv preprint arXiv:1412.6980 (2014)

27. Kourou, K., Exarchos, T.P., Exarchos, K.P., Karamouzis, M.V., Fotiadis, D.I.: Machine learning applications in cancer prognosis and prediction. Comput. Struct. Biotechnol. J.**13**, 8–17 (2015). https://doi.org/10.1016/J.CSBJ.2014.11.005
28. Liu, Z., et al.: Swin transformer: hierarchical vision transformer using shifted windows. In: Proceedings of the IEEE/CVF International Conference on Computer Vision, pp. 10012–10022 (2021)
29. Miere, A., et al.: Deep learning-based classification of inherited retinal diseases using fundus autofluorescence. J. Clin. Med. **9** (2020). https://doi.org/10.3390/jcm9103303
30. Nilsson, N.J.: The Quest for Artificial Intelligence. Cambridge University Press, Cambridge (2009). https://doi.org/10.1017/CBO9780511819346
31. Ovchinnikov, S.: Similarity relations, fuzzy partitions, and fuzzy orderings. Fuzzy Sets Syst. **40** (1991). https://doi.org/10.1016/0165-0114(91)90048-U
32. Pekala, M., Joshi, N., Liu, T.Y., Bressler, N.M., DeBuc, D.C., Burlina, P.: Deep learning based retinal oct segmentation. Comput. Biol. Med. **114**, 103445 (2019). https://doi.org/10.1016/J.COMPBIOMED.2019.103445
33. Plis, S.M., et al.: Deep learning for neuroimaging: a validation study. Front. Neurosci. (2014). https://doi.org/10.3389/fnins.2014.00229
34. Pratt, H., Coenen, F., Broadbent, D.M., Harding, S.P., Zheng, Y.: Convolutional neural networks for diabetic retinopathy, vol. 90 (2016). https://doi.org/10.1016/j.procs.2016.07.014
35. Rudin, C.: Stop explaining black box machine learning models for high stakes decisions and use interpretable models instead (2019). https://doi.org/10.1038/s42256-019-0048-x
36. Samhan, B., Crampton, T., Ruane, R.: The trajectory of it in healthcare at HICSS: a literature review, analysis, and future directions. Commun. Assoc. Inf. Syst. **43** (2018). https://doi.org/10.17705/1CAIS.04341
37. Sarabakha, A., Kayacan, E.: Online deep fuzzy learning for control of nonlinear systems using expert knowledge. IEEE Trans. Fuzzy Syst. **28**(7), 1492–1503 (2019)
38. Schiffman, J.S., Patel, N.B., Cruz, R.A., Tang, R.A.: Optical coherence tomography for the radiologist. Neuroimaging Clin. **25**(3), 367–382 (2015)
39. Simonyan, K., Vedaldi, A., Zisserman, A.: Deep inside convolutional networks: visualising image classification models and saliency maps. arXiv preprint arXiv:1312.6034 (2013)
40. Soleymani, A., Asl, A.A.S., Yeganejou, M., Dick, S., Tavakoli, M., Li, X.: Surgical skill evaluation from robot-assisted surgery recordings. In: 2021 International Symposium on Medical Robotics (ISMR), pp. 1–6. IEEE (2021)
41. Soomro, T., Talks, J.: The use of optical coherence tomography angiography for detecting choroidal neovascularization, compared to standard multimodal imaging. Eye **32**(4), 661–672 (2018)
42. Struyf, A., Hubert, M., Rousseeuw, P.: Clustering in an object-oriented environment. J. Stat. Softw. **1**, 1–30 (1997)
43. Sun, C.T., Jang, J.S.: A neuro-fuzzy classifier and its applications. In: Proceedings of the [Proceedings 1993] Second IEEE International Conference on Fuzzy Systems, pp. 94–98. IEEE (1993)
44. Szegedy, C., et al.: Going deeper with convolutions, vol. 07-12-June-2015 (2015). https://doi.org/10.1109/CVPR.2015.7298594
45. Wang, D., Wang, L.: On oct image classification via deep learning. IEEE Photonics J. **11**(5), 1–14 (2019)

46. Wang, H., Ji, Y., Song, K., Sun, M., Lv, P., Zhang, T.: VIT-P: classification of genitourinary syndrome of menopause from oct images based on vision transformer models. IEEE Trans. Instrum. Meas. **70**, 1–14 (2021). https://doi.org/10.1109/TIM.2021.3122121
47. Wen, H., et al.: Towards more efficient ophthalmic disease classification and lesion location via convolution transformer. Comput. Methods Programs Biomed. **220**, 106832 (2022)
48. Yang, J., et al.: Medmnist v2 - a large-scale lightweight benchmark for 2D and 3D biomedical image classification. Sci. Data **10**, 41 (2023)
49. Yeganejou, M.: Interpretable deep convolutional fuzzy networks (2019)
50. Yeganejou, M., Dick, S.: Classification via deep fuzzy c-means clustering, vol. 2018-July (2018). https://doi.org/10.1109/FUZZ-IEEE.2018.8491461
51. Yeganejou, M., Dick, S.: Improved deep fuzzy clustering for accurate and interpretable classifiers, vol. 2019-June (2019). https://doi.org/10.1109/FUZZ-IEEE.2019.8858809
52. Yeganejou, M., Dick, S., Miller, J.: Interpretable deep convolutional fuzzy classifier. IEEE Trans. Fuzzy Syst. **28**(7), 1407–1419 (2019)
53. Yeganejou, M., Kluzinski, R., Dick, S., Miller, J.: An end-to-end trainable deep convolutional neuro-fuzzy classifier. In: 2022 IEEE International Conference on Fuzzy Systems (FUZZ-IEEE), pp. 1–7. IEEE (2022)
54. Zhang, Z., Huang, M., Liu, S., Xiao, B., Durrani, T.S.: Fuzzy multilayer clustering and fuzzy label regularization for unsupervised person reidentification. IEEE Trans. Fuzzy Syst. **28** (2020). https://doi.org/10.1109/TFUZZ.2019.2914626
55. Zheng, Y.J., Chen, S.Y., Xue, Y., Xue, J.Y.: A pythagorean-type fuzzy deep denoising autoencoder for industrial accident early warning. IEEE Trans. Fuzzy Syst. **25** (2017). https://doi.org/10.1109/TFUZZ.2017.2738605
56. Zwaan, L., Singh, H.: The challenges in defining and measuring diagnostic error. Diagnosis **2** (2015). https://doi.org/10.1515/dx-2014-0069

Fuzzy Inference System-Based Collision Avoidance of Unmanned Aerial Vehicles Optimized Using Genetic Algorithm

Shyam Rauniyar(✉) and Donghoon Kim

University of Cincinnati, Cincinnati, OH, USA
rauniysm@mail.uc.edu, Donghoon.Kim@uc.edu

Abstract. Fixed-wing Unmanned Aerial Vehicle (UAVs) cannot hover in potential collision scenarios as flying lower than critical speeds results in the stall of the aircraft. Hovering is also not an optimal solution for Collision Avoidance (CA) as it increases mission time and is innately fuel inefficient. Thus, an algorithm that ensures the continuous motion of UAVs during CA while maintaining maximum mission performance is needed. This work proposes a decentralized Fuzzy Inference System (FIS)-based resolution algorithm that avoids potential stall speeds during CA and is optimized using a genetic algorithm to minimize path deviation during a given mission. The mathematical model of a fixed-wing UAV in this work considers coordinated turn condition. They are assumed to be equipped with range-bearing sensors to observe potential intruders in a three-dimensional environment. The UAV guidance model uses a PD control of commanded airspeed, bank angle, and flight path angle obtained for desired path variables - airspeed, heading, and altitude which in turn are obtained using lookahead point generated for given waypoints. The optimized FIS mimics the pilot behavior during collision scenarios to provide modulation parameters for desired path variables and achieves CA while ensuring minimum path deviation. FIS optimization was conducted using a pairwise conflict scenario and tested for several pairwise and multiple UAV scenarios. Multiple simulations conducted within a given functional space with random conflict scenarios of multiple aircraft yielded effective CA with minimal collisions and overall path deviation; thus, validating the effectiveness of the proposed CA algorithm.

1 Introduction

Global air traffic is expected to grow annually at 3.4% [1], and global advanced air mobility of automated aircraft at lower altitudes is expected to grow at a compounded annual growth rate of around 22% [2]. This growth would also include Unmanned Aerial Vehicles (UAVs) for surveillance, military, transportation, cargo, and many more applications. Thus, potential conflicts and collisions among the aircraft will be inevitable. Collision Avoidance (CA) algorithms are

K. Cohen et al. (Eds.): NAFIPS 2023, LNNS 751, pp. 13–24, 2023.
https://doi.org/10.1007/978-3-031-46778-3_2

vital for all UAVs. However, the formulation of CA algorithms is even more challenging for fixed-wing UAVs. Unlike rotary-wing UAVs, fixed-wing UAVs can't stop and hover or even slow down below stall speeds in potential collision scenarios. Hovering in itself is also a non-optimal solution for CA as it increases mission time and is innately fuel-inefficient. Moreover, fixed-wing UAVs carry many advantages over rotary-wing UAVs like greater speed and endurance, higher payload capacity for the same endurance, lower maintenance, lower noise levels, etc. which could be essential for many emergency, reconnaissance, and cargo missions. Thus, this work proposes to tackle CA for fixed-wing UAVs that can ensure continuous motion of the aircraft over its stall speeds while avoiding collision with minimum path deviation. Such an algorithm that ensures continuous motion could also be an efficient solution to CA for rotary-wing UAVs which can avoid hover resolutions for many potential collision scenarios.

Aircraft CA System (ACAS) II is the most common method of CA in crewed aircraft. It is implemented as Traffic CAS (TCAS) II which provides traffic advisory for warning and resolution advisory of appropriate CA maneuvers to the pilots [3]. However, TCAS II has several limitations including the provision of resolutions in the vertical planes only as climb, descent, or level flight, that too for specific encounter geometries, and is deterministic [4]. There have been a vast number of research and algorithms proposed on the CA of UAVs as well. Some propose geometric methods like heading changes [5], velocity modulation in horizontal plane [6], while some propose concepts of potential fields which are either modified or reformulated with UAV's physical constraints for better real-life performance [7]. Many algorithms also modulate or optimize the path planning-based approaches of UAVs using different methods like using 3 Dimensional (3D) vector field histogram [8], combining differential game problem with tree-based path planning [9], or utilizing reinforcement learning for UAV guidance [10]. All of the above algorithms are numerically inundated, inducing some computational effort into the system, and are not easily comprehensible. Analytical formulations like speed approach [11] and the use of buffered Voronoi Cells have been proposed but are not intuitive. Humans are innately capable to avoid collision intuitively [12, 13].

Humans are innately capable to avoid collision intuitively [12, 13]. Pilots can perform CA maneuvers manually based on their visual perception and flight instrument information of incoming static or dynamic obstacles. The Fuzzy Inference System (FIS) is one of the best tools that emulate the human way of decision-making with the linguistic characterization of numerical variables. A research paper has proposed a fuzzy-based aircraft CA system that is capable of generating an alert of a potential midair collision while taking control if no preventive action is taken within a specified time [14]. Cook et al. used a fuzzy logic-based approach to help mitigate the risk of collisions between aircraft using separation assurance and CA techniques [15]. However, this work utilizes FIS to generate modulation to path variables based on the relative state of the UAV in consideration with respect to the nearest intruder or obstacle point. The FIS is optimized using Genetic Algorithm (GA) to achieve CA while ensuring minimum path deviation. The UAV in a certain path to its desired waypoint is controlled to deviate from the path minimally during potential collision using the optimized FIS. The algorithm is also applied to quadcopters to examine its effectiveness in rotary-wing UAVs as well. The effectiveness of the algorithm is validated through multiple simulations conducted within a given functional space with random conflict scenarios of multiple aircraft.

2 System Modelling

UAVs are considered to operate in a north-east-down inertial frame (N). Henceforth, an own-ship refers to a UAV in consideration that is running the CA algorithm, and intruders are other UAVs in the airspace. An own-ship's body frame is designated as O-frame and that of an intruder as I-frame. A UAV position in Nframe and its attitude represented by 3-2-1 Euler angles in O-frame are denoted as $\boldsymbol{\rho} = [\rho_\mathrm{n},\ \rho_\mathrm{e},\ \rho_\mathrm{d}]^\top$ and ${}^O\boldsymbol{\Lambda} = [\psi,\ \theta,\ \phi]^\top$, respectively. For simplicity, one omits N-frame expression. It should be noted that altitude (h) is measured in the opposite direction to ρ_d. That is, $h = -\rho_\mathrm{d}$. Also, the state of a UAV can be elaborately defined with airspeed (ν_a), ground speed (ν_g), angle of attack (α), side-slip (β), course angle (χ), and flight path angle (γ) as shown in Fig. 1.

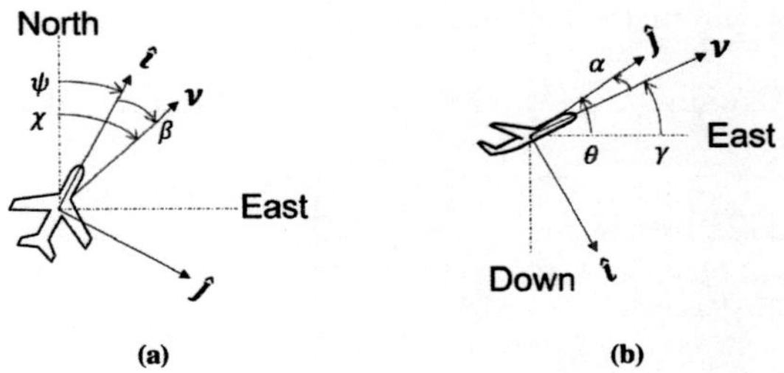

Fig. 1. UAV States in (a) horizontal plane and (b) vertical plane

Here, UAVs are controlled via a PD control of ν_a, γ, and bank angle (ϕ). The subsequent kinematic guidance model for the UAV is given by [16]

$$\dot{\rho}_\mathrm{n} = \nu_\mathrm{g} \cos\chi \cos\gamma, \tag{1}$$

$$\dot{\rho}_\mathrm{e} = \nu_\mathrm{g} \sin\chi \cos\gamma, \tag{2}$$

$$\dot{\rho}_\mathrm{d} = -\nu_\mathrm{g} \sin\gamma, \tag{3}$$

$$\dot{\nu}_\mathrm{a} = k_{\nu_\mathrm{a}} \left(\nu_\mathrm{a}^\mathrm{c} - \nu_\mathrm{a}\right), \tag{4}$$

$$\dot{\gamma} = k_\gamma \left(\gamma^\mathrm{c} - \gamma\right), \tag{5}$$

$$\dot{\chi} = \frac{g \cos\left(\chi - \psi\right)}{\nu_\mathrm{g}} \tan\phi, \tag{6}$$

$$\ddot{\phi} = k_\phi \left(\phi^\mathrm{c} - \phi\right) + k_{\dot{\phi}} \left(-\dot{\phi}\right), \tag{7}$$

where ${}^{G}\boldsymbol{\nu}_{\mathrm{g}} = [\nu_{\mathrm{g}},\ 0,\ 0]^{\top}$ is the ground speed vector, k_{γ}, $k_{\nu_{\mathrm{a}}}$, $k_{P\phi}$, and $k_{D\phi}$ are gain values tuned to achieve smoother flight maneuvers, and g is the acceleration due to gravity. Note that in a wind-less condition, $\nu_{\mathrm{a}} = \nu_{\mathrm{g}}$, $\psi = \chi$, and $\theta = \gamma$. For simplicity, UAVs are considered to be in coordinated-turn condition with zero side-slip and zero angles of attack, where course angle (χ) is related to bank angle (ϕ) as expressed in Eq. (6).

Thus, the UAV state space vector considered here is $\mathbf{x} = \left[\boldsymbol{\rho}^{\top}, \nu_{\mathrm{a}}, \chi, \gamma, \phi, \dot{\phi}\right]^{\top}$. The command vector $\mathbf{u}^{\mathrm{c}} = [\nu_{\mathrm{a}}^{\mathrm{c}},\ \gamma^{\mathrm{c}},\ \phi^{\mathrm{c}}]^{\top}$ is obtained from the desired airspeed $(\nu_{\mathrm{a}}^{\mathrm{d}})$, desired course angle (χ^{d}), and desired altitude (h^{d}). These desired path variables are obtained from the waypoint follower function discussed in detail in Sect. 3.1. It should be noted that the desired path remains unmodulated when no potential collisions are detected. These commanded variables are obtained via modulation of desired path using the proposed FIS explained in Sect. 3.3 when a potential collision is detected as discussed in Sect. 3.2. Commanded bank angle (ϕ^{c}) and commanded flight path angle (γ^{c}) are calculated via the following expressions that relate them to the desired course (χ^{d}) and desired altitude (h^{d}), respectively:

$$\frac{\cos(\chi - \psi)}{\nu_{\mathrm{g}}} \tan \phi^{\mathrm{c}} = k_{\chi}\left(\chi^{\mathrm{d}} - \chi\right), \tag{8}$$

$$\nu_{\mathrm{g}} \sin(\gamma^{\mathrm{c}}) = \min\left(\max\left(k_h\left(h^{\mathrm{d}} - h\right), -\nu_{\mathrm{g}}\right), \nu_{\mathrm{g}}\right). \tag{9}$$

Here, k_h and k_{χ} are gain values tuned for smoother bank and flight path angles.

3 Proposed Method

3.1 Waypoint Follower

The waypoint follower function is a path generator that calculates the desired variables which fully defines a trajectory for the UAV in a mission to a waypoint. This function takes inspiration from pure pursuit control [9] by considering the lookahead point $(\boldsymbol{\rho}^{\mathrm{t}})$ and lookahead distance (d^{t}) when calculating the desired variables. For a given waypoint $(\boldsymbol{\rho}^{w})$ in N-frame, its relative position with respect to the own-ship is given by

$$\boldsymbol{\rho}_{w/o} = \boldsymbol{\rho}_{w} - \boldsymbol{\rho}_{o}. \tag{10}$$

Then, the lookahead point is obtained by

$$\boldsymbol{\rho}^{\mathrm{t}} = \boldsymbol{\rho} + d^{\mathrm{t}} \hat{\boldsymbol{\rho}}_{w/o} = \left[\rho_{\mathrm{n}}^{\mathrm{t}},\ \rho_{\mathrm{e}}^{\mathrm{t}},\ \rho_{\mathrm{d}}^{\mathrm{t}}\right]^{\top}, \tag{11}$$

where $\hat{\boldsymbol{\rho}}_{w/o}$ is the unit vector in the direction of $\boldsymbol{\rho}_{w/o}$.

Now, the desired course (χ^{d}), relative desired heading ($\Delta\chi^{\mathrm{d}}$), desired altitude (h^{d}), and relative desired altitude (Δh^{d}) are

$$\chi^{\mathrm{d}} = \chi^{\mathrm{t}} = \tan^{-1}\left(\frac{\rho_{\mathrm{e}}^{\mathrm{t}} - \rho_{\mathrm{e}}}{\rho_{\mathrm{n}}^{\mathrm{t}} - \rho_{\mathrm{n}}}\right), \tag{12}$$

$$\Delta\chi^{\mathrm{d}} = \chi^{\mathrm{d}} - \psi, \tag{13}$$

$$\Delta h^{\mathrm{d}} = -\left(\rho_{\mathrm{d}}^{\mathrm{t}} - \rho_{\mathrm{d}}\right), \tag{14}$$

$$h^{\mathrm{d}} = -\rho_{\mathrm{d}}^{\mathrm{t}}. \tag{15}$$

Lastly, desired airspeed ($\nu_{\mathrm{a}}^{\mathrm{d}}$) is determined based on the mission requisite. For this research, it is assumed to be equal to an optimal airspeed ($\nu_{\mathrm{a}}^{\mathrm{opt}}$) during the cruise and at lower speeds when near waypoints.

3.2 Collision Detection (CD)

Horizontal and vertical views of potentially colliding UAVs are depicted in Fig. 2. The intruder is detected with the intruder's 3D coordinates (${}^{O}\boldsymbol{\rho}_{i/o}$). The relative velocity and relative separation between the own-ship and the intruder are important for CD. These are required for implementing CD logic provided by TCAS II to determine whether the intruder is in a potential collision course with the own-ship. Firstly, the relative velocity of the intruder with respect to the own-ship as observed in N-frame can be determined using the transport theorem, here expressed in O-frame as

$${}^{O}_{N}\boldsymbol{\nu}_{i/o} = {}^{O}_{N}\dot{\boldsymbol{\rho}}_{i/o} = {}^{O}_{O}\dot{\boldsymbol{\rho}}_{i/o} + {}^{O}\omega_{O/N} \times {}^{O}\boldsymbol{\rho}_{i/o}. \tag{16}$$

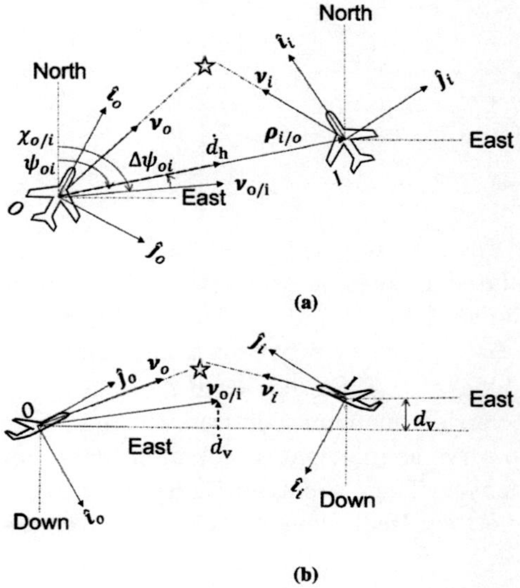

Fig. 2. Collision scenario: (a) horizontal view and (b) vertical view

Here, ${}^{O}_{O}\dot{\boldsymbol{\rho}}_{i/o}$ is determined with successive range-bearing measurements, and the angular velocity in O-frame (${}^{O}\omega_{O/N}$) from an inertial measurement unit.

Now, the horizontal and vertical separations and closure rates can be determined from the relative velocity and relative separation when expressed in the OI-frame. The OI-frame is introduced by rotating the down axis ($\hat{\boldsymbol{k}}_n$) in N-frame about ψ_{oi}. The horizontal separation (d_h), vertical separation (d_v), horizontal closure rate ($\dot{d}_\mathrm{v}$), and vertical closure rate ($\dot{d}_\mathrm{v}$) are obtained as follows:

$$ {}^{OI}\boldsymbol{\rho}_{i/o} = C_N^{OI}\, C_O^{N}\, {}^{O}\boldsymbol{\rho}_{i/o} = [d_\mathrm{h},\ 0,\ d_\mathrm{v}]^\top, \tag{17} $$

$$ {}^{OI}_{N}\boldsymbol{\nu}_{o/i} = -C_N^{OI}\, C_O^{N}\, {}^{O}_{N}\dot{\boldsymbol{\rho}}_{i/o} = \left[\dot{d}_\mathrm{h},\ \dot{d}_\mathrm{y},\ \dot{d}_\mathrm{v}\right]^\top. \tag{18} $$

The CD is confirmed if the time to the Closest Point of Approach (CPA) in both the horizontal range tau (τ_h) and vertical tau (τ_v) direction are lesser than the CPA thresholds (τ_th). These are calculated as follows [17]:

$$ \tau_\mathrm{h} = \dot{d}_\mathrm{h}/d_\mathrm{h}, \tag{19} $$

$$ \tau_\mathrm{v} = \dot{d}_\mathrm{v}/d_\mathrm{v}. \tag{20} $$

The thresholds are chosen using TCAS II logics based on the altitude of operation and the sensitivity level [3]. However, TCAS is for large aircraft and since small UAVs are considered to operate below 1000 ft, the thresholds opted for are further reduced by a factor of 2/3. Likewise, the relative course of the own-ship with respect to the intruder ($\chi_{o/i}$) and the direction of the intruder with respect to inertial North (ψ_{oi}) are important relative variables for CA. They are needed to determine the direction of the intruder with respect to the relative course ($\Delta\psi_{oi}$) which serves as an input for the proposed FIS, and $\Delta\psi_{oi}$ is calculated as

$$ \Delta\psi_{oi} = \chi_{o/i} - \psi_{oi}. \tag{21} $$

3.3 CA Using FIS

Figure 3 depicts the FIS tree structure that is designed to infer from the relative variables calculated in prior sections to provide outputs that modulate the desired path variables. The χ^d and $\dot{d}_\mathrm{h}^\mathrm{d}$ blocks are FISs that yield modulation parameters ($\Delta\chi_\mathrm{ca}$ & $\Delta\nu_\mathrm{ca,h}$) of respective horizontal maneuver desired variables. Likewise, h^d and $\dot{d}_\mathrm{v}^\mathrm{d}$ blocks are FISs that yield modulation parameters (Δh_ca & $\Delta\nu_\mathrm{ca,v}$) of respective vertical maneuver desired variables. Finally, the outputs of $\Delta v_\mathrm{ca,v}$ and Δh_ca also serve as information for intensities required in each direction of the total maneuver. These are again aggregated to determine weights for horizontal (w_h) and vertical (w_v) maneuvers via the Weights FIS block.

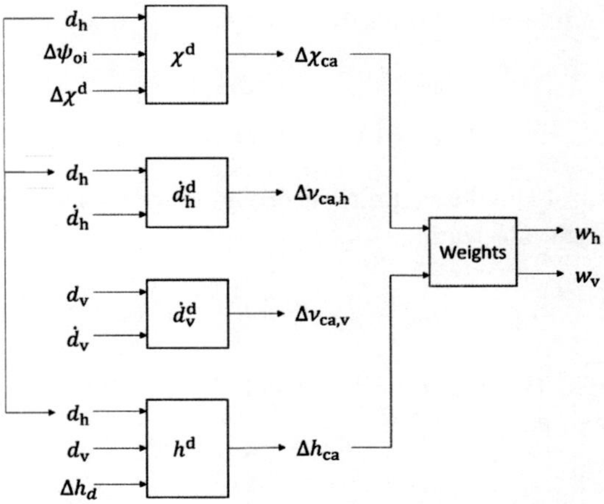

Fig. 3. FIS tree for command modulation

Each of the inputs and outputs is fuzzified into three membership functions (MFs) — two trapezoidal at each end and one triangular in the middle. For instance:

- $\Delta\chi^d$ is fuzzified as Negative/Left (trapezoidal MF), Zero/Center (triangular MF), and Positive/Right (trapezoidal MF) signifying whether the lookahead point is towards the left, center, or right of the current heading respectively,
- h is fuzzified as Low, Medium, and High, $\Delta\psi_{oi}$ as Negative/Left, Zero/Front, and Position/Right, and v as Negative/Below, Zero/Front, and Positive/Over.

The MFs for all inputs and outputs are considered to be symmetrical about the midpoint. The extremes of each variable of the FIS are chosen based on the UAV limits and/or sensor limits. For instance:

- h and v cannot be more than the volume set 3D range-bearing sensor limits, as they can only be measured once the intruder is detected,
- $\Delta\chi_{ca}$ is limited by the yaw rate limits of the UAV,
- Δh_{ca} is limited by the climb rate limits of the UAV,
- $\Delta\nu_{ca,h}$ and $\Delta\nu_{ca,v}$ by UAV acceleration limits.

Finally, the rule-base is developed based on basic human intuition toward potential collision scenarios. For instance:

- If the desired lookahead point direction ($\Delta\chi^d$) is towards left, and the nearest intruder is towards the right of relative motion ($\Delta\psi_{oi}$), the modulation of course angle ($\Delta\chi_{ca}$) is not required and thus is Zero,
- If the nearest intruder is approaching Over (Positive) current own-ship position (v) and the horizontal separation (h) is Below, the modulation of lookahead point's altitude (Δh_{ca} should be Negative to instill a decent motion.

The desired heading and the desired altitude are modulated as follows:

$$\chi^{\mathrm{c}} = \chi^{\mathrm{d}} + w_{\mathrm{h}}\ \Delta\chi_{\mathrm{ca}}, \tag{22}$$

$$h^{\mathrm{c}} = h^{\mathrm{d}} + w_{\mathrm{v}}\ \Delta h_{\mathrm{ca}}. \tag{23}$$

The modulation of the desired airspeed begins with the determination of the desired velocity in the OI-frame as follows:

$${}^{OI}\left[d^{\mathrm{jd}}_{\mathrm{h}},\ d^{\mathrm{jd}}_{\mathrm{y}},\ d^{\mathrm{jd}}_{\mathrm{v}}\right]^{\top} = C^{OI}_{N} C^{N}_{O}\,{}^{O}\left[0,\ 0,\ \nu^{\mathrm{d}}_{\mathrm{a}}\right]^{\top}. \tag{24}$$

Then, the desired horizontal and vertical closure rate components obtained above are modulated as follows:

$$\nu^{\mathrm{c}}_{\mathrm{h}} = d^{\mathrm{jd}}_{\mathrm{h}} + w_{\mathrm{h}}\ \Delta\nu_{\mathrm{ca,h}}, \tag{25}$$

$$\nu^{\mathrm{c}}_{\mathrm{v}} = d^{\mathrm{jd}}_{\mathrm{v}} + w_{\mathrm{v}}\ \Delta\nu_{\mathrm{ca,v}}. \tag{26}$$

To avoid extreme deviation from the desired path, one sets $\nu^{\mathrm{d}}_{y} = d^{\mathrm{jd}}_{y}$. The commanded airspeed ($\nu^{\mathrm{c}}_{\mathrm{a}}$) is obtained from the norm of modulated velocity given below:

$$\nu^{\mathrm{c}}_{\mathrm{a}} = \left(C^{N}_{OI}\right)^{\top}\left[\nu^{\mathrm{c}}_{\mathrm{h}},\ \nu^{\mathrm{d}}_{y},\ \nu^{\mathrm{c}}_{\mathrm{v}}\right]^{\top}. \tag{27}$$

3.4 FIS Optimization Using a GA

The GA is an evolutionary optimization algorithm that tunes the parameters of a system based on the maximization of a fitness function via careful selection, crossover, and mutation. The rule-base and the MFs together make up the knowledge base of the FIS and comprise 510 parameters to be completely defined. The computational effort of the optimization of the proposed FIS is significantly reduced by considering constant rules chosen based on human intuition, retaining the symmetry of MFs of each FIS with consideration of only two tunable parameters (a and b), and limits are chosen to correspond to aircraft limits. The total number of tunable parameters considered dropped to 26. The fitness function is formulated to minimize the path deviation by reducing the total distance traveled (σ) for each UAV during CA and also penalizing the collisions and undesired trajectories (δ) as follows:

$$\min F = \sum_{i=1}^{n} \sigma_i + \delta. \tag{28}$$

4 Simulation Results

The parameters and limits of UAVs for simulations are tabulated in Table 1.

Table 1. Simulation parameters and UAV limits

Gain Parameters	Values	UAV Limits	Values	Range-bearing Sensor Limits	Values
d^t	13 m	v_a^{opt}	70 kts	Range Limit	0.4 nm
k_h	13	v_a^{max}	80 kts	Azimuth Limit	±60 deg
k_γ	0.39	v_a^{min}	45 kts	Elevation Limit	±45 deg
k_ψ	0.6	γ^{max}	45 deg		
k_{ν_a}	1	γ^{min}	−45 deg		
$k_{P\phi}$	3402.97	ϕ^{max}	45 deg		
$k_{D\phi}$	116.67	ϕ^{min}	−45 deg		

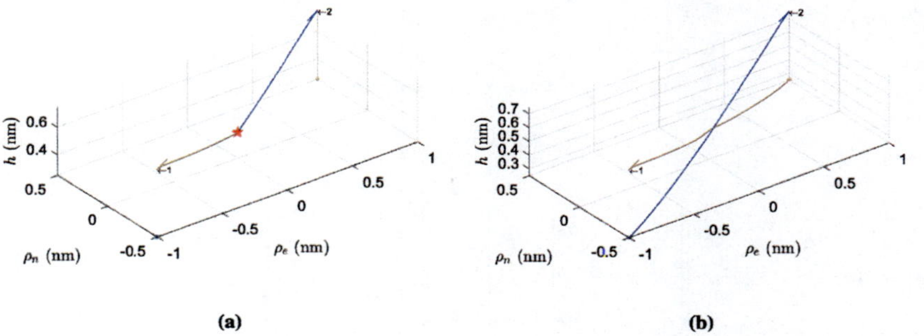

Fig. 4. Training case (a) without CA and (b) with CA using optimized FIS

A pairwise conflict scenario shown in Fig. 4a is used for tuning the FIS parameters using the GA. Here, the UAVs are heading directly toward each other from opposite ends while descending to their respective waypoints at the other end. Once optimized, this pairwise conflict is resolved as shown in Fig. 4b. Initially, without the CA, the states of the UAV 1 follow smooth trends as shown in Fig. 5a. However, there are spikes and changes in the states of the UAV 1 when the collision is detected, and maneuvers are performed for avoidance as shown in Fig. 5b.

The algorithm performs effectively for several test scenarios, one of which is depicted in Fig. 6a. This scenario demands CA such that there is some climb and/or descent of UAVs. The algorithm here performs well enough for the UAVs as they avoid collisions. Figure 6b shows how the states of UAV 1 deviate from the original path when encountering and avoiding the potential collision. Similarly, multiple UAVs were simulated in a closed space shown in Fig. 7 where all individual UAVs consider all detected UAVs and choose the nearest UAV based on which the CA decision is calculated. It can be seen that some UAVs can

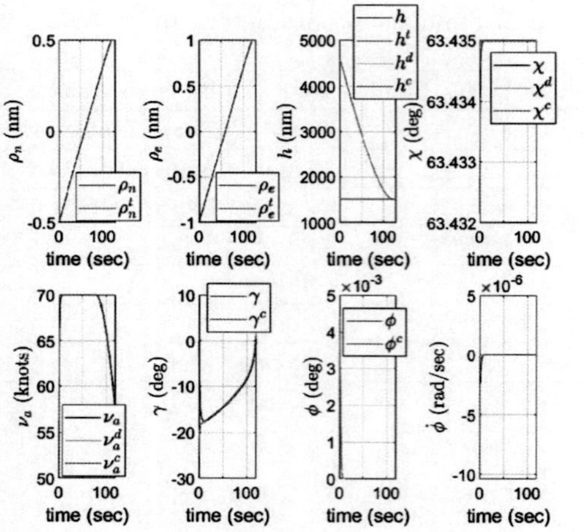

(a)

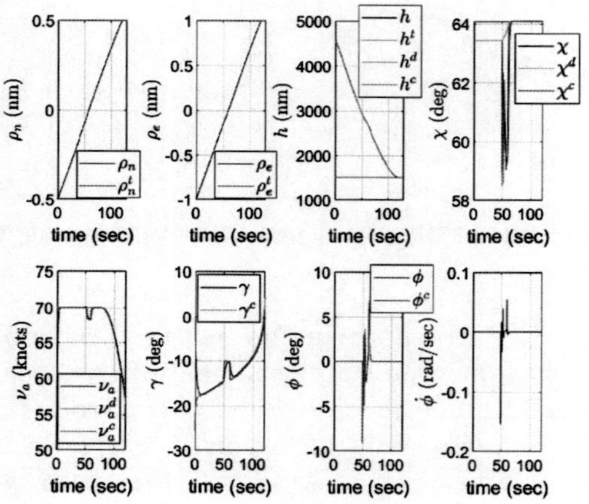

(b)

Fig. 5. UAV 1 for training case (a) without CA and (b) with CA using optimized FIS

avoid successive collisions, thus confirming the effectiveness of the algorithm in multi-UAV conflict scenarios as well. However, there were instances where the lack of detection of UAVs due to limited sensor vision resulted in a collision. These instances were resolved when intruders were detected with a greater field of view.

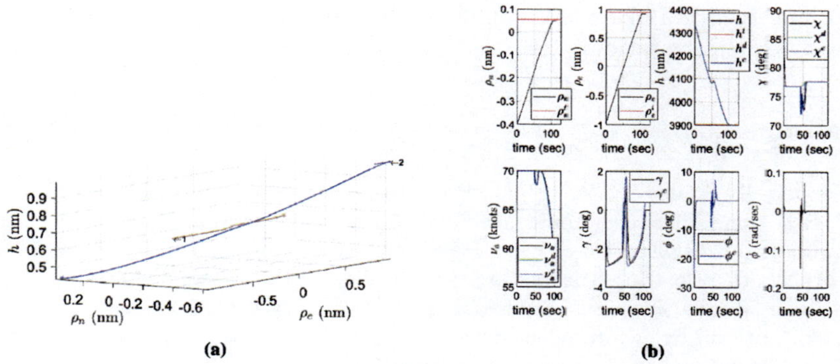

Fig. 6. Pairwise conflict test case with CA: (a) visualization and (b) UAV 1 states

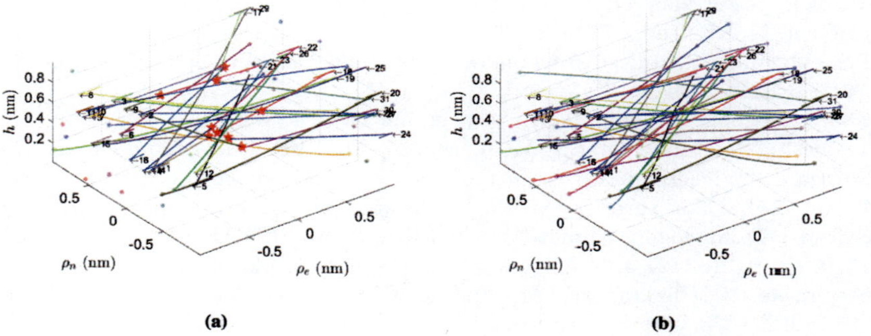

Fig. 7. Multiple UAVs test case (a) without CA and (b) with CA using optimized FIS

5 Conclusion

The work proposed an optimized 3 Dimensional (3D) Collision Avoidance Fuzzy Inference System (3D-CAFIS) that combined horizontal and vertical maneuvers to perform Collision Avoidance (CA) among unmanned aerial vehicles in potential collision scenarios. The 3D-CAFIS generates modulation parameters that are easy to understand and interpret. These parameters are applied to which modulate the desired path variables when a potential collision is detected for CA. The FIS is optimized using a genetic algorithm to improve the 3D maneuvers during CA while maintaining a path toward the target with for minimum deviation during CA. The algorithm proposed performed well enough to avoid collision in various test scenarios despite the use of a simple pairwise conflict training scenario. The results could still be further optimized considering better fitness function. It should be noted that the algorithm is not claimed to perform better than other collision avoidance algorithms, rather is proposed as another method that has simple-to-understand FIS logic such that further tuning and optimization are easier to achieve. Further research will focus on effectiveness during uncertainties like wind, application to quadcopters, and validation with hardware experimentation.

References

1. Cooper, T., Smiley, J., Porter, C., Precourt, C.: Olyver Wyman (2018)
2. P. Research, Advanced Aerial Mobility Market (By Mode of Operation: Piloted, Autonomous; By End-Use: Cargo, Passenger; By Propulsion Type: Parallel Hybrid, Turboshaft, Electric, Turboelectric) - Global Industry Analysis, Size, Share, Growth, Trends, Regional Outlook, and Forecast 2023-2032. Technical report. 2110 (2022). https://www.precedenceresearch.com/advanced-aerial-mobility-market
3. F.A. Administration. Introduction to tcas ii version 7.1 (2011)
4. Manfredi, G., Jestin, Y.: In: 2016 IEEE/AIAA 35th Digital Avionics Systems Conference (DASC), pp. 1–9. IEEE (2016)
5. Trapani, A., Erzberger, H., Dunbar, W.: In: AIAA Guidance, Navigation, and Control Conference, p. 5746 (2009)
6. Peng, L., Lin, Y.: In: 2010 Chinese Control and Decision Conference, pp. 4416–4421. IEEE (2010)
7. Zhu, L., Cheng, X., Yuan, F.G.: Measurement **77**, 40 (2016)
8. Vanneste, S., Bellekens, B., Weyn, M.: In: MORSE 2014 Model-Driven Robot Software Engineering: Proceedings of the 1st International Workshop on Model-Driven Robot Software Engineering co-located with International Conference on Software Technologies: Applications and Foundations (STAF 2014), York, UK, vol. 1319, pp. 91–102, 21 July 2014/Assmann, Uwe [edit.] (2014)
9. Stepanyan, V., Chakrabarty, A., Ippolito, C.A.: In: AIAA Scitech 2021 Forum, p. 1575 (2021)
10. Zhao, Y., Guo, J., Bai, C., Zheng, H.: Complexity **2021**, 1 (2021)
11. Berdonosov, V., Zivotova, A., Zhuravlev, D., Naing, Z.H.: In: 2018 International Multi-Conference on Industrial Engineering and Modern Technologies (FarEastCon), pp. 1–6. IEEE (2018)
12. Aiyama, Y., Ishiwatari, Y., Seki, T.: In: Mechatronics for Safety, Security and Dependability in a New Era, pp. 177–180. Elsevier (2007)
13. Hawkins, K., Tsiotras, P.: In: 2018 IEEE Conference on Decision and Control (CDC), pp. 6301–6306. IEEE (2018)
14. Younas, I., Ahmed, Z., Mohyud-Din, S.T.: In: Proceedings of the 9th WSEAS International Conference on Applied Informatics and Communications (World Scientific and Engineering Academy and Society (WSEAS), Stevens Point, Wisconsin, USA, 2009), AIC'09, pp. 344–350 (2009)
15. Cook, B., Arnett, T., Cohen, K.: Modern fuzzy control systems and its applications, pp. 225–256 (2017)
16. Beard, R.W., McLain, T.W.: Small Unmanned Aircraft: Theory and Practice, Princeton University Press, Princeton (2012)
17. Munoz, C., Narkawicz, A., Chamberlain, J.: In: AIAA Guidance, Navigation, and Control (GNC) Conference 2013, p. 4622 (2013)

Air Traffic Control Using Fuzzy Logic

David Mulligan(✉) and Kelly Cohen

Department of Aerospace Engineering, University of Cincinnati, Cincinnati, OH, USA
mulligdd@mail.uc.edu, cohenky@ucmail.uc.edu

Abstract. Air traffic control (ATC) is a crucial aspect of safe air travel. There have been dozens of documented incidents that compromised the safety of passengers, crew, and bystanders–many of which are due to faulty decision-making from the ATC personnel. A potential way to mitigate the risk of these incidents is through Fuzzy Logic; by implementing a robust and explainable fuzzy inference system, ATC would receive expert recommendations from enormous amounts of historical data in a timely manner. This paper explores the benefits of implementing a fuzzy inference system to monitor air traffic.

1 Introduction

Air travel has quickly become a quintessential part of the modern era. Business professionals can travel across the globe for work, companies can transport large shipments between various locations, and people can visit distant loved ones or reach vacation destinations with ease; even more recently, unmanned air travel has allowed for the rapid transport of goods and services. While all these applications of flight are incredibly beneficial for society, they also come with a great risk–massive, complex vehicles flying at high speeds have the potential for catastrophic damage if managed improperly.

In an FAA human factors analysis on air traffic, it was determined that skill-based errors by Air Traffic Control personnel were the most common occurrences in the investigation of aerial accidents and incidents; more specifically, these mistakes were present in approximately 82% of events reviewed [1]. Air Traffic Control is an extremely difficult and fast-paced job in which safety-critical decisions must be made in rapid succession. For this reason, the search for improved decision making techniques in this industry is of the utmost importance. One possible technique which has been proven to be effective and accurate in various other fields is fuzzy logic. The following pages will examine the feasibility of using fuzzy logic to assist with Air Traffic Control.

2 Literature Review

2.1 Air Traffic Control

Air Traffic Control (ATC) refers to the team of specialists that monitor and direct traffic passing through their designated airspace. These specialists must balance a variety of motivations–passenger safety, bystander safety, industry standards/requirements, efficiency, and many others–and make difficult decisions regarding flight paths of each

K. Cohen et al. (Eds.): NAFIPS 2023, LNNS 751, pp. 25–35, 2023.
https://doi.org/10.1007/978-3-031-46778-3_3

aircraft in the airspace [2]. While the ultimate driving factor is obviously safety, it is also important for ATC to ensure aircraft reach their destination in a timely and resource-efficient manner; this includes making efficient use of personnel, runways, and the surrounding airspace.

Typically, when an aerial incident–event causing damage or minor injuries–or accident–catastrophic event resulting in major injuries/casualties–occurs, the public immediately assigns blame to the pilots. However, a pilot simply doesn't have the ability or resources to account for every variable (especially those occurring outside their own aircraft) while flying. ATC towers attempt to fill in those gaps and monitor variables that the pilot cannot. For example, low visibility due to darkness or adverse weather conditions may prevent a pilot from seeing another aircraft on the runway, while ATC specialists are aware of the position, speed, and flight path of all aircraft in the area. This means that, while pilots certainly have a responsibility to prevent aircraft-related incidents, ATC must be held accountable as well [1].

2.1.1 Consensus

One characteristic of Air Traffic Control that renders decision-making so difficult is the idea of consensus–a decision-making process in which all involved parties work towards an unanimously approved solution. This doesn't necessarily mean every single ATC specialist readily agrees to the presented solution; rather, there can be varying degrees (or "levels") of agreement among a team of specialists that still constitutes a consensus. These levels can range from a definitive agreement to someone who disagrees, but is willing to stand down and not prevent the decision from going through. In a typical consensus-based team environment, everyone's opinion is equally respected and the team will not move forward until all participants agree at minimum to support the decision; this differs from most group decision-making processes, in that a consensus-based team will not accept the idea of "majority rules" [3].

As with many other safety-critical environments, having a consensus-based team is a great quality for ATC teams to have. That being said, making all decisions through unanimous support has its drawbacks–the most important of these relates to a lack of time efficiency. While a majority-based team could simply vote on possible solutions and implement the most popular decision, consensus might require lengthy discussions to reach a solution upon which all participants can agree; in emergency ATC situations such as avoiding collisions and rerouting takeoff/landing assignments, there might not be time for these in-depth discussions. This is where Fuzzy Logic would come into play–analyzing copious amount of historical data and expert knowledge to provide trustworthy recommendations to guide the team's decision in a timely fashion.

2.2 Fuzzy Logic

Fuzzy logic refers to the search for robust, approximate solutions that–due to innate uncertainty in the world–may fall between the absolute "true" and "false" values of traditional logic [4]. This approach reasons that finding such a solution is more cost- and time-effective than using massive amounts of computing power to find a perfectly true answer; keep in mind that this is only true when some tolerance is allowed (i.e., it isn't necessary to get an absolutely correct solution 100% of the time).

There are two main concepts in fuzzy logic applications: linguistic variables and a fuzzy if-then rule. A linguistic variable is simply a parameter expressed in natural language, and a fuzzy if-then rule is a cause/effect statement containing these linguistic variables. Fuzzy logic works by applying fuzzy if-then rules to data sets, thereby compressing the data into only what is needed for decision-making [4]. This differs from traditional logic in that fuzzy logic can use linguistic variables to determine a solution's degree of truthfulness, while a lack of such variables in traditional logic forces it to be binary (either true or false).

3 Methodology

3.1 Simulation Setup

The binary and fuzzy systems are written in MATLAB. Each system is inserted in an existing Air Traffic Control simulation written by MathWorks, in which a centrally-located radar obtains aircraft flight data and predicts their trajectories [5]. This simulation has been modified to include data relevant to this paper such as aircraft type and desired flight path. These parameters will be discussed more thoroughly in the next section.

The MathWorks ATC simulation requires additional software beyond a standard MATLAB license. The "Sensor Fusion and Tracking Toolbox" [6] provides tools and functions that allow MATLAB to design, simulate, and analyze outputs from various data collection methods. This specific example utilizes the "Tracking Scenario" function to simulate an airspace, and then designs all required specifications of a centralized radar using the "Fusion Radar Sensor" function.

3.2 Defining Scenarios

This paper investigates various scenarios that Air Traffic Control might encounter related to an aircraft's safe airspace. A massive number of factors must be considered when dictating whether an aircraft is in safe airspace–aircraft type, weight, altitude, lateral position, proximity to airports, presence of other airborne entities, etc. To concisely analyze our binary and fuzzy systems, we focus on aircraft weighing less than 19,000 lb and residing in Class B airspace–generally up to 10,000 feet mean sea level (MSL). Such cases meet the criteria outlined in Sect. 3.2.3 of the Aeronautical Information Manual (AIM), and therefore have clearly defined ATC clearance and separation standards [2]. These standards–which are shown graphically in Fig. 1–are as follows:

- Minimum vertical separation of +/– 500 feet.
- Minimum lateral separation of 1.5 miles (7,920 feet) [2].

The "disk" shown in Fig. 1 defines the safe/legal airspace that must surround an aircraft at all times. When planning air traffic, ATC personnel work to ensure that this airspace is unobstructed for the entire flight; this means that a properly planned flight path guarantees an aircraft will be safe inside its disk–assuming all other aerial objects

Fig. 1. Visual Representation of an Aircraft's Reserved Airspace [7]

remain in their defined airspace as well. Therefore, it is appropriate to simulate flight path deviation based on the extent to which an aircraft moves within its disk (or leaves the disk entirely). Three such flight paths are defined to simulate small, moderate, and large deviations from the planned path. See Table 1 below.

Table 1. Description of ATC Scenarios

Scenario	Description	Max Deviation
Small deviation	Aircraft flies very close to the flight path for the entire simulation, and does not leave the predefined airspace	Vertical: 200 feet Lateral: 0.5 miles
Moderate deviation	Aircraft gradually deviates from the flight path, until it reaches the edge of the predefined airspace	Vertical: 500 feet Lateral: 1.5 miles
Large deviation	Aircraft experiences a major deviation from the predefined airspace	Vertical: 750 feet Lateral: 2 miles

For testing purposes, characteristics of three different AIM Class B qualified aircraft are used–the Airbus A380, Boeing 757, and Embraer 120. These aircraft have the following weight and range, which serve as inputs for the fuzzy inference system (Table 2):

Table 2. Aircraft Characteristics [8]

Aircraft	Weight (lb)	Range (NM)
Airbus A380	1,234,600	8,000
Boeing 757	220,000	3,929
Embraer 120	25,353	550

Every ATC scenario is simulated twice for each aircraft–once with pilot communication and once without. This results in 18 test cases, which are defined below. Each case is assigned a "desired outcome", the recommendation that a decision making system should provide based on the given information. This information comes from a combination of FAA requirements and ATC data. [2] (Table 3).

Table 3. Parameters for Each Test Case

Case #	Aircraft	ATC Scenario	Pilot Communication?	Desired Outcome
1	Airbus A380	Small deviation	Yes	Do Not Intervene
2	Airbus A380	Moderate deviation	Yes	Do Not Intervene
3	Airbus A380	Large deviation	Yes	Intervene
4	Airbus A380	Small deviation	No	Do Not Intervene
5	Airbus A380	Moderate deviation	No	Intervene
6	Airbus A380	Large deviation	No	Intervene
7	Boeing 757	Small deviation	Yes	Do Not Intervene
8	Boeing 757	Moderate deviation	Yes	Do Not Intervene
9	Boeing 757	Large deviation	Yes	Do Not Intervene
10	Boeing 757	Small deviation	No	Do Not Intervene
11	Boeing 757	Moderate deviation	No	Intervene
12	Boeing 757	Large deviation	No	Intervene
13	Embraer 120	Small deviation	Yes	Do Not Intervene
14	Embraer 120	Moderate deviation	Yes	Do Not Intervene
15	Embraer 120	Large deviation	Yes	Do Not Intervene
16	Embraer 120	Small deviation	No	Do Not Intervene
17	Embraer 120	Moderate deviation	No	Intervene
18	Embraer 120	Large deviation	No	Intervene

3.3 Binary System

Each of the cases described above is first evaluated using a binary decision making system–either the aircraft resides in safe/legal airspace, or it does not. The formula for this system is very straightforward; if the aircraft violates either AIM Class B clearance standard defined in Sect. 3.1 above, then the system recognizes that it has left its "disk" and recommends that ATC intervenes. Figure 2 below contains a flowchart that depicts the binary decision making process.

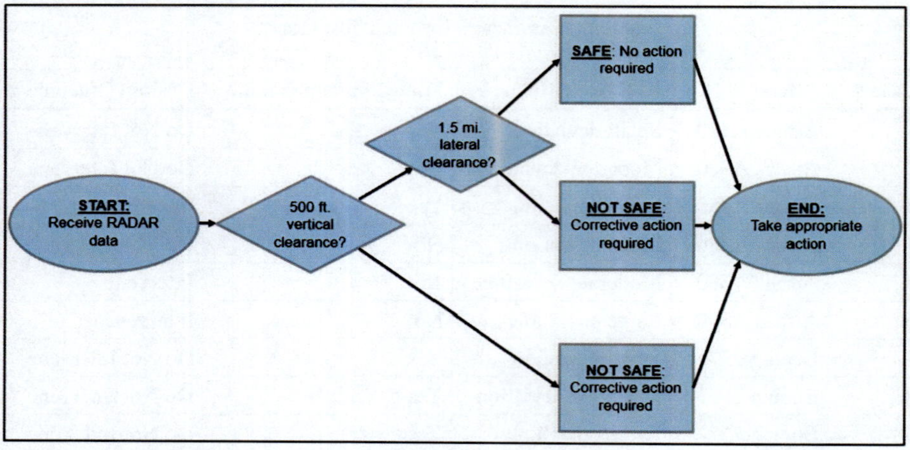

Fig. 2. Flow Diagram for Binary ATC Decision-Making System

3.4 Fuzzy Inference System

Next, each test case is simulated using a cascading Mamdani fuzzy inference system. With this method, the inputs of lateral and vertical clearance are defined by their respective degrees of membership to three fuzzy sets–"small", "medium", and "large"–rather than by a numerical value. Two additional fuzzy inputs–aircraft weight and range–are used to gauge potential impact level of an uncontrolled aircraft, and a binary input regarding the pilot communication status aids in the decision making. Figure 3 below contains a flowchart that depicts the fuzzy decision making process.

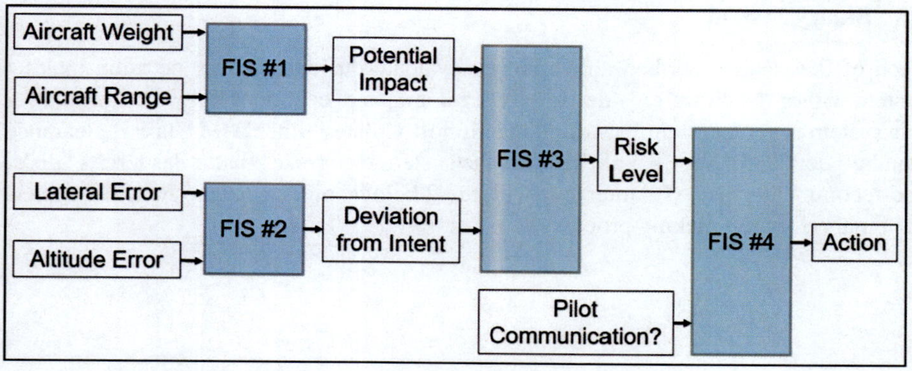

Fig. 3. Flow Diagram for Fuzzy ATC Decision-Making System

Within these fuzzy inference systems, a Mamdani-based engine translates each set of inputs into the desired output using membership functions and rules. The membership functions for all four fuzzy inputs are plotted in Figs. 4 below. Pilot communication status is binary (either there is pilot communication or not), so no membership function is required.

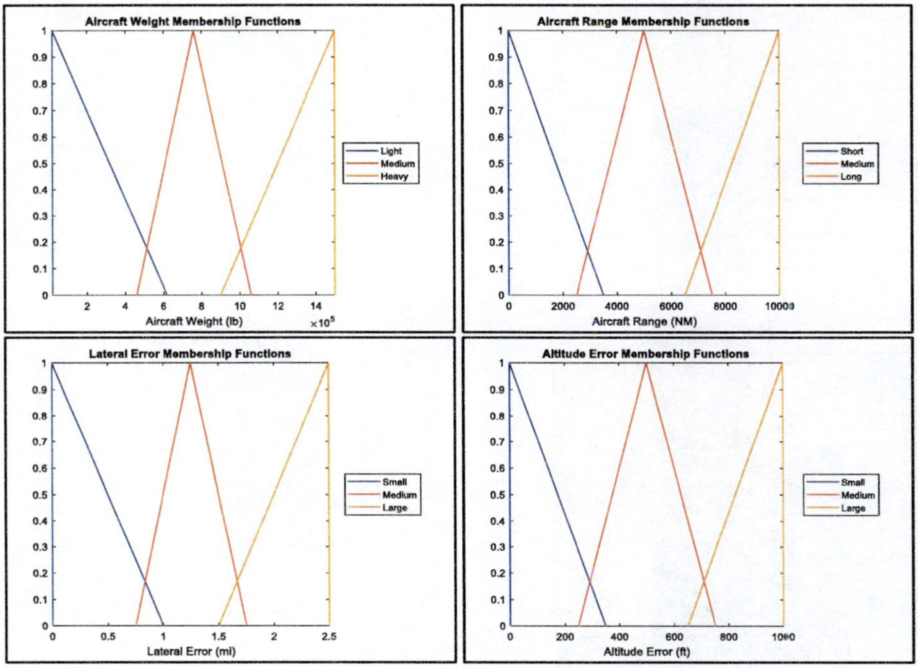

Fig. 4. Membership Functions for Fuzzy Inputs

Once these inputs have been fuzzified in their respective inference systems, they are applied to a rule base which determines the output. The rule bases for FIS #1–3 consist of nine rules, one for each fuzzy input combination; the rule base for FIS#4 has six rules, as the pilot communication input is treated as binary. See Fig. 5 below for the rule base.

FIS #4 produces the final output of this cascading inference system–a recommendation for ATC personnel based on all inputs. The system either recommends that ATC does not intervene, monitors the aircraft, or takes corrective action (intervenes).

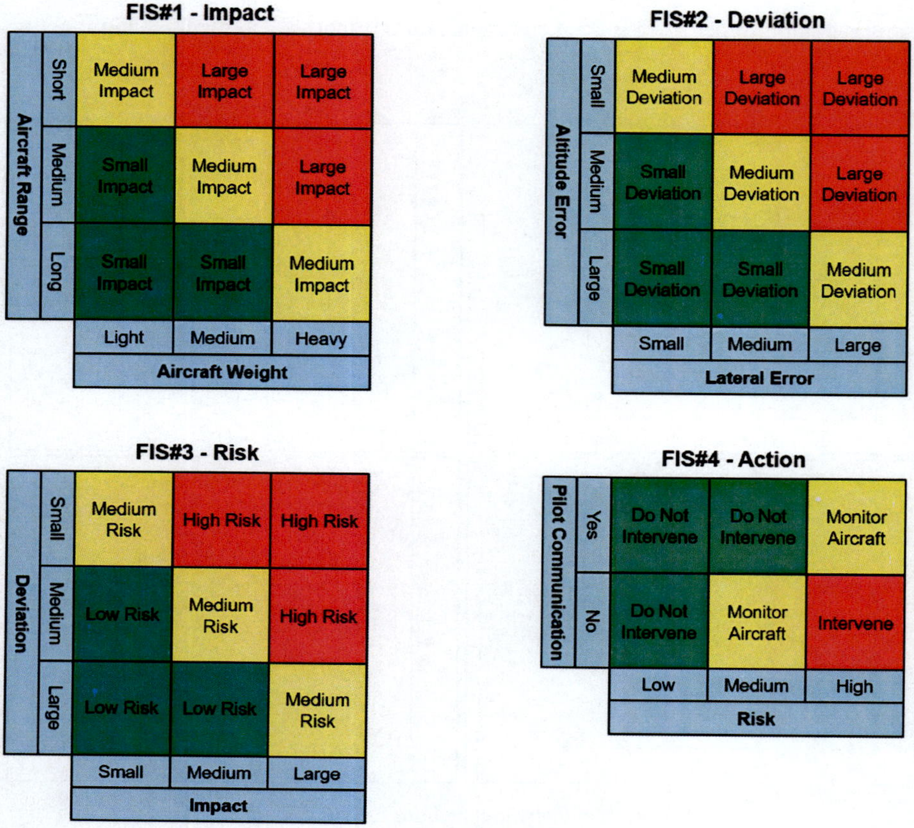

Fig. 5. Rule Base for Each FIS

4 Results and Discussion

Each test case was simulated in MATLAB, and the recommendations of both systems were recorded. See Table 4 below.

Rather than looking at each case individually, let us measure the overall performance of the two systems using the following classification metrics:

- Accuracy–the percentage of total cases that were predicted correctly.
- Precision (Intervene)–the percentage of cases recommending intervention that were predicted correctly.
- Precision (Do Not Intervene)–the percentage of cases not recommending intervention that were predicted correctly.

All three of these classification metrics can be derived from a confusion matrix, which divides the results into four categories–correctly-predicted interventions, incorrectly-predicted interventions, correctly-predicted non-interventions, and incorrectly-predicted

Table 4. Test Results for Binary and Fuzzy Systems

Case	Desired Outcome	Binary System Recommendation	Fuzzy System Recommendation
1	Do Not Intervene	Do Not Intervene	Do Not Intervene
2	Do Not Intervene	Intervene	Monitor Aircraft
3	Intervene	Intervene	Monitor Aircraft
4	Do Not Intervene	Do Not Intervene	Intervene
5	Intervene	Intervene	Intervene
6	Intervene	Intervene	Intervene
7	Do Not Intervene	Do Not Intervene	Do Not Intervene
8	Do Not Intervene	Intervene	Do Not Intervene
9	Do Not Intervene	Intervene	Do Not Intervene
10	Do Not Intervene	Do Not Intervene	Do Not Intervene
11	Intervene	Intervene	Intervene
12	Intervene	Intervene	Intervene
13	Do Not Intervene	Do Not Intervene	Do Not Intervene
14	Do Not Intervene	Intervene	Do Not Intervene
15	Do Not Intervene	Intervene	Do Not Intervene
16	Do Not Intervene	Do Not Intervene	Do Not Intervene
17	Intervene	Intervene	Intervene
18	Intervene	Intervene	Intervene

non-interventions. Tables 5 and 6 below contain a confusion matrix with accuracy/precision values for the binary and fuzzy system, respectively. Note that for the fuzzy system, a recommendation to "Monitor Aircraft" is treated as a non-intervention.

Table 5. Results and Performance Metrics for Binary System

		Actual	
		Intervene	**Do Not Intervene**
Recommended	**Intervene**	5	7
	Do Not Intervene	0	6

Accuracy	61%
Precision (Intervene)	42%
Precision (Do Not Intervene)	100%

Looking at both sets of results, a fuzzy inference system is slightly more accurate (recommends the correct action most frequently) than a binary system. It is also considerably more precise when recommending that ATC intervene–i.e., it is very rare for the system to falsely recommend an intervention. However, the fuzzy system has worse precision when recommending that ATC does not intervene; this means the fuzzy system

Table 6. Results and Performance Metrics for Fuzzy Inference System

		Actual	
		Intervene	Do Not Intervene
Recommended	Intervene	6	1
	Do Not Intervene	1	10

Accuracy	89%
Precision (Intervene)	86%
Precision (Do Not Intervene)	91%

will falsely recommend that intervention is unnecessary more frequently than the binary system.

To understand exactly what this means, it is important to analyze the cost of each outcome. If ATC intervenes when it should not, the result is a series of corrective actions that may be untimely and expensive; however, no additional safety risks are presented by this "false positive". On the other hand, incorrectly recommending that ATC doesn't intervene can have disastrous impacts including loss of aircraft and loss of life. Taking this cost analysis into consideration, the fuzzy inference system is slightly less precise when it comes to the outcome with higher stakes. The fuzzy system's overall accuracy still makes it a viable tool, but this risk is important to note.

5 Conclusion

This paper investigated the potential benefits of using fuzzy logic to improve Air Traffic Control operations. Results show that, compared to a standard binary system based on FAA guidelines [2], the fuzzy inference system is over 45% more accurate for the simulated data set. The fuzzy system is also significantly more precise when recommending non-intervention, although it is slightly less precise when recommending intervention. Since the fuzzy inference system was developed using the Mamdani approach, it is limited to the rules and membership functions chosen by the authors. Future work with this project should adapt a genetic algorithm to better optimize the fuzzy system design, which could result in even more accurate and precise results. Either way, it is clear that Air Traffic Control is a very promising application for fuzzy logic.

References

1. Pape, A., Wiegmann, D., Shappell, S.: Air Traffic Control (ATC) related accidents and incidents: a human factors analysis. In: Proceedings of the 11th International Symposium on Aviation Psychology (2001). https://www.faa.gov/about/initiatives/maintenance_hf/library/documents/media/human_factors_maintenance/air_traffic_control_(atc)_related_accidents_and_incidents.a_human_factors_analysis.pdf
2. U.S. Department of Transportation, Federal Aviation Administration: Aeronautical Information Manual: Official Guide to Basic Flight Information and ATC Procedures. Federal Aviation Administration (2022). https://www.faa.gov/air_traffic/publications/atpubs/aim_html/
3. Consensus-Based Decision-Making Processes. The Consensus Council, Inc (2018). http://www.csh.org/wp-content/uploads/2018/07/38-National-Partner-Recommendation-Consensus-Decision-Making-Process-incl-Modified-Consensus.pdf

4. Zadeh, L.: Soft computing and fuzzy logic. In: IEEE Software, pp. 48–56 (1994). http://projectsweb.cs.washington.edu/research/projects/multimedia5/JiaWu/review/Cite2.pdf
5. Air Traffic Control. The MathWorks, Inc (2022). https://www.mathworks.com/help/fusion/ug/air-traffic-control.html
6. Sensor Fusion and Tracking Toolbox. The MathWorks, Inc (2022). https://www.mathworks.com/help/fusion/index.html?s_tid=srchtitle_Sensor%20Fusion%20and%20Tracking%20Toolbox_
7. D'Arcy, J., Della Rocco, P., Air Traffic Control Specialist Decision Making and Strategic Planning–A Field Survey. Federal Aviation Administration (2001). https://rosap.ntl.bts.gov/view/dot/16683/dot_16683_DS1.pdf
8. Aircraft Technical Data & Specifications. VerticalScope, Inc (2023). https://www.airliners.net/aircraft-data

Data Driven Level Set Fuzzy Classification

Fernando Gomide[1(✉)] and Ronald Yager[2]

[1] University of Campinas, Sao Paulo, Brazil
gomide@unicamp.br
[2] Iona College, New Rochelle, USA
yager@panix.com

Abstract. The paper addresses the structure of fuzzy rule-based classifiers from the point of view of a function relating membership grades of inputs with rule outputs and a discriminant function. Rule-based level set based models are employed to produce classifiers using data to find the output functions of the fuzzy rules. A simple and effective formulation consists of estimating the parameters of the output function of each rule using input and output class data. The data driven method gives an easy and efficient mechanism to produce rule-based fuzzy classifiers. Performance evaluation is done using the data sets and classifiers available in scikit-learn, currently a reference in machine learning. The results suggest that data driven level set fuzzy classifiers performance competes closely or surpasses state of the art classifiers.

1 Introduction

Classification is a fundamental supervised machine learning process that means predicting the class of given data points. Classes can be targets or categories. For example, spam filter algorithms aim to classify emails as either spam or not spam. Process health diagnosis systems attempt to classify the process states into normal or faulty. Common classification algorithms include nearest neighbors, support vector machines, Gaussian process classifiers, decision trees, neural networks, naive Bayes, quadratic discriminant analysis, fuzzy rule-based systems, fuzzy trees, and neural fuzzy networks.

Like regression problems, classification problems may be viewed as learning a mapping from the input data space to an output space. While regression uses learning algorithms to learn a mapping to produce a continuous output, classification learning algorithms learn a mapping with finite number of categories as output. Ultimately, the objective is to find decision boundaries that divide the input data space into different regions, each of which associated with a labeled category. The idea is to assign class labels to input data, depending on which side of the decision boundary they stand.

Classification can be encompassed into three major approaches. Discriminative deterministic approach directly develops a deterministic mapping between

K. Cohen et al. (Eds.): NAFIPS 2023, LNNS 751, pp. 36–43, 2023.
https://doi.org/10.1007/978-3-031-46778-3_4

the input space and labels via a parameterized function. Examples of the discriminative deterministic approach include neural networks, fuzzy rule-based classifiers, and decision trees. Discriminative probabilistic approach models the probability of a data point to belong to class via a parameterized conditional probability density function. Examples of discrininative probabilistic include Gaussian process classifiers. Generative probabilistic approach models the joint distribution of input data space and class labels by specifying the prior distribution and the class-dependent probability distribution. An example of generative probabilistic is the naive Bayes classifier.

Fuzzy rule-based modeling paradigms either compute the output of the linguistic model as the centroid of the union of the consequents of the active rules [1], or as the sum the consequents fuzzy sets and the centroid of the sum [2], or as the weighted average of the consequents functions of the active rules [3]. Another basic paradigm to calculate the output of a fuzzy model relies upon the Yager level set method [4]. The level set is a modeling technique whose output is found using a function that maps rule activation levels into a value of the output space. Level set models output the weighted sum of the output function of each fuzzy rule. Level set modeling has been generalized to produce the output functions of the fuzzy rules using data. The data driven level set method has been shown to be very efficient in regression, and in applications such as nonlinear system modeling [5], and time series forecasting [6].

This paper addresses data driven level set fuzzy rule-based classification. The purpose is to introduce a general data-driven framework for fuzzy classification using the level set process. As shown in the next sections, the framework is a simple, intuitive, computationally efficient, and effective discriminative approach for data classification. Examples concerning classic benchmarks are given using synthetic data sets of scikit-learn, a major public domain machine learning environment. The performance of the data driven level set classifier is compared with several classifiers of scikit-learn to illustrate the nature of decision boundaries of the level set classifier, and classification scores of the classifiers. The results show that the data driven level set classifier performs similarly or better than the nearest neighbors, support vector machines, Gaussian process classifiers, decision tree, neural networks, naive Bayes, quadratic discriminant analysis, and fuzzy decision tree. The paper concludes summarizing its contributions and listing issues for further development.

2 Level Set Fuzzy Classifier

The level set method is a fuzzy rule-based modeling technique whose output maps rule activation levels into values of the output space. In its simplest form, the model output is the weighted sum of the outputs of each fuzzy rule. More precisely, assume a linguistic fuzzy rule-based model whose i-th rule is $\mathcal{R}_i$: if x is $\mathcal{A}_i$ then y is $\mathcal{B}_i$ with fuzzy sets $\mathcal{A}_i$ in $\mathcal{X}$, and $\mathcal{B}_i$ in $\mathcal{Y}$. If an input $x \in \mathcal{X}$ activates rule $\mathcal{R}_i$ with level $\tau_i = \mathcal{A}_i(x) \in [0,1]$, then it scales the corresponding output to $\mathcal{B}_i{}' = \tau_i \wedge \mathcal{B}_i$. The level set $\mathcal{B}_{\tau_i}$ of $\mathcal{B}_i$ is $\mathcal{B}_{\tau_i} = \{y | \tau_i \leq \mathcal{B}_i(y)\}$ the τ_i-cut of

$\mathcal{B}_i$. When the fuzzy set $\mathcal{B}_i'$ needs to be converted into a value, $\mathcal{B}_{\tau_i}$ is the set of the candidates that best represent $\mathcal{B}_i'$. If $\mathcal{B}_i$ is convex, then the level set is the closed interval $\mathcal{B}_{\tau_i} = [y_{il}, y_{iu}]$ and a potential output is the midpoint $m_i = (y_{il} + y_{iu})/2$. The level set method assumes output functions of form $\mathcal{F}_i : [0,1] \to \mathcal{Y}$ for each rule $\mathcal{R}_i$. In particular, when considering midpoint of the level set, the output function of $\mathcal{R}_i$ becomes $\mathcal{F}_i : [0,1] \to m_i$. Notice that the fuzzy set $\mathcal{B}_i$ maps $\mathcal{Y}$ into membership degrees, whereas $\mathcal{F}_i$ maps membership degrees into a value of $\mathcal{Y}$. This value is the mean of the elements of $\mathcal{B}_{\tau_i}$ when considering midpoints. When the $\mathcal{B}_i$ is triangular, $\mathcal{F}_i$ is a line segment, and if in addition it is symmetric, then $\mathcal{F}_i$ is the centroid of $\mathcal{B}_i$. The output of the model is the weighted average of the output functions values of the active rules.

In classification problems, the level set method requires an additional step to map the output y produced by the rule base into a value in a finite set of labels $\mathcal{L} = \{\ell_1, ..., \ell_c\}$, namely a map $C : \mathcal{Y} \to \mathcal{L}$. For instance, in binary classification with labels in $\mathcal{L} = \{0,1\}$, the map can be a threshold function $C : \mathcal{Y} \to \{0,1\}$, a discriminant function such that $Class = 0$ if $C(y) < 0.5$, $Class = 1$ otherwise. In sum, assume a rule-based classifier in the form:

$$\mathcal{R}_i : \text{ if } x \text{ is } \mathcal{A}_i \text{ then } y \text{ is } \mathcal{B}_i \tag{1}$$

where $i = 1, 2, \ldots, N$ and $\mathcal{A}_i$ and $\mathcal{B}_i$ are convex fuzzy sets with membership functions $\mathcal{A}_i : \mathcal{X} \to [0,1]$ and $\mathcal{B}_\rangle : \mathcal{Y} \to [0,1]$. Given an input data $x \in \mathcal{X}$, the level set classifier is as follows

1. Compute the activation degree of each rule $\mathcal{R}_i$, $\tau_i = \mathcal{A}_i(x)$
2. Find the level set for each τ_i, $\mathcal{B}_{\tau_i} = \{y | \tau_i \le \mathcal{B}_i(y)\} = [y_{il}, y_{iu}]$
3. Compute the midpoint of the level set $m_i = (y_{il} + y_{iu})/2$
4. Compute the model output y as $y = \sum_{i=1}^N \tau_i m_i / \sum_{i=1}^N \tau_i$
5. Output $Class = \ell_i$ if $\ell_{i-1} \le C(y) < \ell_i$

Essentially, this is the level set method developed in [4] translated for classification purposes. In its original formulation, the level set is a framework in that allows $\mathcal{F}_i : [0,1] \to \mathcal{Y}_i$ to be computed in advance, a feature that makes level set modeling very efficient computationally speaking. Functions $\mathcal{F}_i$ change only when consequent membership functions change.

3 Data-Driven Level Set Fuzzy Classifier

Likewise the original level set method, the data driven level set classifier requires only a minor modification in the original data driven method developed for regression [7]. The data driven classifier is as follows.

Let $\mathcal{D} = \{(x^k, y^k)\}$, $x^k \in \mathbb{R}^p$, $y^k \in \mathcal{L} = \{\ell_1, ..., \ell_c\}$ be a data set such that $y^k = f(x^k)$, $k = 1, 2, \ldots, K$. The task is to build a fuzzy classifier $C(\mathcal{F})$ that approximates the function f using $\mathcal{D}$.

Building linguistic fuzzy rule-based models from data requires the identification of the model structure, the specification of the number of rules, and the estimation of the antecedent and consequent membership functions. Typically, these steps can be done using clustering algorithms. The paradigm is one cluster - one rule which assigns cluster centers as modal values of the membership functions $\mathcal{A}_i$ of the antecedents. The number N of clusters can be found either using cluster validity measures, or experimentally. Given the membership functions $\mathcal{A}_i$, it remains to estimate consequent membership functions $\mathcal{B}_i$ or, equivalently, to estimate their corresponding output functions $\mathcal{F}_i$, $i = 1, 2, \ldots, N$ from data. The simplest way to do this is to assume $\mathcal{F}_i$ affine:

$$\mathcal{F}_i(\tau_i) = m_i(\tau_i) = v_i\tau_i + w_i \tag{2}$$

Parameters v_i and w_i can be estimated using any appropriate procedure, such as least squares-based learning procedures for instance. Regularized, recursive, or alternative solutions can be derived. Here the classic pseudo inverse-based solution, summarized as follows, is used. Consider the activation degrees $\tau_i^k = \mathcal{A}_i(x^k)$, $i = 1, 2, \ldots, N$, for each data pair (x^k, y^k), and let $s^k = \sum_{i=1}^{N} \tau_i^k$ According to the level set method, the corresponding outputs z^k are

$$z^k = \frac{\tau_1^k(v_1\tau_1^k + w_1)}{s^k} + \ldots + \frac{\tau_N^k(v_N\tau_N^k + w_N)}{s^k} \tag{3}$$

If $\mathbf{d}^k = [(\tau_1^k)^2/s^k, \tau_1^k/s^k, \ldots, (\tau_N^k)^2/s^k, \tau_N^k/s^k]$, and the vector of parameters $\mathbf{u} = [v_1, w_1, \ldots, v_N, w_N]^T$, then expression (3) can be rewritten as

$$z^k = \mathbf{d}^k \cdot \mathbf{u}, \quad k = 1, \ldots, K \tag{4}$$

Let $\mathbf{z} = [z^1, \ldots, z^K]^T$, $\mathbf{D} = [d^{1T}, \ldots, d^{KT}]^T$, and $\mathbf{y} = [y^1, \ldots, y^K]^T$. The collection of equations (4) can be expressed compactly as $\mathbf{z} = \mathbf{D}\mathbf{u}$. The vector of parameters $\mathbf{u}$ is the solution of $min_{\mathbf{u}}\|\mathbf{y} - \mathbf{z}\|^2$, respectively

$$\mathbf{u} = \mathbf{D}^{+}\mathbf{z} \tag{5}$$

where $\mathbf{D}^{+}$ is the Moore-Penrose pseudo inverse of $\mathbf{D}$ [9]. Keeping in mind that $\mathbf{d} = \mathbf{d}(\tau_1, \ldots, \tau_N)$ and that $\tau_i = \mathcal{A}_i(x)$, the model output for input is x is

$$y = \mathbf{d} \cdot \mathbf{u} \tag{6}$$

The steps to develop a data driven level set classifier are:

1. Cluster the data set $\mathcal{D}$ into N clusters accounting for labels in $\mathcal{L}$.
2. Assign membership function $\mathcal{A}_i$ to cluster $i = 1, \ldots, N$.
3. Find consequent parameter vector $\mathbf{u}$ using (5).
4. Compute the model output y using (6)
5. Output $Class = \ell_i$ if $\ell_{i-1} \leq C(y) < \ell_i$

Clustering can be accomplished by any appropriate clustering algorithm such as the fuzzy c-means or its variations, adaptive vector quantization, grid, or knowledge-based partitioning [8]. Given an input datum $x \in \mathcal{X}$, the level set classifier outputs the corresponding class estimate using a discriminant function $C(y)$. In binary classification $\mathcal{L} = \{0, 1\}$ and $C(y)$ is a threshold function such that $Class = 0$ if $C(y) < 0.5$, and $Class = 1$ otherwise.

4 Classifier Comparison

This section compares the level set classifier with the ones of scikit-learn using synthetic data sets. The idea is to illustrate the nature of the decision boundaries and the classification accuracy of the different classifiers. scikit-learn is an open, and widely recognized Python library for machine learning. The same data sets, classifiers, training and testing data split process adopted in [10] are adopted. The scikit-learn classifiers are nearest neighbors (KNNB), linear support vector machine (LSVM), radial basis support vector machine (RSVM), Gaussian process (GPRC), decision tree (DTRE), random forest (RFOR), neural network (MLNN), AdaBoost (ADAB), naive Bayes (NBAY), and quadratic discriminant analysis (QDAN). The fuzzy tree (FTRE) classifier developed within the framework of scikit-learn reported in [11] is also chosen as an alternative fuzzy classifier. The level set fuzzy classifier is denoted LSFC for short.

4.1 Data Sets and Decision Boundaries

The data sets are produced using the make_classification function to create multiclass data sets by allocating each class one or more normally-distributed clusters of points. Function make_classification specializes in introducing noise by way of: correlated, redundant and uninformative features; multiple Gaussian clusters per class; and linear transformations of the feature space of three functions, respectively make_moons, make_circles, and linearly separable [10]. As highlighted in [10], make_moons and make_circles generate 2-dimensional binary classification data that challenge certain algorithms (e.g. centroid-based clustering, linear classification), and include optional Gaussian noise. They also are useful for visualization. Data sets display training data points in solid colors, and testing data points in semi-transparent colors.

4.1.1 Moons Data Function make_moons produces (makes half circles in 2 dimensions) Gaussian data for binary classification, Fig. 1 (left). The classification boundary developed by a six rules level set fuzzy classifier is shown in Fig. 1 (right). The color code is the same used by scikit-learn.

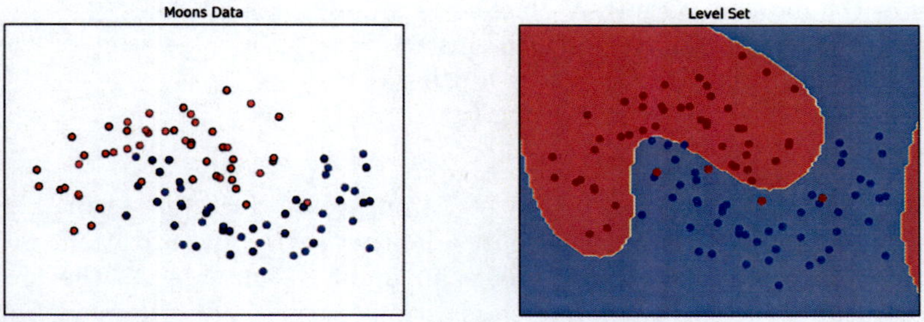

Fig. 1. Moons data set and 6 rules level set fuzzy classifier decision boundary.

4.1.2 Circles Data Function make_circles makes a large circle containing a smaller circle in 2 dimensions Gaussian data with a spherical decision boundary for binary classification, Fig. 2 (left). The classification boundary developed by two rules level set fuzzy classifier is shown in Fig. 2 (right).

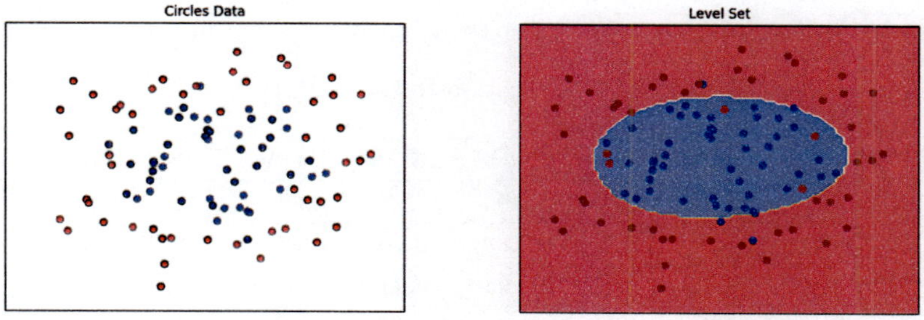

Fig. 2. Circles data set and 2 rules level set fuzzy classifier decision boundary.

4.1.3 Linearly Separable Data

Function linearly_separable makes 2 dimensional linear separable data sets with Gaussian noise, Fig. 3 (left). The classification boundary developed by two rules level set fuzzy classifier is shown in Fig. 3 (right).

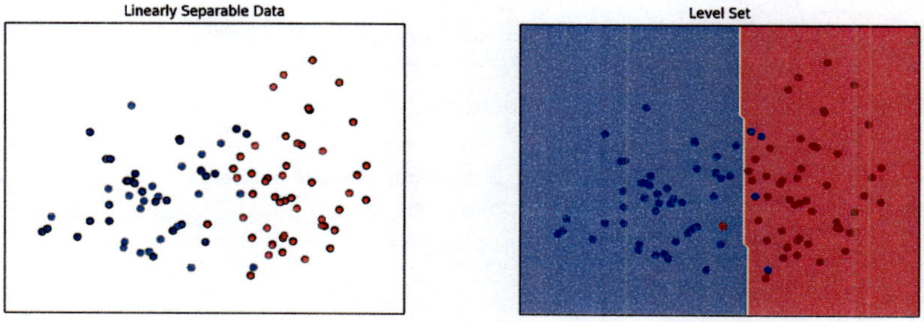

Fig. 3. Linearly separable data set and 2 rules level set fuzzy classifier decision boundary.

4.2 Classification Accuracy

Classifier comparison of scikit-learn uses the function score to compute the classification accuracy, the rate between the number of correctly classified test data and number of test data. The LSFC uses the function accuracy_score of scikit-learn. For classification, score and accuracy_score are both the same, just different ways to calculate the same thing. Table 1 and Table 2 summarize the results.

Table 1. Classification Rate

Input Data	KNNB	LSVM	RSVM	GPRC	DTRE	FTRE	LSFC
Moons	0.97	0.88	0.97	0.97	0.95	0.95	$\mathbf{0.98}^{6}$
Circles	0.93	0.40	0.88	0.90	0.78	0.78	$\mathbf{0.95}^{2}$
Linearly	**0.95**	0.93	**0.95**	0.93	0.93	**0.95**	$\mathbf{0.95}^{2}$

Best classifiers marked bold. Superscripts denote number of fuzzy rules.

Table 2. Classification Rate (continued)

Input Data	RFOR	MLNN	ADAB	NBAY	QDAC	FTRE	LSFC
Moons	0.93	0.90	0.93	0.88	0.85	0.95	$\mathbf{0.98}^{6}$
Circles	0.80	0.90	0.82	0.70	0.72	0.78	$\mathbf{0.95}^{2}$
Linearly	0.93	**0.95**	**0.95**	**0.95**	0.93	**0.95**	$\mathbf{0.95}^{2}$

Best classifiers marked bold. Superscripts denote number of fuzzy rules.

5 Conclusion

The paper has introduced a novel data driven approach to construct rule-based fuzzy classifiers using the level set method. They differ from previous fuzzy classifying paradigms in the way the fuzzy rules are constructed and processed. The level set method uses output functions to map membership grades into values in the domain of the output variables. The level set classifier uses a threshold function to assign a class label to an input data. Comparison using data sets and classifiers of scikit-learn, a major tool for machine learning, shows that data driven level set fuzzy classifiers are effective, and compete closely or surpass state of the art classifiers. Future work shall address model interpretability issues, recursive adaptive level set modeling, and applications in high dimensional spaces.

Acknowledgements. The first author is grateful to the Brazilian National Council for Scientific and Technological Development for grant 302467/2019-0. The authors also tank the reviewers for their helpful comments.

References

1. Mamdani, E.: Application of fuzzy algorithms for control of simple dynamic plant. Proc. IEEE **121**(12), 1585–1588 (1974)
2. Kosko, B.: Fuzzy Engineering. Prentice Hall, Upper Saddle River, New Jersey (1997)
3. Sugeno, M., Takagi, T.: A new approach to design of fuzzy controller. In: Wang, P. (ed.) Advances in Fuzzy Sets, Possibility Theory and Applications, pp. 325–334. Plenum Press, New York (1983)
4. Yager, R.: An alternative procedure for the calculation of fuzzy logic controller values. J. Jpn. Soc. Fuzzy Theory Syst. **3**(4), 736–746 (1991)

5. Maciel, L., Ballini, R., Gomide, F.: Data driven level set method in fuzzy modeling and forecasting. In: Dick, S., Kreinovich, V., Lingras, P. (eds.) Applications of Fuzzy Techniques. NAFIPS 2022. LNNS, vol. 500, pp. 125–134. Springer, Cham (2023). https://doi.org/10.1007/978-3-031-16038-7_14
6. Maciel, L., Ballini, R., Gomide, F., Yager, R.: Forecasting cryptocurrencies prices using data driven level set fuzzy models. Expert Syst. Appl. **210**, 118387 (2022)
7. Leite, D., Gomide, F., Yager, R.: Data driven fuzzy modeling using level sets. In: Proceedings of the FUZZ-IEEE 2022, Padova, Italy, 18–23 July 2022
8. Jang, R., Sun, C., Mizutani, E.: Neuro-fuzzy and Soft Computing: A Computational Approach to Learning and Machine Intelligence. Prentice-Hall, Upper Saddle River, New Jersey (1997)
9. Serre, D.: Matrices: Theory and Applications. Springer, New York (2010). https://doi.org/10.1007/978-1-4419-7683-3
10. scikit-learn 1.2.1 (2022). https://scikit-learn.org/stable/
11. fuzzytree 0.1.4 (2021). https://balins.github.io/fuzzytree/. Accessed February 2022

Equivalence Between 1-D Takagi-Sugeno Fuzzy Systems with Triangular Membership Functions and Neural Networks with ReLU Activation

Barnabas Bede[1], Vladik Kreinovich[2](✉), and Peter Toth[1]

[1] DigiPen Institute of Technology, 9931 Wilows Rd. NE, Redmond, WA 98012, USA
{bbede,petert}@digipen.edu

[2] Department of Computer Science, University of Texas at El Paso, 500 W University Ave, El Paso, TX 79968, USA
vladik@utep.edu

Abstract. In the present paper we establish the equivalence between a neural network with one input and one hidden layer with ReLU activation and a Takagi-Sugeno fuzzy system with triangular membership functions. The results verify, both from theoretical and practical point of view, that we can convert a neural network into a TS fuzzy system and make it more explainable via this method. Possible extensions to multiple dimensions and multiple layers are also discussed.

Keywords: Fuzzy systems · Neural networks · Takagi-Sugeno fuzzy systems

1 Introduction

Since fuzzy sets were introduced in [12] as mathematical models for modeling under uncertainty, they were widely utilized in control applications [7] and were used in conjunction with adaptive and learning algorithms. Takagi-Sugeno fuzzy systems combine nonlinearity of the antecedents with the linearity of consequences [10], making them very successful in modeling and control applications [9].

Neural networks [2] are widely used in applications that are very similar to those on fuzzy systems. Also, Adaptive Network based Fuzzy Inference Systems (ANFIS) can combine the learning of neural networks with fuzzy systems [6].

Deep learning [3] is a hot topic in science, with current research focusing on various applications, from image generation [4] to large language models [13]. Explainable AI [11] is a paradigm that has been developed based on the need to explain the inferences of various ML models. One of the approaches to explainable AI is explainable fuzzy AI [8].

In view of the above discussion, we can see that there is a strong connection between fuzzy systems and neural networks. In fact, it is know that Radial Basis

K. Cohen et al. (Eds.): NAFIPS 2023, LNNS 751, pp. 44–56, 2023.
https://doi.org/10.1007/978-3-031-46778-3_5

Function (RBF) Networks and Fuzzy inference systems with Gaussian membership functions are equivalent [5]. Also, inter-approximation between Neural Networks with sigmoid activation and TSK fuzzy systems with difference of sigmoid membership functions has been recently investigated [1].

In the present paper we intend to further investigate this connection, and we prove that a one dimensional neural network with ReLU activation is equivalent to a TS fuzzy system with triangular membership functions.

1.1 What We Plan to Do

In many practical situations, we know that a quantity y depends on the quantity x. In practice, the values of each quantity is bounded by a certain interval $[\underline{x}, \overline{x}]$; for example:

- time duration is bounded by the lifetime of the Universe,
- speed is limited by the speed of light, etc.

So, the desired dependence can be described by a function $y = F(x)$ from the interval $[\underline{x}, \overline{x}]$ to real numbers.

Such a function can be describe both in fuzzy terms and by using a neural network. It is known that both fuzzy systems and neural networks are universal approximators, so both can represent any continuous function with any given accuracy. In particular, this means that every function described by fuzzy rules can be approximated by a neural network, and, vice versa, every function described by a neural network can be described by fuzzy rules.

In this text, we will show that for functions of one variable, there is a stronger result: that there is actually the exact equality between these two classes of functions. To be more precise, we will prove the equivalence between:

- one of the most widely used type of fuzzy systems – Takagi-Sugeno systems with triangular membership functions and constant consequences, and
- one of the most widely use type of neural networks – neural networks that use the activation function $s(x) = \max(0, x)$; this function is known as Rectified Linear Unit, ReLU for short.

Specifically, we will prove that on each interval $[\underline{x}, \overline{x}]$:

- every Takagi-Sugeno fuzzy system with triangular membership functions and constant consequences can be represented as a ReLU-based neural network (actually, as a 1-hidden-layer ReLU-based neural network), and that
- every ReLU-based neural network can be represented as a Takagi-Sugeno system with triangular membership function with constant consequences.

To formulate and prove this equivalence, let us recall the main definitions.

1.2 Fuzzy Systems: Brief Reminder

Definition 1. *By a* membership function *on an interval* $[\underline{x}, \overline{x}]$, *we mean a function from this interval to* $[0, 1]$.

Definition 2. *Let* $[\underline{x}, \overline{x}]$ *be an interval, and let* $a \le b \le c$ *be real numbers from this interval for which* $a \ne c$. *By a* triangular membership function $m_{a,b,c}(x)$, *we mean the following function from the interval* $[\underline{x}, \overline{x}]$ *to the interval* $[0, 1]$:

- $m_{a,b,c}(x) = 0$ *for* $x < a$;
- $m_{a,b,c}(a) = 0$ *if* $a < b$;
- $m_{a,b,c}(x) = \dfrac{x-a}{b-a}$ *for* $a < x < b$;
- $m_{a,b,c}(b) = 1$;
- $m_{a,b,c}(x) = \dfrac{c-x}{c-b}$ *for* $b < x < c$;
- $m_{a,b,c}(c) = 0$ *if* $b < c$; *and*
- $m_{a,b,c}(x) = 0$ *for* $x > c$.

Comments. When $a < b < c$, we have the usual triangular membership function:

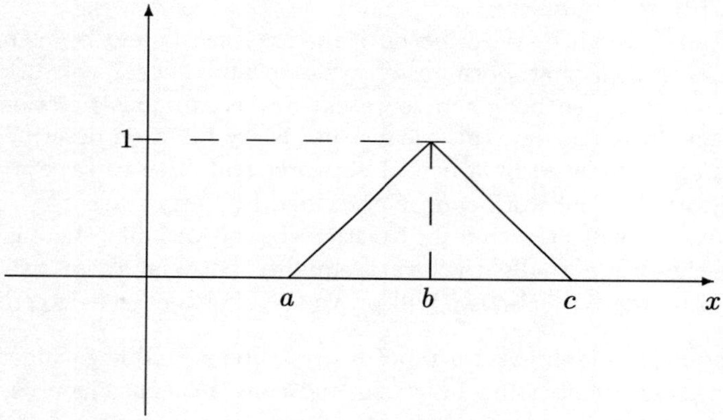

When $a = b < c$, we have the following function:

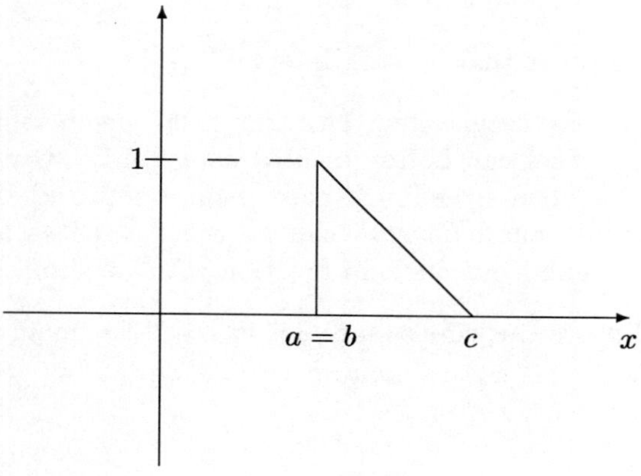

Finally, when $a < b = c$, we have the following function:

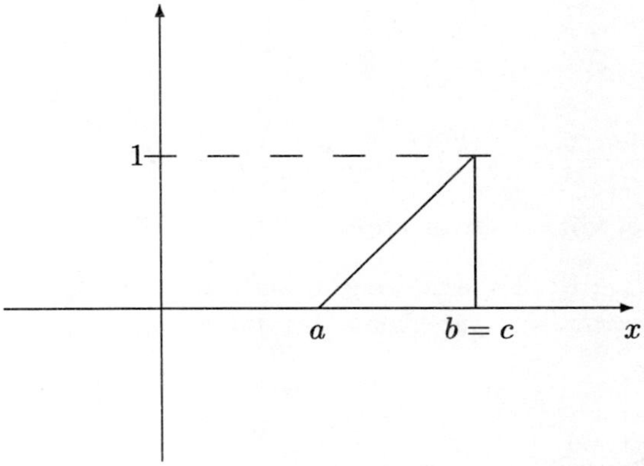

Definition 3.

- *By a* Takagi-Sugeno (TS) system with constant consequences *(or simply* TS system, *for short), we means a finite sequence of statements*

$$\text{“}if\ m_i(x)\ then\ y_i\text{”},\ i = 0, 1, \ldots, n,$$

where:

  - $m_i(x)$ *are membership functions for which, for each* $x \in [\underline{x}, \overline{x}]$, *at least one value* $m_i(x)$ *is different from 0, and*
  - y_i *are real numbers.*
- *By the* output *of the TS system, we mean the function*

$$F(x) = \frac{\sum_{i=0}^{n} m_i(x) \cdot y_i}{\sum_{i=0}^{n} m_i(x)}.$$

Definition 4. *Let* $X = (x_0, \ldots, x_n)$ *be a sequence of real numbers for which*

$$\underline{x} = x_0 < x_1 < \ldots < x_{n-1} < x_n = \overline{x},$$

and let $Y = (y_0, \ldots, y_n)$ *be another sequence of real numbers. By the* TS system with triangular membership functions, *we mean the following sequence of rules:*

- *if* $m_{x_0,x_0,x_1}(x)$ *then* $y = y_0$;
- *if* $m_{x_0,x_1,x_2}(x)$ *then* $y = y_1$;
- ...
- *if* $m_{x_{i-1},x_i,x_{i+1}}(x)$ *then* $y = y_i$;
- ...
- *if* $m_{x_{n-1},x_{n-1},x_n}(x)$ *then* $y = y_{n-1}$;
- *if* $m_{x_{n-1},x_n,x_n}(x)$ *then* $y = y_n$.

Comment. Here, the sum of these membership functions is 1 for all x, so the denominator in the formula for the outputs is equal to 1, and thus, the output is equal to

$$F(x) = m_{x_0,x_0,x_1}(x) \cdot y_0 + m_{x_0,x_1,x_2}(x) \cdot y_1 + \ldots + m_{x_{i-1},x_i,x_{i+1}}(x) \cdot y_i + \ldots +$$
$$m_{x_{n-1},x_{n-1},x_n}(x) \cdot y_{n-1} + m_{x_{n-1},x_n,x_n}(x) \cdot y_n.$$

1.3 Neural Networks: Brief Reminder

Definition 5. *By a* ReLU-based neuron *with n inputs $x_1, \ldots, x_n$, we mean a function that, given these inputs, computes the value*

$$y = s(w_1 \cdot x_1 + \ldots + w_n \cdot x_n + w_0)$$

for some values w_i, where $s(x) \stackrel{\text{def}}{=} \max(0, x)$.

Definition 6. *By the* output of a 1-hidden-layer ReLU-based neural network, *we mean a linear combination of outputs of several ReLU-based neurons, i.e., a function*

$$F(x) = \sum_{k=1}^{K} W_k \cdot s(w_{k,1} \cdot x_1 + \ldots + w_{k,n} \cdot x_n + w_{k,0}) + W_0.$$

Definition 7. *By a* ReLU-based neural networks *with n inputs and L layers, we mean a finite set of neurons divided into subsets called* layers*:*

- *neurons from the first layer process the inputs $x_1, \ldots, x_n$ and generate outputs $y_1^{(1)}, \ldots, y_{n_1}^{(1)}$;*
- *neurons from the second later process the values $y_1^{(1)}, \ldots, y_{n_1}^{(1)}$ as inputs and generate outputs $y_1^{(2)}, \ldots, y_{n_2}^{(2)}$;*
- ...
- *neurons from the i-th layer process the values $y_1^{(i-1)}, \ldots, y_{n_{i-1}}^{(i-1)}$ as inputs and generate outputs $y_1^{(i)}, \ldots, y_{n_i}^{(i)}$ given as*

$$y_j^{(i)} = s\left(\sum_{k=1}^{n_{i-1}} W_{jk}^{(i-1)} \cdot y_k^{(i-1)} + W_{j0}^{(i-1)}\right), j = 1, \ldots, n_i;$$

- ...
- *neurons from the $(L-1)$-st layer process the values $y_1^{(L-2)}, \ldots, y_{n_{L-2}}^{(L-2)}$ as inputs and generate outputs $y_1^{(L-1)}, \ldots, y_{n_{L-1}}^{(L-1)}$;*
- *finally, at the last layer, we generate a linear combination of the outputs of the previous layer:*

$$F(x) = \sum_{k=1}^{n_{L-1}} W_k^{(L-1)} \cdot y_k^{(L-1)} + W_0^{(L-1)}.$$

In the present paper we consider a regression network with a one-dimensional output, therefore the last layer has no activation function, or, equivalently, this layer has the identity function as its activation. It is easy to see that the results can easily be extended to a situation where there is an additional function applied to the last layer, such as a sigmoid or softmax.

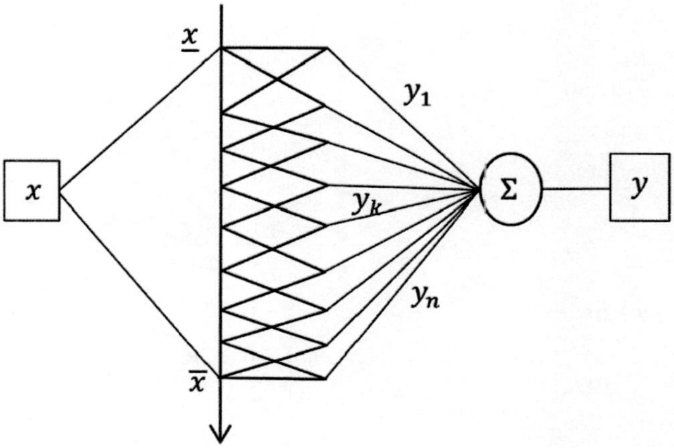

Fig. 1. The structure of a TS-system

2 Main Results

2.1 Equivalence Between TS Fuzzy Systems with Triangular Membership and Neural Networks with ReLU Activation

Theorem.

- *For every TS system with triangular membership functions, there exists a 1-hidden-layer ReLU-based neural network that produces the same output function $F(x)$.*
- *For every ReLU-based neural network, there exists a TS system with triangular membership functions that produces the same output $F(x)$.*

Proof. 1°. Let us first prove that for every TS system with triangular membership functions, there exists a 1-hidden-layer ReLU-based neural network that produces the same output function.

For this purpose, we will prove that each of the membership functions used in such a TS system can be represented as a linear combination of ReLU-based neurons. Since the output of a TS system is a linear combination of these membership functions, this implies that this output is a linear combination of ReLU-based neurons – and thus, the output of a 1-hidden-layer ReLU-based neural network.

Indeed, one can check that for i between 1 and $n-1$, we have

$$m_{x_{i-1},x_i,x_{i+1}}(x) = \frac{s(x-x_{i-1}) - s(x-x_i)}{x_i - x_{i-1}} - \frac{s(x-x_i) - s(x-x_{i+1})}{x_{i+1} - x_i}.$$

We can check this equality by considering all four possible locations of x with respect to the corresponding points x_j:

- $x \leq x_{i-1}$,
- $x_{i-1} \leq x \leq x_i$,
- $x_i \leq x \leq x_{i+1}$, and
- $x \geq x_{i+1}$.

For $i = 0$, we have

$$m_{x_0,x_0,x_1}(x) = 1 - \frac{s(x-x_0) - s(x-x_1)}{x_1 - x_0},$$

and for $i = n$, we have

$$m_{x_{n-1},x_n,x_n}(x) = \frac{s(x-x_{n-1}) - s(x-x_m)}{x_n - x_{n-1}}.$$

2°. Let us now prove that for every ReLU-based neural network, there exists a TS system with triangular membership functions that produces the same output $F(x)$.

Indeed, the activation function $s(x) = \max(0, x)$ is:

- continuous and
- piecewise linear – in the sense that the real line can be divided into finitely many intervals – finite or infinite – on each of which the function is linear.

By definition, the output of a ReLU-based neural network is a composition of activation function and linear functions. Linear functions are also continuous and piecewise linear. It is known that the composition of continuous functions is continuous, and that the composition of piecewise linear functions is also piecewise linear. Thus, the output function of a ReLU-based neural network is continuous and piece-wise linear. In other words, there exists values

$$x_1 < x_2 < \ldots < x_{n-1}$$

for which $x_0 \stackrel{\text{def}}{=} \underline{x} < x_1$ and $x_{n-1} < x_n \stackrel{\text{def}}{=} \overline{x}$ for which the output function $F(x)$ is linear – i.e., has the form $F(x) = a_i \cdot x + b_i$ for some a_i and b_i – on each interval $[x_i, x_{i+1}]$.

To represent this function as TS system, we use the same membership functions as in the first part of the proof, and take $y_i = F(x_i)$. On each of the intervals $[x_i, x_{i+1}]$, each of the above membership functions is linear, so the resulting TS-system output $T(x)$ – which is a linear combination of the membership functions

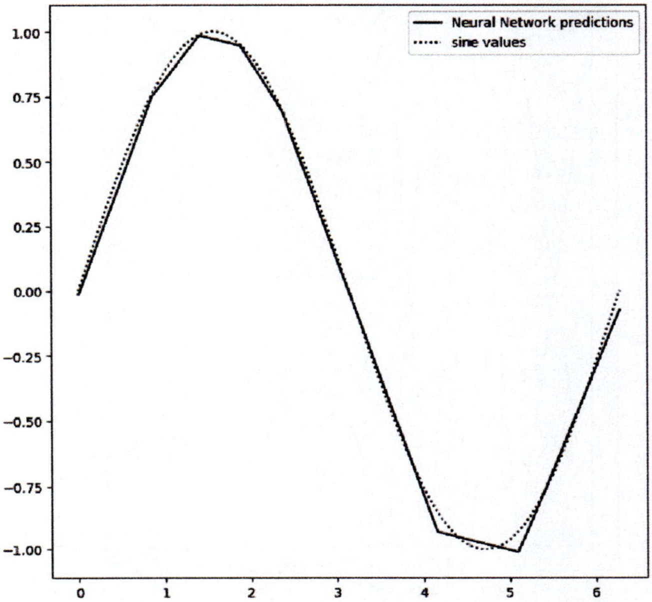

Fig. 2. Neural Network approximation of the sine function

– is also linear. For $x = x_i$, only one membership function is non-zero: the i-th one, so for the TS-system output $T(x)$, we have $T(x_i) = y_i$. Thus, $T(x_i) = F(x_i)$ for all i, and since both functions are linear on each interval $[x_i, x_{i+1}]$, these two functions coincide.

The theorem is proven.

Comment. From the proof, it follows that the transition from a TS system to a ReLU-based neural network does not increase the complexity: we start with a TS-system with $n+1$ rules, and we end up with a 1-hidden-layer neural network with $n+1$ neurons.

Vice versa, suppose that we start with a 1-hidden-layer ReLU-based neural network with $n+1$ neurons, with outputs $\max(0, a_i \cdot x + b_i)$. For each neuron, the break-point between two linear parts is at the value $x = -b_i/c_i$ at which $a_i \cdot x + b_i = 0$. So, we have $n+1$ point at which the function stops being linear. Thus, we can represent its output function by $n+1$ rules.

The structure of the TS-system constructed by the previous theorem can be illustrated as in Fig. 1. This structure receives one input but we can generalize it to a TS layer receiving the output from multiple inputs with different weights, similar to the input of a neural network (Fig. 3).

2.2 Experimental Results

We are going to demonstrate the connection between Neural Networks and the Takagi-Sugeno system through experimentation.

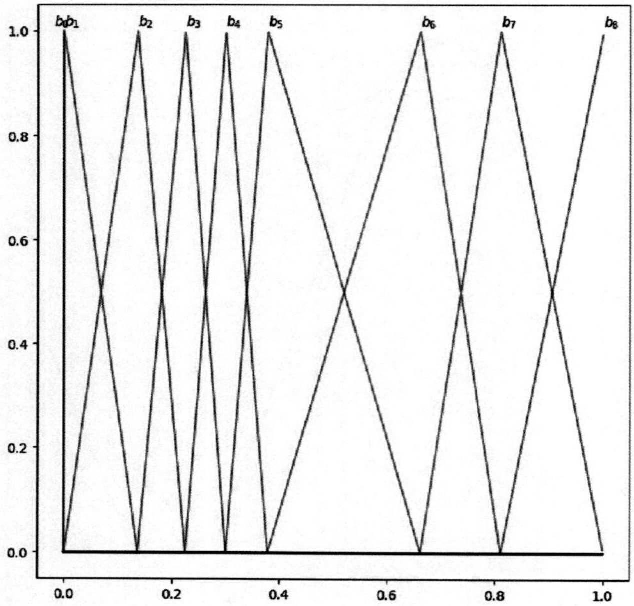

Fig. 3. TS membership functions based on b_i parameters

We will use *keras* and *tensorflow* libraries to train a neural network to approximate a specific function. We will visualize the specific function and its neural network approximation. Using the parameters of the trained network and using the results in Theorem, we will find the parameters of the TS fuzzy system and visualize the fuzzy system that we obtain.

Let us choose $f(x) = sin(x)$ as the function to approximate and limit its domain to $0 - 2\pi$. Train a neural network with one hidden layer of 9 nodes, set acceptable minimal loss and achieve high accuracy. The training will use metrics of Mean Squared Error for loss and Root Mean Squared Error for accuracy. ReLU activation for the single hidden layer and identity function for the single output. By tuning the parameters, we will find that ADAM optimizer and 0.01 learning rate can yield with very good approximation: 0.001 loss and 0.03 RMSE (Fig. 2).

Now we concentrate of the output of a TS system, specifically to the numerator of the function

$$TS(x) = \frac{\sum_{i=0}^{n} m_i(x) \cdot y_i}{\sum_{i=0}^{n} m_i(x)}.$$

since the denominator will be 1 for any given x.

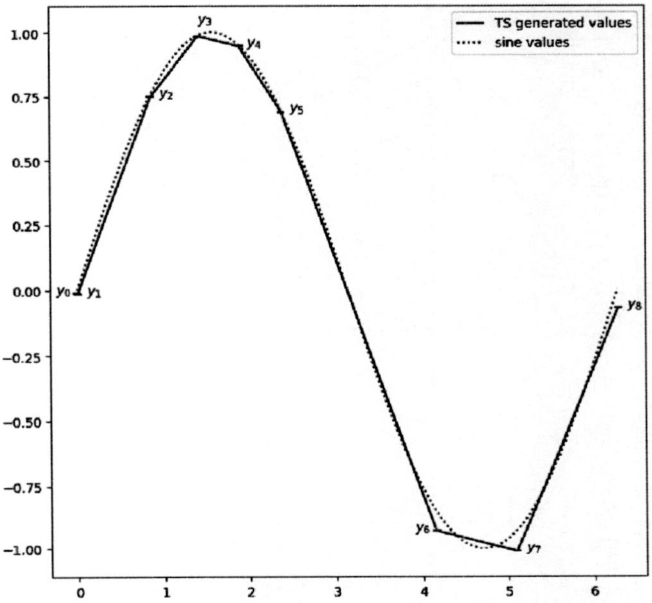

Fig. 4. Takagi-Sugeno system approximation of the sine function

For every single node, we calculate

$$b_i = -\frac{c_i}{a_i}$$

where c_i are input biases and a_i are input weights. We then sort by b_i values so we will have the base values in order to set up the m_i triangular membership functions (Fig. 3).

We also pass the b_i values to the trained NN to calculate y_i:

$$y_i = NN(b_i)$$

Now we use these parameters and the *fuzzylogic* library to calculate and visualize TS system generated values. Observe that the TS system approximates the function the very same way as the Neural Network does (Fig. 4).

This method allows us to extract a fuzzy rule base from a neural network built using existing efficient libraries such as keras and tensorflow, allowing both efficient learning algorithms to be used in conjunction with fuzzy systems, and improves explainability of neural networks.

2.3 Possible Extensions to Multiple Inputs and Multiple Layers

The previous theorem shows that each layer of a neural network can be written as a TS fuzzy system. The results of the previous section are assuming a single

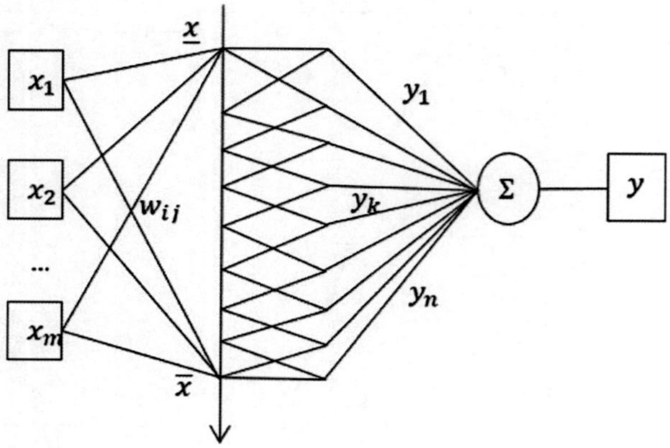

Fig. 5. The structure of a TS-system with multiple inputs

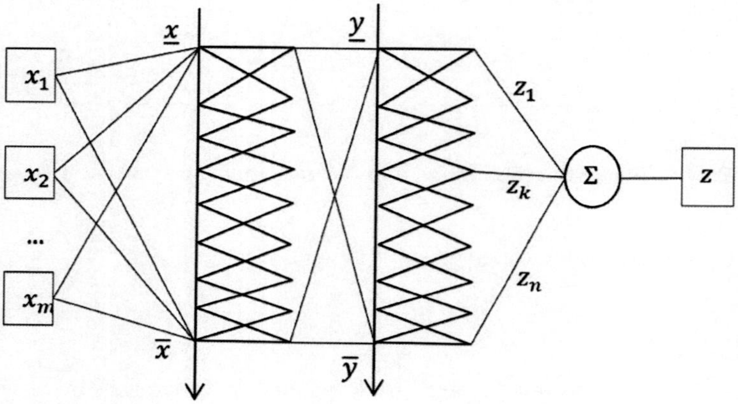

Fig. 6. The structure of a TS-system with multiple inputs

input. To generalize this structure to the case of multiple inputs we can use the same general idea as in neural networks, where the input layer is connected to the hidden layer through a weight matrix $W = [w_{ij}]_{i=1,...,n,j=1,...,m}$. In Fig. 5 we illustrate this structure. The equation for such a system is

$$F(x) = \sum_{i=1}^{n} y_i \cdot m_i \left(\sum_{j=1}^{m} w_{ij} x_j \right).$$

This fuzzy system can be described by using the rule base

$$\text{if } \sum\nolimits_{j=1}^{m} w_{ij} x_j \text{ is } m_i \text{ then output } = y_i, i = 1, ..., n.$$

We can rewrite the rules in a vector form by using the matrix W and organizing the inputs into a vector $\mathbf{x}$. Then the input of the TS layer can be written as $W\mathbf{x}$

as a matrix multiplication. The fuzzy rules can be rewritten as

$$\text{if } (W\mathbf{x})_\mathrm{i} \text{ is } m_i \text{ then } o = y_i,\ i = 1, ..., n.$$

where $(W\mathbf{x})_\mathrm{i}$ represents the i-th component of the vector $W\mathbf{x}$ and o represents the output of the network.

If we have a network with multiple layers, then it can be transformed into a series of TS fuzzy system connected as layers, similar to a neural network. In Fig. 6 we illustrate the structure of such a network. The output of the first TS layer serves as input of the second TS layer.

For an interpretation of the system in Fig. 6 we can consider it as a fuzzy syllogism. The output of the first layer is input of the second layer. This structure can be interpreted as rules of the form

$$\text{if } (W\mathbf{x})_i \text{ is } m_i^{(1)} \text{ then if } (o^{(1)})_j, \text{ is } m_j^{(2)} \text{ then } o^{(2)} = z_j,\ i = 1, ..., n,\ j = 1, ..., m,$$

with $(o^{(1)})_j$ denoting the component j of, the output of the first layer, and $o^{(2)}$ being the output of the second TS layer.

The mathematical expression for such a network is

$$o^{(2)} = \sum_{j=1}^{n} m_j^{(2)}(o^1)_j \cdot z_j$$

where

$$o^{(1)} = \sum_{i=1}^{n} m_j^{(1)}(W\mathbf{x})_i \cdot y_i$$

Of course a better understanding of the TS network structure and the fuzzy rule base with syllogisms will be subject of further research.

3 Conclusions and Further Research

We have shown that a one-dimensional neural network with ReLU activation is equivalent to a TS fuzzy system with triangular membership functions. The proposed approach improves interpretability of neural networks and also, allows efficient optimization algorithms to be used in the context of fuzzy systems. We have also discussed possible extensions to the multi-dimensional case. For further research we plan to investigate the proposed multi-layer and multi-dimensional architectures, both from theoretical and practical points of view. Also, we plan to investigate different architectures such as extensions to TSK fuzzy systems with linear consequences. Another area for future research is extensions to the multi-input, multi-output case, and combinations between neural and fuzzy architectures.

References

1. Bede, B.: Fuzzy systems with sigmoid-based membership functions as interpretable neural networks. In: Kearfott, R.B., Batyrshin, I., Reformat, M., Ceberio, M., Kreinovich, V. (eds.) IFSA/NAFIPS 2019. AISC, vol. 1000, pp. 157–166. Springer, Cham (2019). https://doi.org/10.1007/978-3-030-21920-8_15
2. Bishop, C.M.: Neural networks and their applications. Rev. Sci. Instrum. **65**(6), 1803–1832 (1994)
3. Goodfellow, I., Bengio, Y., Courville, A.: Deep Learning. MIT Press, Cambridge (2016)
4. Goodfellow, I., et al.: Generative adversarial networks. Commun. ACM **63**(11), 139–144 (2020)
5. Hunt, K.J., Haas, R., Murray-Smith, R.: Extending the functional equivalence of radial basis function networks and fuzzy inference systems. IEEE Trans. Neural Netw. **7**(3), 776–781 (1996)
6. Jang, J.S.: ANFIS: adaptive-network-based fuzzy inference system. IEEE Trans. Syst. Man Cybern. **23**(3), 665–685 (1993)
7. Mamdani, E.H., Assilian, S.: An experiment in linguistic synthesis with a fuzzy logic controller. Int. J. Man Mach. Stud. **7**(1), 1–13 (1975)
8. Mencar, C., Alonso, J.M.: Paving the way to explainable artificial intelligence with fuzzy modeling. In: Fullér, R., Giove, S., Masulli, F. (eds.) WILF 2018. LNCS (LNAI), vol. 11291, pp. 215–227. Springer, Cham (2019). https://doi.org/10.1007/978-3-030-12544-8_17
9. Sugeno, M., Kang, G.: Structure identification of fuzzy model. Fuzzy Sets Syst. **28**(1), 15–33 (1988)
10. Takagi, T., Sugeno, M.: Fuzzy identification of systems and its applications to modeling and control. IEEE Trans. Syst. Man Cybern. **8**(1), 116–132 (1985)
11. Xu, F., Uszkoreit, H., Du, Y., Fan, W., Zhao, D., Zhu, J.: Explainable AI: a brief survey on history, research areas, approaches and challenges. In: Tang, J., Kan, M.-Y., Zhao, D., Li, S., Zan, H. (eds.) NLPCC 2019. LNCS (LNAI), vol. 11839, pp. 563–574. Springer, Cham (2019). https://doi.org/10.1007/978-3-030-32236-6_51
12. Zadeh, L.A.: Fuzzy sets. Inf. Control **8**(3), 338–353 (1965)
13. Zhang, S., et al.: OPT: open pre-trained transformer language models. arXiv preprint arXiv:2205.01068 (2022)

Fuzzy Logic-Aided Inverse Kinematics Control for Redundant Manipulators

Anirudh Chhabra(✉), Sathya Karthikeyan, Daegyun Choi, and Donghoon Kim

University of Cincinnati, Cincinnati, OH, USA
{chhabrad,karthist,choidg}@mail.uc.edu, Donghoon.Kim@uc.edu

Abstract. Redundant robotic systems offer flexibility during control and offer the advantage of achieving multiple tasks such as trajectory tracking, collision avoidance, joint limit avoidance, singularity avoidance, etc. However, it is difficult to achieve all these tasks without active prioritization as the environmental conditions can change and certain behaviors may be undesirable. This work proposes a fuzzy logic-aided inverse kinematics control technique that aims to actively prioritize these secondary tasks to ensure that minimal control effort is required for achieving all the tasks while operating the manipulator. The proposed control design leverages the advantages of fuzzy inference systems in order to properly assign weights to the secondary accelerations obtained using the null-space optimization technique. Through simulations, it is shown that the proposed controller is able to achieve all the secondary tasks while maintaining the primary tracking control.

1 Introduction

Redundant robotic manipulators are controlled by designing a controller based on a unique solution to the Inverse Kinematics (IK) problem. A very popular, generic description of this problem is provided by Siciliano [1]. While operating robotic platforms, usually the desired task is not just to track a point or trajectory but also to ensure that it avoids collisions and stays within its mechanical limits or otherwise risks damage to the environment or the platform itself. The addition of sub-tasks while obtaining the IK is known as the Closed-Loop IK (CLIK) control technique. Here, several of the sub-tasks such as collision avoidance or singularity avoidance are modeled as objective functions and then an appropriate correction acceleration is obtained within the null space of the robot's Jacobian to obtain a set of joint accelerations and velocities that allow the end-effector to continue with its primary task of tracking the given point. Existing literature has focused on this problem for years. Conventional methods such as Sliding Mode Control (SMC) and Model Predictive Control (MPC) have been applied to this problem but do not offer flexibility in modeling multiple sub-tasks. Nicolis et al. present a combination of SMC and MPC methods to control a redundant robot and experimentally validate it on a 7-Degrees-Of-Freedom (DOF) prototype ABB YuMi robot arm [2]. Di Vito et al. present a

K. Cohen et al. (Eds.): NAFIPS 2023, LNNS 751, pp. 57–68, 2023.
https://doi.org/10.1007/978-3-031-46778-3_6

review of the various damped least square approaches to the control of redundant manipulators in order to ensure singularity avoidance while maintaining low tracking errors [3]. However, this approach doesn't guarantee a non-singular pose while ensuring low errors. Optimization techniques such as the Fruit-Fly technique [4], Newton's method [5], and multi-start algorithm [6] have been proposed and work well in achieving the required objectives but all optimization approaches suffer from the problem of computational time and are not desirable for real-time application. Model-free techniques such as Neural Networks (NNs) are very popular as they do not require robot dynamic modeling and can work without it thereby simplifying the design of the controller [7]. Some of the work that considers NNs include the use of a radial basis function NN [8] and recurrent NNs [9,10]. However, these methods are not scalable to a new problem and need to be retrained for changes within the dynamic model. Furthermore, the black-box nature of NNs doesn't provide any meaningful information regarding the operation of the controller, thereby reducing its trustworthiness in sensitive applications such as space robotics. Another paper considers applying a single-neuron PID model to the control law instead of a static PID model [11]. Fuzzy-based approaches have been studied in application to the problem of IK solvers, however, only focus on gain adaptation for PID and SMC control and estimating uncertainty adaptively. None of the above literature considers the active task prioritization of multiple secondary tasks in the null space of the robot's Jacobian. Chiacchio et al. provide a general framework of multiple-task augmentation and prioritization for the CLIK control scheme [12]. Furthermore, he highlights the multiple ways to approach redundancy resolution through various case studies. However, this study doesn't focus on task prioritization but on adding new tasks and the difference in modeling these tasks. Fiore et al. propose the extended Saturation in the Null Space (eSNS) optimization approach that is able to successfully tackle the generalized IK control problem for redundancy resolution by handling multiple tasks [13]. However, the approach suffers from computational efficiency and therefore offers the FastSNS and OptSNS methods that focus on speed and optimality separately. As discussed in the above literature, there is a lack of a holistic approach that is able to adaptively tackle the redundancy resolution problem while ensuring optimality and computational efficiency at the same time.

As mentioned above, it is difficult to obtain the inverse kinematic solution of a redundant robotic platform analytically and numerical methods can produce undesirable behaviors due to a poor choice of the inverse kinematic solution. CLIK control offers the distinct advantage of providing a unique solution using null-space optimization to achieve multiple tasks such as trajectory tracking in the task space, joint-limit avoidance, collision avoidance, singularity avoidance, etc. However, as mentioned in the literature, these multiple tasks must be appropriately prioritized using weights to obtain safe, smooth, and stable control. For instance, the controller must be able to give higher priority to collision avoidance compared to trajectory tracking when close to an obstacle in the environment. Therefore, this paper proposes a Fuzzy Logic-aided IK (FLIK) controller that adaptively prioritizes multiple tasks, thereby providing an inverse kinematic solution that fully utilizes the capabilities of the redundant DOF while ensuring the safe and stable operation of the robotic platform. Since Fuzzy Inference Systems (FISs) generally require expert knowledge, this research considers the use of a Genetic Algorithm (GA) to optimize the knowledge base of the designed FISs. Since the optimization process is offline, the FLIK control technique offers scalability and interpretability of the control method while ensuring computational efficiency, thereby leveraging the advantages of Genetic Fuzzy System (GFS)-based interpretable artificial intelligence. Therefore, the following paper focuses on the design of the FLIK control technique applied to a simple 3-DOF planar manipulator which aims to track a specified 2-Dimensional (2D) reference trajectory. The paper primarily focuses on three control tasks applied to the manipulator: trajectory tracking, collision avoidance, and joint limit avoidance. The controller is validated in simulations and compared with the traditional CLIK control method to validate the improvement in performance. The paper is structured as follows: Sect. 2 describes the design of the FLIK controller; Sect. 3 presents the results of the simulation study conducted with a brief description of the manipulator considered and the corresponding simulation environment; and Sect. 4 concludes the work based on the obtained results and highlights the future direction for this study.

2 Fuzzy Logic-Aided Inverse Kinematics (FLIK) Control

The control of redundant manipulator platforms is based on the solution of IK. This is known as redundancy resolution. To obtain a unique solution, certain sub-tasks (other than trajectory tracking), deemed important for the operation of the platform, are selected and corresponding secondary accelerations are obtained within the null space of the platform's Jacobian. This allows the platform to assume a different internal configuration while maintaining the end-effector position. The first step is to define the IK relationship as follows:

$$\dot{\boldsymbol{x}}_e = J(\boldsymbol{q})\dot{\boldsymbol{q}}, \tag{1}$$

where $\boldsymbol{x}_e$ denotes the end-effector position, $\boldsymbol{q}$ is the vector of generalized coordinates that define the robotic system, and J_e is the corresponding Jacobian matrix of the platform. On differentiating Eq. (1), one can obtain the relationship between the end-effector acceleration and the generalized coordinates as

$$\ddot{\boldsymbol{x}}_e = \dot{J}\dot{\boldsymbol{q}} + J\ddot{\boldsymbol{q}}, \tag{2}$$

$$\ddot{\boldsymbol{q}} = J^{\dagger}\left(\ddot{\boldsymbol{x}}_e - \dot{J}\dot{\boldsymbol{q}}\right), \tag{3}$$

where $J^{\dagger}$ denotes the Moore-Penrose pseudo-inverse of the Jacobian and is represented as $J^{\dagger}(\boldsymbol{q}) = J^T\left(JJ^T\right)^{-1}$. Note that the manipulator dynamics are represented in their general form using the generalized coordinates (joints) $\boldsymbol{q}$ as follows:

$$M(\boldsymbol{q})\ddot{\boldsymbol{q}} + C(\boldsymbol{q}, \dot{\boldsymbol{q}})\dot{\boldsymbol{q}} + G(\boldsymbol{q}) = Q, \tag{4}$$

where M denotes the inertia matrix of the platform, C matrix represents the nonlinear Centrifugal and Coriolis interactions, G is the gravity vector, and Q denotes the vector of applied torques on each joint. Based on the above information, a simple PD tracking control law can be written as follows:

$$\ddot{\boldsymbol{q}}_d = J^{\dagger}\left[\ddot{\boldsymbol{x}}_d + K_D(\dot{\boldsymbol{x}}_d - \dot{\boldsymbol{x}}_e) + K_P(\boldsymbol{x}_d - \boldsymbol{x}_e) - \dot{J}\dot{\boldsymbol{q}}\right], \tag{5}$$

where $\boldsymbol{x}_d$ is the desired end-effector trajectory and $\boldsymbol{q}_d$ is the corresponding joint-space trajectory. Note that K_P and K_D are positive definite gain matrices. However, when considering additional sub-tasks, secondary accelerations in the null space of the Jacobian are added to the controller design as

$$\ddot{\boldsymbol{q}} = \ddot{\boldsymbol{q}}_P + \ddot{\boldsymbol{q}}_S, \tag{6}$$

where $\ddot{\boldsymbol{q}}_P$ is the primary task of trajectory tracking and is defined as in Eq. (5). The secondary task accelerations $\ddot{\boldsymbol{q}}_S$ are obtained through a process of null-space optimization and the corresponding control law can be expressed as [14,15]

$$\ddot{\boldsymbol{q}}_S = (I - J_e^{\dagger}J_e)(\dot{\Phi}_N + K_N\dot{\boldsymbol{e}}_N) - (J_e^{\dagger}\dot{J}_eJ_e^{\dagger} + \dot{J}^{\dagger})J_e(\Phi_N - \dot{\boldsymbol{q}}). \tag{7}$$

Note that such a definition of the control law ensures that null-space optimization generates a null-space velocity Φ_N that allows the manipulator to change its internal configuration in order to achieve the sub-tasks while maintaining the primary tracking control. Furthermore, this control law is stable and its Lyapunov stability can be proven by defining the Lyapunov candidate as $V = \frac{1}{2}||\dot{\boldsymbol{e}}_N||$. The corresponding null-space error is denoted by

$$e_N = \left(I - J_e^{\dagger}J_e\right)\left(\Phi_N - \dot{\boldsymbol{q}}\right). \tag{8}$$

The null space velocity Φ_N can be obtained as the negative gradient of a properly selected objective function (ϕ) defined to model the sub-tasks [12,14].

However, when multiple sub-tasks are considered, just adding the multiple secondary accelerations doesn't produce acceptable results as these sub-tasks need to be properly prioritized. Static prioritization is often helpful but cannot offer a scalable solution when many sub-tasks are considered. Therefore, certain weighting factors are defined to properly prioritize the secondary accelerations originating from these multiple sub-tasks and the new total acceleration for n tasks can be represented in a simple general form as follows:

$$\ddot{\boldsymbol{q}} = K_1\ddot{\boldsymbol{q}}_1 + K_2\ddot{\boldsymbol{q}}_2 + K_3\ddot{\boldsymbol{q}}_3 + \cdots + K_n\ddot{\boldsymbol{q}}_n. \tag{9}$$

Here, K_i are positive definite weighting matrices assumed to only contain non-zero diagonal elements. These diagonal elements provide individual weights to joint accelerations across each DOF.

2.1 Description of Sub-tasks

This study considers two sub-tasks - collision avoidance and joint-limit avoidance - in addition to the primary task of trajectory tracking. As mentioned above, the choice of the objective function depends on the sub-task and will be defined in this section. Note that the null-space joint velocities (Φ_k) for the kth task can be obtained as follows:

$$\Phi_k = -\frac{\partial \phi_k}{\partial \boldsymbol{q}}. \tag{10}$$

2.1.1 Joint Limit Avoidance Generally, a given robotic manipulator's joints are constrained by certain angular limits due to the physical structure of the manipulator. When designing a controller, these limits must be taken into consideration to avoid damage to the manipulator structure. Therefore, a proper choice for the objective function is to ensure that the joints stay close to the center of the angular range. A corresponding objective function (ϕ_1) can then be defined as

$$\phi_1 = \frac{1}{m}\sum_{i=1}^{m}\left[\frac{q_i - \bar{q}_i}{\max(q_i) - \min(q_i)}\right]^2, \tag{11}$$

where m denotes the DOF of the manipulator and $\bar{q}_i$ represents the center of the angular range of ith joint q_i.

2.1.2 Collision Avoidance While operating a robotic manipulator, generally the requirement to avoid collisions with obstacles in the environment is important and this requirement must be carefully defined in the controller design. For this purpose, a simple objective function that minimizes the inverse square of the distance between the jth point on the manipulator (defined as critical points) and the kth obstacle can be defined as follows

$$\phi_2 = \sum_{k=1}^{N_2}\sum_{j=1}^{N_1}\frac{1}{d_{jk}^2}, \tag{12}$$

where N_1 and N_2 are the total numbers of critical points and obstacles, respectively.

2.2 Fuzzy Inference System (FIS)

Since the elements of K_i need to be updated adaptively based on the system state information, environment, and task requirements, this work considers the use of FISs to update the values of K_i based on the available information. Note that the values of secondary accelerations are still obtained using null-space optimization as explained in Eq. (7). A generic representation of a FIS used to obtain the elements of a given weighting matrix, say corresponding to the collision avoidance sub-task, is shown in Fig. 1. As shown in the figure, the internal parameters of the FISs are optimized using a GA to form a GFS.

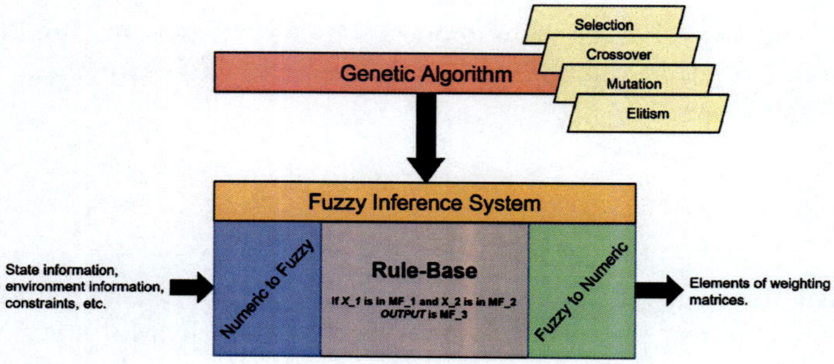

Fig. 1. Generic representation of the proposed FIS

In this study, two different FIS structures are defined for each sub-task. Each FIS is then used for all the i joints as the nature of the relationship between each joint's motion and the corresponding weights is similar. These FISs are shown in Fig. 2. Note that for obtaining a simple FIS model that can be easily optimized, all inputs and corresponding outputs are normalized between 0 and 1. Figure 3 shows the general description of each fuzzy variable's (MFs) and the parameters that need to be optimized for defining the MFs. As shown in the figure, only two parameters are optimized for each FIS denoted by $\sigma(i)$ whereas the trailing and leading parameters are fixed as per the range of input data. Since the inputs are normalized, the MF ranges are set to be between 0 and 1. Therefore, a total of $2 \times 6 = 12$ parameters need to be optimized for the proper definition of the system's MFs. Based on these MFs, the fuzzy rules can be defined as shown in Table 1. Based on this, a total of 30 parameters are selected for optimization using GA. Note that since the FISs only output a value between 0 and 1, a scaling factor equal to the weights used for the CLIK control is multiplied to the FIS output for obtaining the proper weights for null-space optimization of sub-tasks. Using this method, one can obtain smaller and better weights, thereby allowing reduction of control effort and improvement in performance. This is also good for the optimization process as the outputs of FISs are bounded within a small range and minimize the computational effort.

Once the MFs and rules of the FISs are defined, the GA is used to optimize the elements of vector σ. The GA initializes a random population of potential solutions and optimizes them using a series of operations such as crossover, mutation, selection, and elitism. However, a proper fitness function needs to be

Table 1. FIS rules

		Joint Angle		
		S	**M**	**L**
Angular Rate	**S**	$\sigma(7)$	$\sigma(10)$	$\sigma(13)$
	M	$\sigma(8)$	$\sigma(11)$	$\sigma(14)$
	L	$\sigma(9)$	$\sigma(12)$	$\sigma(15)$

		Relative position		
		S	**M**	**L**
Relative velocity	**S**	$\sigma(22)$	$\sigma(25)$	$\sigma(28)$
	M	$\sigma(23)$	$\sigma(26)$	$\sigma(29)$
	L	$\sigma(24)$	$\sigma(27)$	$\sigma(30)$

Fig. 2. Structure of the FISs

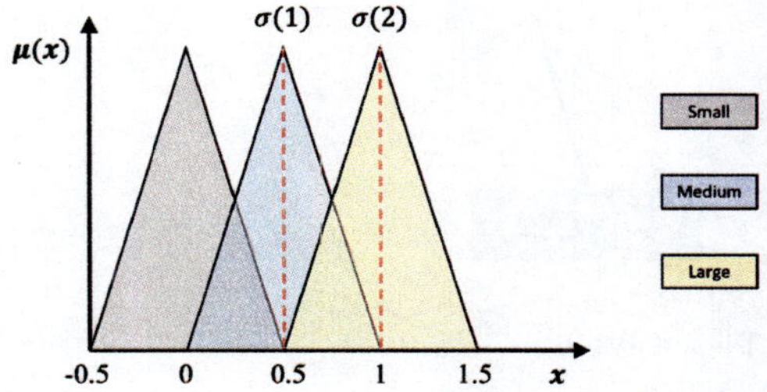

Fig. 3. Description of the FIS MFs

defined such that the GA can appropriately define the fitness of each solution. This work defines the following fitness function in order to minimize the total control effort,

$$\mathcal{J} = \int_{t=0}^{t_F} \boldsymbol{\tau}^{\mathrm{T}}(t)\boldsymbol{\tau}(t)\ \mathrm{d}t + \rho, \tag{13}$$

where $\tau(t)$ is the control effort exerted by the system at a given time step and ρ is defined as a penalty in case the controller generates a motion that violates the specified constraints, i.e., doesn't achieve all the desired tasks. The penalty is usually defined as a large number ($\rho \gg \int \boldsymbol{\tau}^{\mathrm{T}}(t)\boldsymbol{\tau}(t)\ \mathrm{d}t$) such that it is much bigger than the non-penalty fitness value and therefore reduces the fitness of the undesirable solutions enough to avoid them.

3 Simulation Study

3.1 Description

This work considers a simple 3-DOF planar manipulator that aims to follow a 2D trajectory in the presence of a point obstacle. Figure 4 shows the structure of the manipulator considered where each joint angle is defined in the local frame. Each link of the manipulator is assumed to be a thin rod of length 1 m and a point mass located at the end of the rod (1 kg each). The limits applicable on the manipulator's joints are specified as

$$0 < \theta_1 < \pi, \quad -\pi < \theta_2 < \pi, \quad -\pi < \theta_3 < \pi. \tag{14}$$

Two trajectories are defined for the manipulator system. One is used for training and the other for testing the scalability of the proposed FLIK control method.

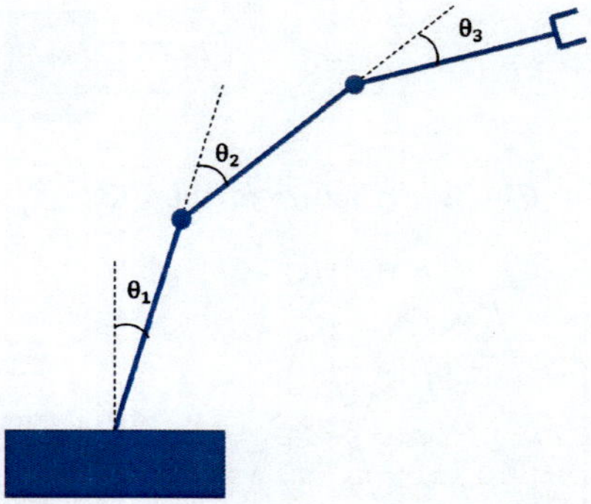

Fig. 4. 3-DOF planar manipulator robot considered for simulation study

In both cases, the results are compared with the CLIK controller. The general reference trajectory is defined as follows:

$$x_{\mathrm{d}}(t) = c_{\mathrm{x}} + a\sin(bt), \tag{15}$$
$$y_{\mathrm{d}}(t) = c_{\mathrm{y}} + a\cos(bt), \tag{16}$$

where $(c_{\mathrm{x}}, c_{\mathrm{y}})$ is used to represent the offset of the trajectory from the origin. Therefore, the resulting trajectory is a circle of radius a located at $(c_{\mathrm{x}}, c_{\mathrm{y}})$. Furthermore, as shown in the figure, the base of the manipulator is located at the origin $(0, 0)$. The initial conditions for both the training and testing scenarios are shown in Table 2. Note that for training, the GA is run for 1000 generations with a crossover rate of 80% and a mutation rate of 30%. Also, in each generation, 10% of the fittest solutions are considered elite and preserved.

Table 2. Initial conditions for the simulation study

	Description	Parameter	Value	Unit
Training	Manipulator Pose	$(\theta_1, \theta_2, \theta_3)$	$(\pi/2, -\pi/2, \pi/2)$	rad
	Obstacle Location	$(o_{\mathrm{x}}, o_{\mathrm{y}})$	(1.26, 0.59)	m
	Reference trajectory center	$(c_{\mathrm{x}}, c_{\mathrm{y}})$	(1.00, 1.50)	m
Testing	Manipulator Pose	$(\theta_1, \theta_2, \theta_3)$	$(\pi/2, 0.5, 0)$	rad
	Obstacle Location	$(o_{\mathrm{x}}, o_{\mathrm{y}})$	(–0.25, 0.8)	m
	Reference trajectory center	$(c_{\mathrm{x}}, c_{\mathrm{y}})$	(–1.00, 1.50)	m

3.2 Results

Once the training and testing scenarios are properly defined, one can obtain the results from the simulation study. In the training scenario selected, one can see the comparison of the manipulator trajectories generated by both the controllers in Fig. 5. It can be seen that the CLIK controller is able to manage

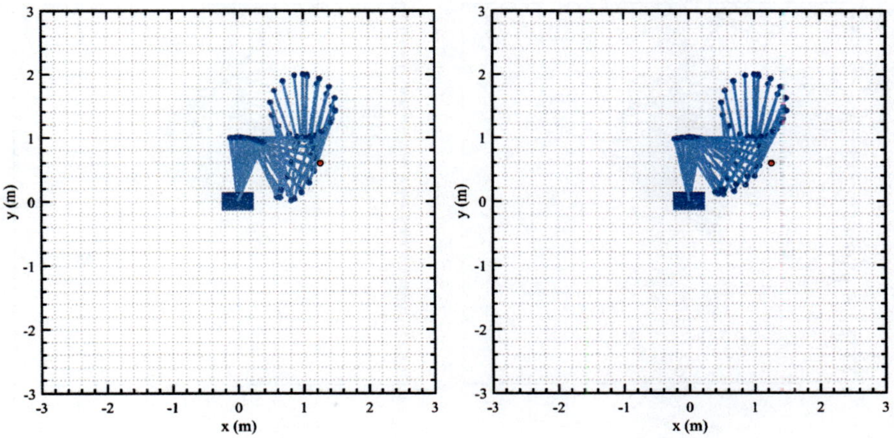

Fig. 5. Training scenario: CLIK (left) and FLIK (right) trajectories

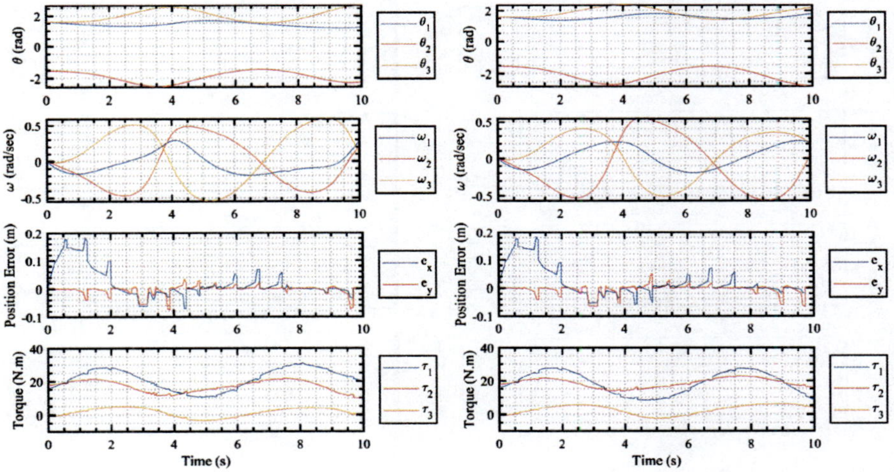

Fig. 6. Training scenario: CLIK (left) and FLIK (right). This figure compares the state histories and the corresponding errors and torques

multiple tasks after careful tuning of control gains and weights for secondary tasks in this environment. As shown in the figure, the FLIK controller is also able to generate adaptive motion in accordance with the applied constraints. Even though the Root-Mean-Square Error (RMSE) for both cases is similar, i.e., 0.0585 m for FLIK vs 0.0533 m for CLIK, the total control effort in the case of CLIK ($\mathcal{J} = 7986.39$) is higher compared to FLIK ($\mathcal{J} = 7570.08$). This difference in the total control effort can also be seen in the torque plots in Fig. 6 as the peaks for CLIK control torque are much higher than the torques generated by FLIK control.

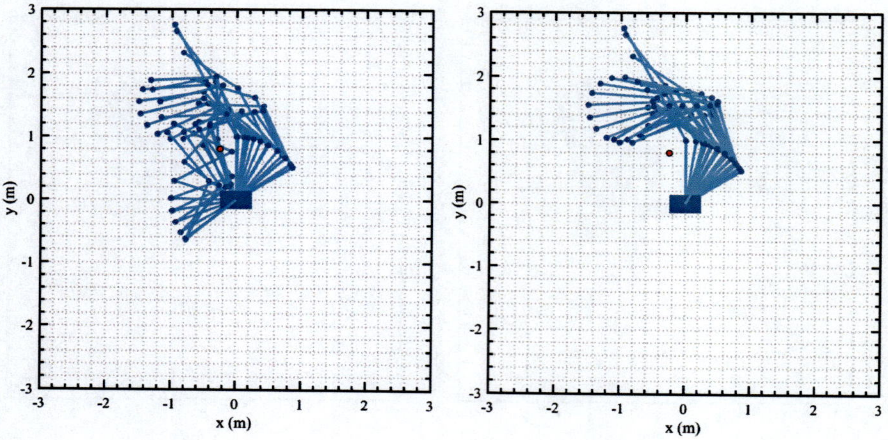

Fig. 7. Testing scenario: CLIK (left) and FLIK (right) trajectories.

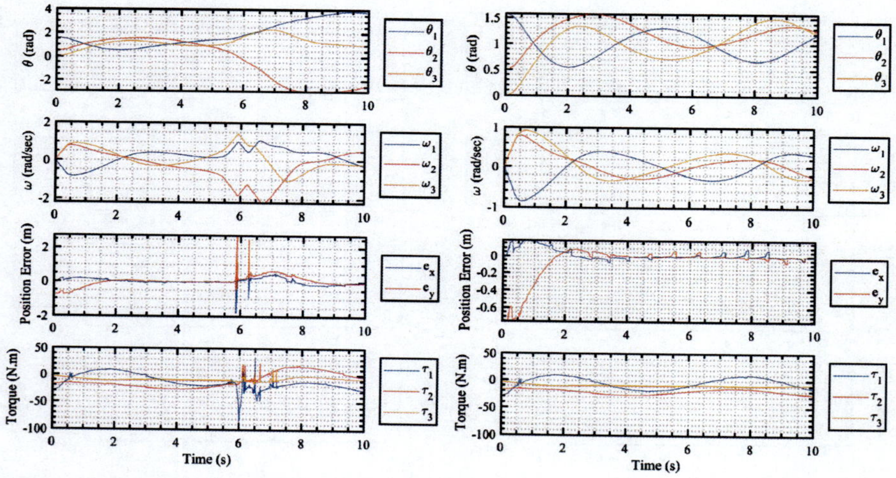

Fig. 8. Testing scenario: CLIK (left) and FLIK (right). This figure compares the state histories and the corresponding errors and torques.

Once the environment is changed, the FLIK controller clearly highlights the advantages of adaptability and scalability as it is able to adapt much better than the CLIK controller. As shown in Fig. 7, the FLIK controller is able to satisfy all constraints, whereas the CLIK controller isn't able to satisfy the collision avoidance constraint. The difference in performance is further evident from the total control effort comparison (FLIK: $\mathcal{J}$ = 6206.98, CLIK: $\mathcal{J}$ = 7052.43). Furthermore, in the testing scenario, the tracking RMSE of the FLIK controller (0.3915 m) is even lesser than the RMSE of CLIK controller (0.5272 m) thereby validating the performance improvement (Fig. 8).

It can be clearly seen from the results that the proposed FLIK control technique offers several advantages over the CLIK controller, such as scalability and active prioritization of manipulator tasks. These are highly necessary when aiming to achieve multiple tasks during manipulator operation. Overall, this work successfully highlights the several advantages of FISs in robotic system applications and how they can be used to enhance conventional techniques.

4 Conclusion

This work proposes a novel control algorithm for redundant robotic systems. Redundant manipulators are often used for several applications for their advantages such as flexibility in completing a task or achieving multiple tasks. These tasks often include trajectory tracking, collision avoidance, avoidance of joint limits, etc. However, the solution to the inverse kinematics problem is difficult and when multiple tasks are desired, it is difficult to prioritize them. In this work, a fuzzy logic-aided inverse kinematic control technique is proposed that aims to overcome these issues and adaptively satisfy all tasks and offer scalability in the controller design. Furthermore, the controller parameters are optimized by a genetic algorithm in order to obtain high accuracy and performance. The performance of the proposed controller is then validated through simulations of a 3-Degrees-of-Freedom planar manipulator in different environments. From the results, it can be clearly seen that the proposed controller offers an improvement in performance over the closed-loop inverse kinematics control technique. In the future, this work will be extended to generalize the proposed controller and improve the controller design.

References

1. Siciliano, B.: J. Intell. Robot. Syst. **3**(3), 201–212 (1990). https://doi.org/10.1007/bf00126069
2. Nicolis, D., Allevi, F., Rocco, P.: IEEE Trans. Robot. **36**(4), 1348 (2020). https://doi.org/10.1109/TRO.2020.2974092
3. Di Vito, D., Natale, C., Antonelli, G.: IFAC-PapersOnLine **50**(1), 6869 (2017). https://doi.org/10.1016/j.ifacol.2017.08.1209, https://www.sciencedirect.com/science/article/pii/S2405896317317159. 20th IFAC World Congress
4. Shi, J., et al.: Math. Probl. Eng. **2020**, 6315675 (2020). https://doi.org/10.1155/2020/6315675
5. Safeea, M., Béarée, R., Neto, P.: Collision avoidance of redundant robotic manipulators using Newton's method. J. Intell. Robot. Syst. 673–681 (2020). https://doi.org/10.1007/s10846-020-01159-3
6. Tringali, A., Cocuzza, S.: Robotics **9**(3), 61 (2020). https://doi.org/10.3390/robotics9030061
7. Liu, Z., Peng, K., Han, L., Guan, S.: Iran. J. Sci. Technol. Trans. Mech. Eng. (2023). https://doi.org/10.1007/s40997-023-00596-3
8. Rani, M., Ruchika, N.K.: Procedia Comput. Sci. **125**, 50 (2018). https://doi.org/10.1016/j.procs.2017.12.009, https://www.sciencedirect.com/science/article/pii/S1877050917327734. The 6th International Conference on Smart Computing and Communications

9. Xu, Z., Li, S., Zhou, X., Yan, W., Cheng, T., Huang, D.: Neurocomputing **329**, 255 (2019). https://doi.org/10.1016/j.neucom.2018.11.001, https://www.sciencedirect.com/science/article/pii/S0925231218313122
10. Zhang, Z., Zheng, L., Chen, Z., Kong, L., Karimi, H.R.: IEEE Trans. Neural Netw. Learn. Syst. **32**(3), 1052 (2021). https://doi.org/10.1109/TNNLS.2020.2980038
11. Zhang, H., Jin, H., Liu, Z., Liu, Y., Zhu, Y., Zhao, J.: IEEE Trans. Ind. Inf. **16**(1), 28 (2020). https://doi.org/10.1109/TII.2019.2917392
12. Chiacchio, P., Chiaverini, S., Sciavicco, L., Siciliano, B.: Int. J. Robot. Res. **10**(4), 410 (1991). https://doi.org/10.1177/027836499101000409
13. Fiore, M.D., Meli, G., Ziese, A., Siciliano, B., Natale, C.: IEEE Trans. Robot. 1–20 (2023). https://doi.org/10.1109/TRO.2022.3232266
14. Hsu, P., Hauser, J., Sastry, S.: In: 1988 American Control Conference, pp. 2135–2139 (1988). https://doi.org/10.23919/ACC.1988.4790077
15. Chhabra, A., Kim, D.: J. Aerosp. Inf. Syst. **19**(7), 480 (2022). https://doi.org/10.2514/1.I011008

Interval Sequence: Choosing a Sequence of the Investment

Marina T. Mizukoshi[1(✉)], Tiago M. da Costa[2], Yurilev Chalco-Cano[2], and Weldon A. Lodwick[3]

[1] Universidade Federal de Goiás, IME, Goiânia, GO, Brazil
tuyako@ufg.br

[2] Universidad Tarapacá, Arica, Chile
ychalco@academicos.uta.cl

[3] University of Colorado, Larimer Street, Denver, USA
Weldon.Lodwick@ucdenver.edu

Abstract. An approach for an investment portfolio is obtained using the constraint interval sequence. We consider interval rates in the formulation of our problem. Interval rates can arise since it will depend on the market. Moreover, we are looking at decisions under the collection of various interest rates the various scenarios and market values. In particular, intervals rate can be used to model the range from pessimistic and to optimistic scenarios. Once an interval solution is obtained, the investor is able to determine the range of possible returns given the range of rates.

1 Introduction

The interval theory can be used to study the generalized uncertainty problems. In general, problems about financial investment decisions present uncertainties about the future since it is not possible to predict the outcome of financial market, such as interest rate variation risk, credit risk, market risk, operational risk and liquidity risk. Thus, it is interesting consider such uncertainties as being variables in a decision making process.

According to the Markowitz theory [8], if there exists a diversification where the money is applied, then the risks are lower. In order to propose an investment portfolio, one needs to consider the type of investor profile [1]: (a) **traditional investor**: is one who wants to have no or very low risk in his investments. (b) **moderate investor**: means the people are willing accepts to take a certain amount of risk in his investments, seeking a little more profitability, but they prefer not to take risks in very daring applications; (c) **agressive investor**: is the one that accepts to take more risks in order to have more profitability. This profile is even willing to lose part of the amount invested.

In this study, the interval theory is applied in choosing of a portfolio of investment. Moreover, we are looking at decisions under the collection of various interest rates the various scenarios and market values. In particular, intervals rate can be used to model the range from pessimistic and to optimistic scenarios.

K. Cohen et al. (Eds.): NAFIPS 2023, LNNS 751, pp. 69–80, 2023.
https://doi.org/10.1007/978-3-031-46778-3_7

Once an interval solution is obtained, the investor is able to determine the range of possible returns given the range of rates. Firstly, some basic notion presented in [10], such as constraint interval representation (CI), interval independency and total dependency according to CI, and constraint interval extension representation, are recalled in Sect. 2. In Sect. 3, the concept of constraint interval distance is introduced. Then the notion of constraint interval sequences is presented in Sect. 4. Section 5 presents an application of these mathematical tools in an investor portfolio problem without considering details about interval expected valued and interval covariance (see [6]), which are intrinsic in decision making theory. Lastly, final considerations are made.

2 Preliminaries

First we consider some definitions [10] that are important to understand the concept for constrained interval (CI) function. According Warmus, Sunaga and Moore

Definition 1. The **Standard Interval Representation** (SI) of an interval $[a, b]$, $(a \leq b)$, is the ordered pair of its endpoints (a, b).

A **Constraint Interval representation (CI)** is a representation of an interval $[x] = [\underline{x}, \overline{x}]$, in the representation space $\mathcal{F}$ via space mapping/function $\mathcal{R} : \mathbb{IR} \to \mathcal{F}$ given by

$$\mathcal{R}([x]) = \mathcal{R}_{[x]} : [0, 1] \to \mathbb{R}$$

such that

$$\mathcal{R}_{[x]}(\lambda_x) = \underline{x} + \lambda_x (\overline{x} - \underline{x}) = \underline{x} + \lambda_x w_x$$

for all $\lambda_x \in [0, 1]$, where $w_x = \overline{x} - \underline{x}$ and the representation space $\mathcal{F}$ is the space of bounded real-valued functions $f : [0, 1] \to \mathbb{R}$ defined on the squares (or, in general case, hypercubes) $[0, 1]$. Denote by $\mathbb{IR}$ the space of all intervals.

Definition 2. A CI of a $m-$tuple of intervals $[x] = ([x_1], [x_2], \cdots, [x_m]) \in \mathbb{IU}^m = \mathbb{IU} \times \ldots \times \mathbb{IU} \subseteq \mathbb{IR}^m = \mathbb{IR} \times \ldots \times \mathbb{IR}$, with $[x_i]$ totally independent of $[x_j]$ for all $i \neq j$, it is embedded into $\mathcal{F}^m = \mathcal{F} \times \ldots \times \mathcal{F}$ through the map

$$\mathcal{R}([x]) = \mathcal{R}_{[x]} : [0, 1]^m \to \mathbb{R}^m$$

where

$$\mathcal{R}_{[x]}(\lambda_x) := (\mathcal{R}_{[x_1]}(\lambda_{x_1}), \mathcal{R}_{[x_2]}(\lambda_{x_2}), \cdots, \mathcal{R}_{[x_m]}(\lambda_{x_m})) = \bigwedge_{i=1}^{m} (\mathcal{R}_{[x_i]}(\lambda_{x_i}))$$

for all $\lambda_x = (\lambda_{x_1}, \lambda_{x_2}, \cdots, \lambda_{x_m}) \in [0, 1]^m$ and $w_i = \overline{x}_i - \underline{x}_i \geq 0$, with $\mathcal{R}_{[x_i]}(\lambda_{x_i}) = \underline{x}_i + \lambda_{x_i} w_i, i = 1, \ldots, m$. Moreover, a m-tuple of intervals $[x] = ([x_1], [x_2], \cdots, [x_m]) \in \mathbb{IU}^m \subset \mathbb{IR}^m$, $[x_i]$ are totally dependent for all $i \in \{i_1, \ldots, i_k\} \subset \{1, \ldots, m\}$, it is embedded into $\mathcal{F}$ through the map

$$\mathcal{R}([x]) = \mathcal{R}_{[x]} : [0, 1]^{m-(k-1)} \to \mathbb{R}^m,$$

given by

$$\mathcal{R}_{[x]}(\lambda_x) = \bigwedge_{i=1}^{m} \left(\mathcal{R}_{[x_i]}(\lambda_{x_j})\right),$$

where $\begin{cases} j = i_1, \text{ if } i \in \{i_1, \ldots, i_k\} \\ j = i, \text{ if } i \notin \{i_1, \ldots, i_k\}. \end{cases}$

Definition 3. Given $[x], [y] \in \mathbb{IR}$ and $\circ \in \{+, -. \times, \div\}$, it follows that:

(a) if $[x]$ and $[y]$ are totally independent, then the operation $\circ_{CIRA}$ between $[x]$ and $[y]$ is the function

$$[x] \circ_{CIRA} [y] = \mathcal{R}_{([x]\circ[y])} : [0,1]^2 \to \mathbb{R} \tag{1}$$

such that

$$\mathcal{R}_{([x]\circ[y])}(\lambda_x, \lambda_y) = \mathcal{R}_{[x]}(\lambda_x) \circ \mathcal{R}_{[y]}(\lambda_y) = (\underline{x} + \lambda_x w_x) \circ (\underline{y} + \lambda_y w_y)$$

for all $(\lambda_x, \lambda_y) \in [0,1]^2$.

(b) If $[x]$ and $[y]$ are totally dependent, then the operation $\circ_{CIRA}$ between $[x]$ and $[y]$ is the function

$$[x] \circ_{CIRA} [y] = \mathcal{R}_{([x]\circ[y])} : [0,1] \to \mathbb{R} \tag{2}$$

such that

$$\mathcal{R}_{([x]\circ[y])}(\lambda) = \mathcal{R}_{[x]}(\lambda) \circ \mathcal{R}_{[y]}(\lambda) = (\underline{x} + \lambda w_x) \circ (\underline{y} + \lambda w_y)$$

for all $\lambda \in [0,1]$.

The set of the arithmetic operations $\circ_{CIRA}$, where $\circ \in \{+, -, \times, \div\}$, it is called the **constraint interval representation arithmetic** (CIRA).

This mapping approach allows to define the concept of *extension* of a function $f : \mathbb{R}^n \to \mathbb{R}$, which was firstly considered by Lodwick and Jenkins (see [7]) in a different version from that presented herein. $\mathcal{F}^n, n \geqslant 2$ denotes the Cartersian product of the representation space $\mathcal{F}$.

The results provided by the CIRA are bounded real-valued functions defined on $[0,1]^n$ for some $n \in \mathbb{N}$. That is, the results provided by the CIRA are elements in $\mathcal{F}^n$. Thus, given $\phi \in \mathcal{F}^n$, it follows that Φ attains its infimum and supremum values on $[0,1]^n$. If one wants to obtain an interval result from the CIRA, it can be done by means of the mapping back operator defined as follows.

Definition 4. Given an element $\phi \in \mathcal{F}^n$, the operator **mapping back** is the map $\mathcal{MB} : \mathcal{F}^n \to \mathbb{IR}$ given by

$$\mathcal{MB}(\phi) = \left[\inf_{\lambda_x \in [0,1]^n} \phi(\lambda_x), \sup_{\lambda_x \in [0,1]^n} \phi(\lambda_x)\right] \tag{3}$$

Definition 5 (Constraint Interval Extension). Given a continuous function $f : U \subseteq \mathbb{R}^m \to \mathbb{R}$ and $\hat{f} : \mathbb{R}^m \to \mathbb{R}$ given by

$$\hat{f}(x) = \begin{cases} f(x), & \text{if } x \in U \\ 0, & \text{otherwise} \end{cases},$$

let $\mathbb{U} \subset \mathbb{IR}^m$ be given such that $U \subseteq \mathbb{U}$. For each $[x] \in \mathbb{U}$, let $\phi_{f_{[x]}} : [0,1]^n \to \mathbb{R}$ be given by

$$\phi_{f_{[x]}}(\lambda_x) = (\hat{f} \circ \mathcal{R}_{[x]})(\lambda_x),$$

where $n \leq m$ and $\mathcal{R}_{[x]}$ is the image of $\mathcal{R} : \mathbb{U} \to \mathcal{F}^m$ at $[x]$. That is,

$$\mathcal{R}([x]) = \mathcal{R}_{[x]} : [0,1]^n \to \mathbb{R}^m$$

is such that

$$\begin{aligned} \mathcal{R}_{[x]}(\lambda_x) &= (\mathcal{R}_{[x_1]}(\lambda_{x_1}), \mathcal{R}_{[x_2]}(\lambda_{x_2}), \cdots, \mathcal{R}_{[x_m]}(\lambda_{x_m})) \\ &= (\underline{x}_1 + \lambda_{x_1} w_1, \underline{x}_2 + \lambda_{x_2} w_2, \cdots, \underline{x}_m + \lambda_{x_m} w_m), \end{aligned}$$

for all $\lambda_x = (\lambda_{x_1}, \lambda_{x_2}, \ldots, \lambda_{x_m}) \in [0,1]^m$, $w_i = \overline{x}_i - \underline{x}_i \geq 0$, whenever the coordinates $[x_i]$ of $[x] = ([x_1], \ldots, [x_m])$ are independent (and in this case $n = m$), and if the interval coordinates $[x_i]$ of $[x] = ([x_1], [x_2], \cdots, [x_m])$ are totally dependent for all $i \in \{i_1, \ldots, i_k\} \subseteq \{1, \ldots, m\}$, then $\mathcal{R}_{[x]}$ is the image of

$$\begin{aligned} \mathcal{R}_{[x]}(\lambda_x) &= \left(\mathcal{R}_{[x_1]}(\lambda_{x_1}), \ldots, \mathcal{R}_{[x_{i_1}]}(\lambda_{x_{i_1}}), \mathcal{R}_{[x_{i_2}]}(\lambda_{x_{i_1}}), \ldots, \mathcal{R}_{[x_{i_{k+1}}]}(\lambda_{x_{i_k+1}}), \ldots, \mathcal{R}_{[x_m]}(\lambda_{x_m})\right) \\ &= \bigwedge_{i=1}^{m} \left(\mathcal{R}_{[x_i]}(\lambda_{x_j})\right) \end{aligned}$$

for all $\lambda_x = (\lambda_{x_1}, \ldots, \lambda_{x_{i_k}}, \cdots, \lambda_{x_m}) \in [0,1]^n$, $w_i = \overline{x}_i - \underline{x}_i \geq 0$ and $w_{i_j} = \overline{x}_{i_j} - \underline{x}_{i_j} \geq 0$, where $\begin{cases} j = i_1, \text{ if } i \in \{i_1, \ldots, i_k\} \\ j = i \text{ if } i \notin \{1_1, \ldots, i_k\} \end{cases}$

If $\phi_{f_{[x]}} \in \mathcal{F}$, then the interval function $F^I : \mathbb{IU} \to \mathbb{IR}$ given by

$$F^I([x]) = \mathcal{MB}\left(\phi_{f_{[x]}}\right)$$

is called a **constraint interval function extension** of f. The map $\Phi_f : \mathbb{U} \to \mathcal{F}$ given by $\Phi_f([x]) = \phi_{f_{[x]}}$ is called a **constraint interval function representation of F^I**.

It turns out that $\hat{f}([x])$ is a hypercube on $\mathbb{R}^m$ whereas $f([x])$ it is not always a hypercube. Moreover, it is always true that $f([x]) \subset \hat{f}([x])$. A numerical set, intervals in our case, can be considered as (see [4] and [5]):

1. A *whole* representation of a mathematical uncertainty, which is said to be **ontic**;
2. An representation of an ill-known real number, which is said to be **epistemic**. In this case, the only information about the interval are its lower and upper bounds.

Definition 6. Two intervals $[a] = [\underline{a}, \overline{a}]$, $[b] = [\underline{b}, \overline{b}] \in \mathbb{IR}$ are said to be:

(i) **numerically equal** if $\underline{a} = \underline{b}$ and $\overline{a} = \overline{b}$, and it is denoted by $[a] = [b]$;
(ii) **epistemically equal** if $\mathcal{R}([a]) = \mathcal{R}([b])$.

3 Distances

In general, the use of ontic distances between interval is adopted to deal with calculus involving numerical uncertainties. On the other hand, Mazarandani et al. [9] provided the concept called granular distance, which is able to deal with intervals considering the epistemic point of view. A new concept of distance between intervals called constraint interval distance it is presented herein through CIA. This new concept is able to deal with both ontic and epistemic intervals. Moreover, it is a first step toward introducing a new notion of distances between fuzzy intervals that is discussed in subsequent works. As motivation to the introduction of constraint interval distance, the following scenary is considered: if $[\underline{a}, \overline{a}]$ and $[\underline{b}, \overline{b}]$ are interval representations of the numerical uncertainties a and b, then it is natural to think about the constraint interval representation of the real distance between a and b since this real distance is also a numerical uncertainty.

Given a metric $d : \mathbb{R} \times \mathbb{R} \to \mathbb{R}_+ \cup \{0\}$, where $\mathbb{R}_+ = \{r \in \mathbb{R} : r > 0\}$, consider the function $d_{CI}([\underline{a}, \overline{a}], [\underline{b}, \overline{b}]) : [0, 1] \times [0, 1] \to \mathbb{R}$ given by

$$d_{CI}([\underline{a}, \overline{a}], [\underline{b}, \overline{b}])(\lambda_a, \lambda_b) = d\left(\mathcal{R}_{[\underline{a}, \overline{a}]}(\lambda_a), \mathcal{R}_{[\underline{b}, \overline{b}]}(\lambda_b)\right). \tag{4}$$

This provides **a real-valued representation of** $d(a, b), a \in [\underline{a}, \overline{a}], b \in [\underline{b}, \overline{b}]$ by means of the real-valued $d_{CI}([\underline{a}, \overline{a}], [\underline{b}, \overline{b}])(\lambda_{a_0}, \lambda_{b_0})$ since there exist a pair $(\lambda_{a_0}, \lambda_{b_0}) \in [0, 1] \times [0, 1]$ satisfying

$$a = \mathcal{R}_{[\underline{a}, \overline{a}]}(\lambda_a) = \underline{a} + \lambda_{a_0}(\overline{a} - \underline{a}),$$

$$b = \mathcal{R}_{[\underline{b}, \overline{b}]}(\lambda_{b_0}) = \underline{b} + \lambda_{b_0}(\overline{b} - \underline{b})$$

and, consequently,

$$d_{CI}([\underline{a}, \overline{a_0}], [\underline{b_0}, \overline{b}])(\lambda_a, \lambda_b) = d(a, b).$$

Thus, the function d_{CI} given in (4) is a suitable candidate to be the constraint interval distance representation. Note that the variable λ_a varies in an independent way from the variable λ_b on the interval $[0, 1]$. Consequently, the function given in (4) can define the distance between the intervals $[a] = [\underline{a}, \overline{a}]$ and $[b] = [\underline{b}, \overline{b}]$, whenever $[a]$ and $[b]$ are totally independent. On the other hand, it is known that in many phenomena may have pairs of totally dependents intervals. If $[a] = [\underline{a}, \overline{a}]$ and $[b] = [\underline{b}, \overline{b}]$ are totally dependent, then the constraint interval representation are the functions given, respectively, by

$$\mathcal{R}_{[a]}(\lambda) = \underline{a} + \lambda(\overline{a} - \underline{a})$$

and

$$\mathcal{R}_{[b]}(\lambda) = \underline{b} + \lambda(\overline{b} - \underline{b}).$$

That is, the constraint interval representation of both intervals use the same variable λ in the interval $[0, 1]$. So, the concept of constraint interval distance representation is presented as follows.

Definition 7 (Constraint interval distance representation). Given $[a] = [\underline{a}, \overline{a}], [b] = [\underline{b}, \overline{b}] \in \mathbb{IR}$, the constraint interval distance representation is the function:

(i) $d_{CI}([a],[b]) : [0,1] \times [0,1] \to \mathbb{R}$ given by (4) if $[a]$ and $[b]$ are totally independent;
(ii) $d_{CI}([a],[b]) : [0,1] \to \mathbb{R}$ given by

$$d_{CI}([a],[b])(\lambda) = d\left(\mathcal{R}_{[a]}(\lambda), \mathcal{R}_{[b]}(\lambda)\right). \tag{5}$$

if $[a]$ and $[b]$ are totally dependent.

Remark 1. If $[a]$ and $[b]$ are totally independent, then for each (λ_a, λ_b) fixed on $[0,1] \times [0,1]$, $d_{CI}([a],[b])(\lambda_a, \lambda_b)$ can be interpreted as an real-valued representation of the numerical uncertainty $d(a,b)$. In this sense, $d_{CI}([a],[b])(\lambda_a, \lambda_b)$ can be seen as a decision making for each (λ_a, λ_b) fixed on $[0,1] \times [0,1]$. Given the **constraint interval distance representation** it is possible to obtain an interval

$$D_{CI}([a],[b]) = \left[\min_{(\lambda_a,\lambda_b)\in[0,1]^2} d_{CI}([a],[b])(\lambda_a,\lambda_b), \max_{(\lambda_a,\lambda_b)\in[0,1]^2} d_{CI}([a],[b])(\lambda_a,\lambda_b)\right] \tag{6}$$

for the distance, where the uncertainty of distance between two interval ill-known real numbers is contained. Analogously, if $[\underline{a}, \overline{a}]$ and $[\underline{b}, \overline{b}]$ are totally dependent, the $d_{CI}([a],[b])(\lambda)$ is contained in the interval

$$D_{CI}([a],[b]) = \left[\min_{\lambda\in[0,1]} d_{CI}([a],[b])(\lambda), \max_{\lambda\in[0,1]} d_{CI}([a],[b])(\lambda)\right]. \tag{7}$$

$D_{CI}([a],[b])$ is said to be the ***constraint interval distance between*** $[a]$ ***and*** $[b]$.

Remark 2. The constraint interval distance representation between $[a]$ and $[b]$ is a function where each element of its range can be a decision choice. Thus, the constraint interval distance representation between $[a]$ and $[b]$ allows a choice dependent on the context of the problem at hand. On the other hand, the granular distance (given by Mazandarani et al. [9]) between $[a]$ and $[b]$ is a real number and a decision has already been made. These facts make it clear that the constraint interval distance representation and the granular distance are different concepts.

4 Sequences

This section develops sequences and convergence from the CI framework.

Definition 8. Given an interval-valued sequence $[x_n] : \mathbb{N} \to \mathbb{IR}$, denoted by $\{[x_n]\}$, where the general term is $[\underline{x}_n, \overline{x}_n]$, the constraint interval sequence representation (CISeq) is defined by $\mathcal{R}_{\{[x_n]\}} : [0,1] \to \mathbb{R}$ as follows:

1. $\mathcal{R}_{\{[x_n]\}}(\lambda_n) = \underline{x}_n + \lambda_n w_{[x_n]},\ \lambda_n \in [0,1], \forall n$, if all the intervals in $\{[x_n]\}$ are independent;
2. $\mathcal{R}_{\{[x_i]\}}(\lambda_i) = \begin{cases} \underline{x}_i + \lambda w_{[x_i]}, \text{ if } k \leq i \leq k+l \\ \underline{x}_i + \lambda_i w_{[x_i]}, \text{ if } i < k \text{ or } i > k+l \end{cases}, \lambda,\ \lambda_i \in [0,1]$, if some are independent and other of them the intervals in $\{[x_n]\}$ are dependent;
3. $\mathcal{R}_{\{[x_n]\}}(\lambda_n) = \underline{x}_n + \lambda w_{[x_n]}, \lambda \in [0,1], \forall n$, if all of the intervals in $\{[x_n]\}$ are totally dependent,

where $\mathcal{R} : \mathbb{IR} \to \mathcal{F}$ is the space mapping/function and $\mathcal{F}$ is the space of real bounded function.

Example 1. Consider the interval-valued sequence $\{[x_n]\} = \left\{\left[1 - \frac{1}{n}, 2 + \frac{1}{n}\right]\right\}$. Then, according the Definition 8 of sequence we can consider:

1. The intervals in the sequence are totally dependents and, in this case, the constraint interval sequence representation CISeq is $\mathcal{R}_{[x_n]}(\lambda) = 1 - \frac{1}{n} + \left(1 + \frac{2}{n}\right)\lambda,\ \lambda \in [0,1]$. Thus, a suitable numerical sequence can be set by means of a convenient λ_n in $[0,1]$.
2. The intervals in the sequence are totally independents and, in this case, the constraint interval sequence representation CISeq is $\mathcal{R}_{[x_n]}(\lambda_n) = 1 - \frac{1}{n} + \left(1 + \frac{2}{n}\right)\lambda_n,\ \lambda_n \in [0,1]$. Thus, a suitable numerical sequence can be defined by the selection of an appropriate λ_n^* for each $n \in \mathbb{N}$.
3. Some intervals are dependents and others independents. For instance, let $[x_i]$ be totally dependents if $i = 3, 4, 5$ and let $[x_j]$ be independent if $j \neq i$. In this case, the constraint interval sequence representation CISeq is

$$\mathcal{R}_{[x_i]}(\lambda_i) = \begin{cases} 1 - \frac{1}{n} + \left(1 + \frac{2}{n}\right)\lambda_i, i = 3, 4, 5 \\ 1 - \frac{1}{n} + \left(1 + \frac{2}{n}\right)\lambda_j, j \neq 3, 4, 5. \end{cases}$$

   Thus, an appropriate numerical sequence can be defined by choosing a suitable $\lambda_i \in [0,1]$ for each $i \in \mathbb{N}$.

This example considers that the unique avalaible information is: each term of the sequence of real numbers is something between 1 and 3, which can be done by choosing suitable values on $[0,1]$. Then in order to obtain a numerical sequence we need to do a decision making to find a possible solution for $n \to +\infty$.

Example 2. Let $[x_n] = [n, n+n^2], [z_n] = \left[1 - \frac{1}{n}, 2 + \frac{1}{n}\right]$ be interval-valued sequences. Then, $\mathcal{R}_{\{[x_n]\}}(\lambda_n) = n + \lambda_n n^2, \mathcal{R}_{\{[z_n]\}}(\lambda_n) = 1 - \frac{1}{n} + \lambda_n\left(1 + \frac{2}{n}\right)$ are the CISeq, respectively. Below is the geometric illustration for the totally dependent case.

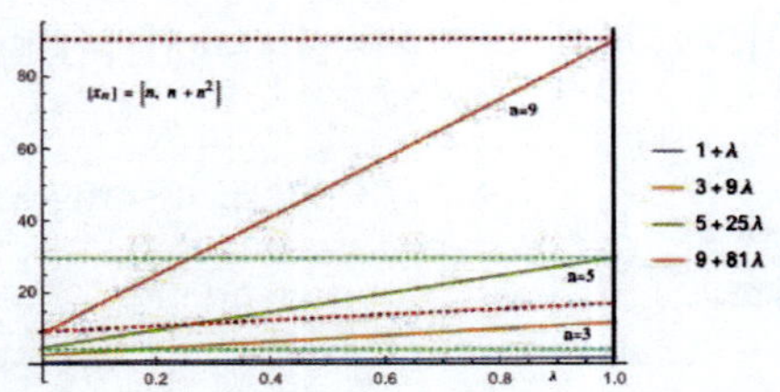

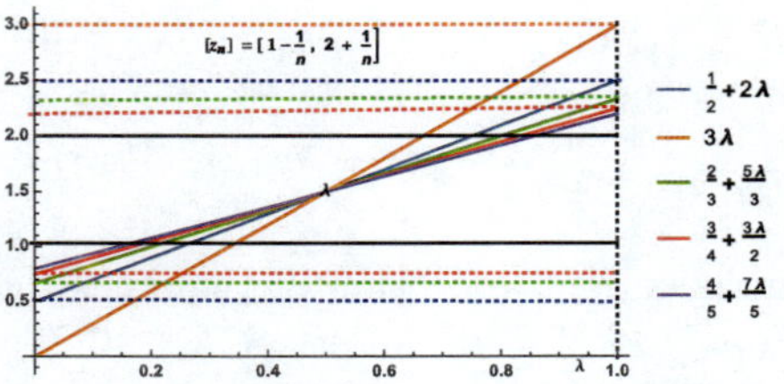

The $d_{CI}([x_n],[x])(\lambda_n)$ can lead to a different interval distance than Pompieiu-Hausdorff since the latter does not capture the epistemic semantic. For example, the d_{CI} distance between $[a] = [-1,1]$ and $[b] = (-1)[-1,1]$ results in $[2,2]$, whereas the Pompieiu-Hausdorff distance between $[a]$ and $[b]$ is 0.

Definition 9 (Convergence). Given an interval-valued sequence $[x_n] : \mathbb{N} \to \mathbb{IR}$, denoted by $\{[x_n]\}$ and whose general term is $[\underline{x}_n, \overline{x}_n]$, the constraint interval sequence representation CISeq $\mathcal{R}_{\{[x_n]\}}(\lambda_n) = \underline{x}_n + \lambda_n w_{x_n}, \lambda_n \in [0,1]$, converges to $\mathcal{R}_{[x]}(\lambda) = \underline{x} + \lambda w_x,\ \lambda \in [0,1]$, if given $\varepsilon > 0, \exists\ n_0|\ \forall n \geq n_0$, follows that $d_{CI}([x_n],[x])(\lambda_n) = d(\mathcal{R}_{\{[x_n]\}}(\lambda_n) - \mathcal{R}_{[x]}(\lambda)) < \varepsilon$ with $w_{[x_n]} \to w_{[x]}$ and choosing a suitable $\{\lambda_{n_j}\}$ convergent subsequence.

Remark:

1) It is possible to choose a convergent subsequence of parameter $\{\lambda_n\}$ such that the interval-valued sequence is not convergent.
2) Given a convergent CISeq, then

$$\lim_{n\to+\infty} \mathcal{R}_{\{[x_n]\}}(\lambda_n) = \underline{x} + \lambda w_x \text{ for some } \lambda_{n_j} \to \lambda.$$

3) When we are saying that the interval-valued sequence converges for an interval we are considering it as a "whole" interval in the ontic point of view or as an element in the interval for what exists a convergent subsequence of parameters λ_n that satisfies a making decision in an ontic point of view.

Definition 8 can be used to obtain a constraint interval representation for $m-$tuples of constraint interval sequences.

Definition 10. Given an interval-valued sequence $[x_n] : \mathbb{N} \to \mathbb{IR}^m, \mathbb{IR}^m = \mathbb{IR} \times \ldots \times \mathbb{IR}$, denoted by $\{[x_n]\}$, where the general term is $([\underline{x}_{n_1}, \overline{x}_{n_1}], \ldots, [\underline{x}_{n_m}, \overline{x}_{n_m}])$, the constraint interval sequence representation CISeq is defined by $\mathcal{R}_{\{[x_n]\}} : [0,1]^n \to \mathbb{R}^m$ as follows:

1. $\mathcal{R}_{\{[x_{n_i}]\}}(\lambda_{n_i}) = \underline{x}_{n_i} + \lambda_{n_i} w_{[x_{n_i}]}, \lambda_n \in [0,1], \forall n$, if all the intervals in $\{[x_n]\}$ are independent for each CISeq representation for $i = 1, \ldots, m$;
2. $\mathcal{R}_{\{[x_{n_i}]\}}(\lambda_i) = \begin{cases} \underline{x}_{n_i} + \lambda w_{[x_{n_i}]}, & \text{if } j \leq i \leq j+l \\ \underline{x}_{n_i} + \lambda_i w_{[x_{n_i}]}, & \text{if } i < j \text{ or } i > j+l \end{cases}, \lambda,\ \lambda_i \in [0,1]$, if some of the intervals in $[x_n]$ are independent and others of them are dependent for each CISeq representation for $i = 1, \ldots, m$;
3. $\mathcal{R}_{\{[x_{n_i}]\}}(\lambda_n) = \underline{x}_{n_i} + \lambda w_{[x_{n_i}]},\ \lambda \in [0,1],\ \forall\ n$, all the intervals in $[x_n]$ are totally dependent for some CISeq representation for $i = 1, \ldots, m$.

where $\mathcal{R} : \mathbb{IR} \to \mathcal{F}$ is the representation mapping/function.

5 Application in Economics

Here we consider the following ordered defined in [3], to define an investment portfolio using constraint interval sequence and the inherent risk that exists in each of them.

Definition 11. Let $\varphi : \mathbb{R}^{2m} \to \mathbb{R}^{2m}$ be a bijection and let $\leq \mathbb{R}^{2m}$ be an order on $\mathbb{R}^{2m}$. The $\varphi, \leq_{\mathbb{R}^{2m}}$-preference order relation $\leq_{\varphi}^{2m}$ on $(\mathbb{IR})^m$ is defined by

$$[A] \leq_{\varphi}^{2m} [B] \tag{8}$$

where $[A] = ([\underline{a}_1, \overline{a}_1], \dots, [\underline{a}_n, \overline{a}_n]), [A] = ([\underline{a}_1, \overline{a}_1], \dots, [\underline{a}_n, \overline{a}_n]), [B] = ([\underline{b}_1, \overline{b}_1], \dots, [\underline{b}_n, \overline{b}_n]) \in (\mathbb{IR})^m$ and $\overline{\varphi} : (\mathbb{IR})^m \to \mathbb{R}^{2m}$ is the injective function given by

$$\overline{\varphi}([\underline{a}_1, \overline{a}_1], \dots, [\underline{a}_n, \overline{a}_n]) = \varphi(\underline{a}_1, \overline{a}_1, \dots, \underline{a}_n, \overline{a}_n). \tag{9}$$

Let $\mathcal{A}^m$ be the particular class of automorphisms $\varphi_1 \times \ldots \times \varphi_m : \mathbb{R}^{2m} \to \mathbb{R}^{2m}$ such that:

$$(\varphi_1 \times \ldots \times \varphi_m)(x_1, x_2, \dots, x_{2m-1}, x_{2m}) = (\varphi_1(x_1, x_2), \dots, \varphi_m(x_{2m-1}, x_{2m})),$$

where the automorphism given by

$$\varphi_i(x_{2i-1}, x_{2i}) = (\alpha_{2i-1}x_{2i-1} + \alpha_{2i}x_{2i}, \beta_{2i-1}x_{2i-1} + \beta_{2i}x_{2i})$$

$\forall\ (x_{2i-1}, x_{2i}) \in \mathbb{R}^2$ and $\forall i \in \{1, \dots, m\}$. Then, $\forall\ [a] = ([a_1], \dots, [a_m]), [b] = ([b_1], \dots, [b_m]) \in (\mathbb{IR})^m$, φ, $\leq_{\mathbb{R}^{2m}}$ $-$preference order relations $\leq_{\varphi}$ on $(\mathbb{IR})^m$ dependents on $\varphi \in \mathcal{A}^m$, is defined by
$([a_1], \dots, [a_m]) \leq_{\varphi} ([b_1], \dots, [b_m]) \Leftrightarrow \overline{\varphi}([a_1], \dots, [a_m]) \leq^{2m} \overline{\varphi}([b_1], \dots, [b_m])$
$\Leftrightarrow \overline{\varphi}_i([a_i]) \leq \overline{\varphi}_i([b_i]) \Leftrightarrow [a_i] \leq_{\varphi_i} [b_i], \forall i \in \{1, \dots, m\}$.

Now, we consider the following data involved in an investment F_1: saving account; F_2: pre-fixed bank deposit certificate; F_3: daily bank deposit certificate; F_4: mortgage letter; F_5: agribusiness credit bills; F_6: brazilian national treasury bonds IPCA (indexed to the Consumer Price Index); F_7: stock exchange; F_8: private social security funds; F_9: debenture; F_{10}: multimarket funds; F_{11}: exchange trade funds (ETF's). These data appear in several types of investments considered in Brazil. Below we describe some of them, taking into account the investor profile.

- **Traditional investor**: the preferred investments are: F_1, F_2, F_3, F_4, F_5
- **The moderate investor**: some preferred investments are: F_4, F_5, F_6, F_8.
- **The agressive investor**: is the one that accepts to take more risks in order to have more profitability. This profile is even willing to lose part of the amount invested. However, this means that he has great chances to recover this amount in the future and still earn more with it. In the investment portfolio of the aggressive investor, most of his applications are composed of variable income products. Some of them are: $F_6, F_7, \dots, F_{11}$.

Three components are important to define the portfolio:

1. Expected Return $= p_1 \cdot r_1 + p_2 \cdot r_2 \dots p_k \cdot r_k$, where p_k reflects the portfolio weight invested in a given asset F_k with r_k expected rates for $k = 1, \dots, 11$.
2. The portfolio risk [11] for two assets is given by

$$PR = \sqrt{p_i^2 \times (D_{F_i})^2 + p_j^2 \times (D_{F_j})^2 + (2 \cdot p_i \cdot_j \cdot \rho_{i,j} \cdot D_{F_i} \cdot D_{F_j})}, i \neq j$$

where: D_{F_i}, D_{F_j} are standard deviations of retruns associated to the assets i and j; $\rho_{i,j}$ correlation coefficient between the returns of assets i and j.
3. There is an index to define the risk of market called sharpe index that is the potential return of an investment compared to its risk

$$SR = \frac{R_X - R_f}{DP}, \tag{10}$$

where R_X, R_f, DP are expected portfolio return, risk-free rate of return and standard deviation (volatibility). According [2] sharpe ratio grading thresholds:
   - $SR < 1$, bad;
   - $SR = [1, 1.99]$, adequate/good;
   - $SR = [2, 2.99]$ very good;
   - $SR > 3$ is excelent.

   The higher the ratio, the greater the return relative to the investment, given the risk taken, and thus the better the investment. The ratio can be used to evaluate a single stock or investment, or an entire portfolio.

If we consider investments with a simple interest, then we have the future amount given by

$$F_k(n_i) = P_k(1 + n_i r_k), \tag{11}$$

where $F_k(n_i)$: future amount, P_k: weight of the value invested in the asset F_k, r_k: interest rate for the assets F_k, $k = 1, \dots, 11$ and n_i is the number of interest periods. In general, the interest rates vary according to what is happening in the world market. Thus $r_k, k = 1, \dots, 11$ can be seen as intervals, instead of real numbers. Hence, $[F_k] : \mathbb{N} \rightarrow \mathbb{IR}$ for $k = 1, \dots, 11$, which are interval valued functions, that represent the simple interest (11) defined by

$$F_k(n_i) = P_k +_{CIA} n_i \cdot_{CIA} [r_k] \cdot_{CIA} P_k, \tag{12}$$

where $w_{r_k} = \overline{r}_k - \underline{r}_k, k = 1, \dots, 11$. and whose CiSeq associated is

$$\mathcal{R}_{\{[F_k(n_i)]\}}(\lambda_{F_k(n_i)}) = P_k + (\underline{r}_k + \lambda_{r_k} w_{r_k}) n_i P_k \tag{13}$$

$\lambda_{r_k} \in [0, 1]$. Now, if we consider the profile of the investor, we can offer some portfolio considering the expected returns provide, the final value is obtained taking into account the profit that each application will give at the end of a suitable period, considering the variation rates and risks involved in the investment. For example, we can consider:

1) Traditional investor: consider two portfolios

$$PA = (P_1\mathcal{R}_{\{[F_1(n_1)]\}}(\lambda_{r_1}), P_2\mathcal{R}_{\{[F_2(n_2)]\}}(\lambda_{r_2}, P_31\mathcal{R}_{\{[F_3(n_3)]\}}(\lambda_{r_3}), P_4\mathcal{R}_{\{[F_4(n_4)]\}}(\lambda_{r_4})),$$

and

$$PB = (P_1\mathcal{R}_{\{[F_1(n_1)]\}}(\lambda_{r_1}), P_2\mathcal{R}_{\{[F_2(n_2)]\}}(\lambda_{r_2}, P_31\mathcal{R}_{\{[F_3(n_3)]\}}(\lambda_{r_3}), P_5\mathcal{R}_{\{[F_5(n_5)]\}}(\lambda_{r_5})),$$

where $P_k, k = 1, 2, 3, 4, 5$ is the amount from the total to be applied in each investment. If the investor wants more profitability, with less risks, then we need to choose a order relation considering the two options for portfolios or more. Note that each sequence in a 4−tuple converge for some value in the representation space and if we want an interval solution after some time n_i then we need minimize and maximize for $\lambda_{r_k} \in [0, 1], k = 1, 2, 3, 4, 5$.

Supposing that there isn't sharpe rate for PA and PB we can choose the weight where the money is applied. The setting rates are as follows: $r_1 = \pm 4,5\%, r_2 = \pm 5,7\%, r_3 = \pm 80\% r_2, r_4 = \pm 6.5\%$ and $r_5 = \pm 6.51\%$. Thus, a simple decision making world be to apply all money considering the greatest interest rate, i.e., r_4. However, this type of investment is more likely to be affected by the financial market than others types. Thus, a portfolio decision can be made taking into account the Definition 11.

2) Agressive investor: here, it's important to consider the sharpe index because this type of investor prefer to apply in stock exchange, brazilian national treasure and ETF's (exchange-traded funds) that have high chances of profitability, even though these investiments also have high level of risk. In general, you need to see what is happening in the world market and perform a decision making with respect to sharp index based on the profile of each investor. We have many options of portfolio to offer and, consequently, it is necessary to consider some criteion to classificate the priorities. This criterion can be obtained from order relation according the Definition 11.

6 Conclusion

In general, the uncertainty in the investment market has a great fluctuation. Consequently, it is not easy to make a decision. For example, there are many oscillations in a day or in an hour in a stock market, while other investments need weeks, months or years. Mathematical modelling can help in a decision making. However, the risks are inherent in the investment market and, consequently, there is no method to ensure which is the best decision making at the beginning of the investment. The investment market problem presented in Sect. 5 illustrates the applicability of the concepts developed in this work. Some additional definitions and ideas are being developed for interval sequences and limit of interval functions from a constraint interval point of view.

References

1. https://economiasc.com/2022/07/26/conservador-moderado-ou-arrojado-analista-de-educacao-financeira-explica-como-saber-seu-perfil-de-investimentos/
2. https://corporatefinanceinstitute.com/resources/risk-management/sharpe-ratio-definition-formula/
3. Costa, T.M., Osuna-Gómez, R., Chalco-Cano, Y.: New preference order relationships and their application to multiobjective interval and fuzzy interval optimization problems (Submitted for Publication)
4. Couso, I., Dubois, D.: Statistical reasoning with set-valued information: ontic vs. epistemic views. Int. J. Approx. Reason. **55**, 1502–1518 (2014)
5. Dubois, D., Prade, H.: Gradualness, uncertainty and bipolarity: making sense of fuzzy sets. Fuzzy Sets Syst. **192**, 3–24 (2012)
6. Jamison, K.D., Lodwick, W.A.: A new approach to interval-valued probability measures, a formal method for consolidating the languages of information deficiency: foundations. Inf. Sci. **507**, 86–107 (2020)
7. Lodwick, W.A., Jenkins, O.A.: Constrained intervals and interval spaces. Soft Comput. **17**, 1393–1402 (2013)
8. Markowitz, H.M.: Portfolio selection. J. Financ. **7**(1), 77–91 (1952)
9. Mazandarani, M., Pariz, N., Kamyad, A.V.: Granular differentiability of fuzzy-number-valued functions. IEEE Trans. Fuzzy Syst. **26**(1), 310–323 (2018)
10. Mizukoshi, M.T., Costa, T.M., Chalco-Cano, Y., Lodwick, W.: A Formalization of Constraint Interval: A Precussor to Fuzzy Interval Analysis (Submitted for Publication)
11. Castro, G.J., Baidya, T.K.N., Aiube, F.A.L.: Using Omega Measure for Performance Assessment of a Real Options Portfolio, Real Options Conference (2008)

Genetic Fuzzy Threat Assessment for Asteroids 2600 Derived Game

Daniel Heitmeyer(✉) and Kelly Cohen

University of Cincinnati, 2600 Clifton Avenue, Cincinnati, OH, USA
{heitmedl,cohenky}@mail.uc.edu

Abstract. A fuzzy logic controller is used to assess the threat of asteroids in a python game derived from Asteroids 2600. The original game has been used extensively for testing and training of autonomous agents using reinforcement learning and serves as a testing ground for the training of a fuzzy system. A system for aiming and firing by planning the trajectory of an asteroid is used in conjunction to eliminate the highest threat asteroid. A genetic algorithm is used to train the fuzzy logic controller's rule base in various custom and random asteroid environments. Using the genetic algorithm, the solution space can be explored to produce a more optimal controller than by hand. Three varying-sized fuzzy controllers are trained and tested on similar environments and training procedures. All controllers outperformed a human operator and produced threat decisions across all asteroids within milliseconds. The smaller systems showed much faster training and evaluation while producing marginally worse results than the largest controller, leading us to use the smallest controller in future work.

1 Introduction

The original arcade game Asteroids 2600 has been used extensively for testing and training of autonomous agents using reinforcement learning [8]. Many deep learning agents have been evaluated on the whole of Atari 2600 games and outperformed the human normalized scores in most games [1]. The Arcade Learning Environment was specifically designed and published openly to provide this test-bed for reinforcement learning where many state of the art models have been trained and compared [2]. These methods however implement black-box methods such as Deep Q-Networks or Neural Networks to achieve this performance [7]. Controllers that attempt to learn to world model of these Atari games have also been made and achieved similarly impressive results [6]. These models also typically achieve this performance with a large amount of computational budget requiring weeks or even months of training time. While methods are being developed to add explainability to these models, utilizing fuzzy logic provides inherit explainability within the fuzzy rule base [3]. The Asteroids game also requires quick decisions with a tolerance for imprecision specifically in the targeting where fuzzy logic can be very beneficial [9].

Similar to Q-Learning methods, a Genetic Algorithm (GA) can be used to reward the behavior of the controller. Benefits for scenario wins, eliminating asteroids, and punishing deaths can be used for the cost or fitness function of these approaches. Previous work has shown the performance of genetic fuzzy systems in simulated air combat

K. Cohen et al. (Eds.): NAFIPS 2023, LNNS 751, pp. 81–90, 2023.
https://doi.org/10.1007/978-3-031-46778-3_8

missions [4]. By employing a tree structure the computation time and solution space are heavily decreased compared to a singular Fuzzy Inference System (FIS). While the tree structure may limit the optimality of the final controller, the benefits of reducing the large solution space are greater when considering the training time using a GA. The FIS and aiming systems developed here are planned to be part of a larger fuzzy decision tree to control all aspects of the asteroids environment.

The kessler-game python package used is based upon the original Atari game and the python arcade package version of asteroids [5]. The game was developed by the Thales Group for use in their Explainable AI challenge. The game includes multiple agents in an asteroid field where the edges of the map wrap to the opposite sides and can be seen in Fig. 1. Notably, previous work dealt with the wrapping of borders by expanding the playing field creating duplicates at each side and corner [8]. Each agent is given control over a ship that can fire bullets, command negative or positive thrusts, and command a turn rate. Custom asteroid and ship scenarios can be created and will be used for the training of the fuzzy system.

2 Methodology

Thirty times per second a list of asteroids and the ship state is read. Asteroid state information is iterated through and the ship is commanded at the end of each time step. Values on a range of -180 to 180 are commanded for the turn rate, -480 to 480 for the thrust, and a boolean is set for deciding to fire a bullet. For targeting and aiming, only the turn rate and firing boolean are used. This is done to reduce the complexity of the scenario while training the rules for the targeting system. This may limit the performance of the system when incorporated into the more complex controller. However, retraining of the rule base within the full system is planned to mitigate these concerns and is left to future work.

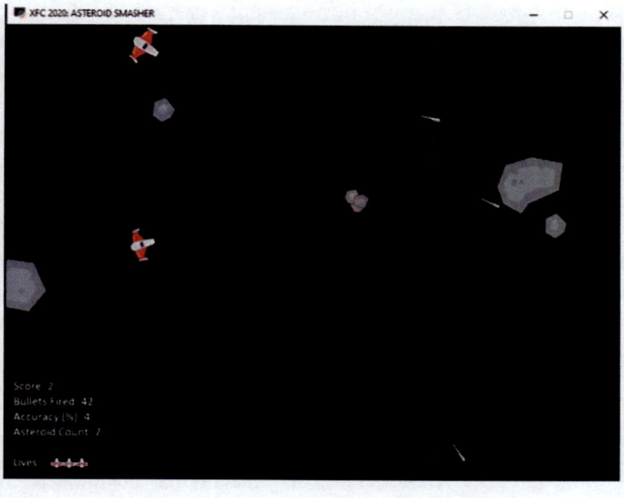

Fig. 1. Kessler Game

2.1 Fuzzy Targeting

Iterating through each asteroid, a threat level is assigned from a FIS. The FIS used is a two-input, one-output system with a number of membership functions and thus varying rule base sizes. Three, five, or seven evenly-spaced and symmetrical triangular membership functions are used for each of the inputs and the output. Increasing the number of membership functions serves to increase the possible complexity of the final FIS trained. The total number of rules increases exponentially as the number of membership functions is increased which causes increased evaluation and training time.

The first input to the FIS used is a normalized distance to the asteroid. This distance in 2-D space is normalized to a percentage of the total map size. This allows for a max radius around the ship to be set where asteroids outside that circle are considered to be equally far away. This can be useful for reducing computation time by only considering asteroids within a certain range of the ship. The radius of the ship and asteroid are also subtracted from the distance to account for the collisions occurring when any part of the ship sprite intersects any part of the asteroid sprite.

The second input is a normalized closure rate of the asteroid towards the ship. This is simply defined as the projection of the asteroid's velocity onto the asteroid's vector from the ship. This can be seen in Fig. 2. This is accomplished with a dot product and can result in either a positive or negative scalar depending on whether the asteroid is moving towards or away. This value is also normalized before being used in the FIS based on the maximum asteroid speed expected. A relative velocity vector and heading could be used instead of this input but would increase the number of inputs to the FIS which is undesirable since the rule base scales exponentially with the number of inputs.

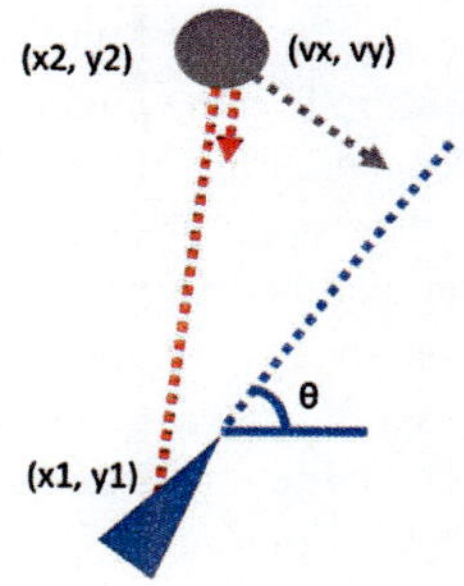

Fig. 2. Sample Asteroid Scenario

The rule base features "AND" rules for each of the permutations of the input membership functions. An example rule created by the system is read as "IF *distance* IS *very close* AND *closure rate* IS *very large* THEN *threat* IS *very large*". These rules in the fuzzy logic system allow for an easily understood decision made by the system. This furthers the explainability of the overall system and allows for trustworthiness in the decisions made by a final trained FIS. For the three by three system this results in nine total "AND" rules whereas the seven by seven system would have forty-nine.

This exponential increase is a large reason why large fuzzy systems benefit from being broken up into fuzzy trees. For a set of inputs, each rule is considered and the output "threat" value is assigned to an asteroid. This output is also normalized from 0 to 1. The threat values of all asteroids considered within the ship's radius are recorded and the asteroid with the largest value is selected to be shot at.

2.2 Aiming

Aiming is handled once a target has been selected from the maximum threat level of all asteroids. The ship position, asteroid position, and asteroid velocity are considered to find the heading at which the ship should fire. The speed of the bullet is explicitly defined in the game as 800 units and is represented with an S. Time is represented with a lowercase t. The future positions of the bullet to be fired are expressed in Eqs. 1 and 2 in reference to Fig. 2. The asteroid's future position is expressed in Eqs. 3 and 4.

$$x_b = x_1 + S * cos(\theta) * t \tag{1}$$

$$y_b = y_1 + S * sin(\theta) * t \tag{2}$$

$$x_{ast} = x_2 + V_x t \tag{3}$$

$$y_{ast} = y_2 + V_y t \tag{4}$$

Solving this systems of equations for $x_b = x_{ast}$ and $y_b = y_{ast}$ gives the desired heading to shoot the asteroid in Eq. 5 where $\Delta x = x_1 - x_2$ and $\Delta y = y_1 - y_2$. A turn rate for the ship is then commanded to reach the desired heading at the next time step.

$$\theta = 2atan(\frac{S\Delta x - \sqrt{2V_y V_x \Delta x \Delta y - V_y^2 \Delta x^2 + S^2 \Delta x^2 + \Delta y^2 (S^2 - V_x^2)}}{-V_y \Delta x - \Delta y (S - V_x)}) \tag{5}$$

Since the ship must turn before shooting, the position of the ship and asteroid during the next time step is used in this calculation. If the ship could not turn fully within one time step, the prediction would be done with a position extrapolated further. The time that the bullet will collide can also be calculated for other uses by reusing Eq. 1. Asteroids that have previously been shot at but the bullet is in the air can be tracked this way. Future threat calculations then can ignore these specific asteroids as they have already been dealt with. Asteroids of larger sizes that can still break up into others are not disregarded however, as they should continue to be shot at to reduce the total number of children asteroids from breaking them apart.

2.3 Genetic Algorithm

A custom GA in python is used to train the rule base of the fuzzy system. Each rule is represented as an integer and has a size totaling either 9, 25, or 49 rules. For the case of three output membership functions the chromosome of the GA has integers up to three which can be read as small, medium, or large in linguistics. The GA is used to search the solution space and optimize the fuzzy rule base. Each chromosome represents an individual in the population which change each generation and is evaluated with a fitness function. This fitness function can be adapted to encourage specific behavior or results from the given controller. Individuals in the population with greater fitness values are selected more often to produce children for the new population. Crossover between two parents is done by swapping two sections of the parent chromosomes to produce two children as seen in Table 1. These children's chromosomes then undergo mutation where each of the integers has a small chance to be changed randomly. After the population has been doubled, a number of the "elite" as determined by the fitness function are sent to the next population and a degree of randomness is used to determine which of the rest should carry onto the next generation.

The fitness function plays a significant role in the final behavior of the system trained. Multiple scenarios were chosen to produce the desired behavior in the final FIS created. The fitness function evaluated over a number of scenarios is defined in Eq. 6. One of the game scenarios included a randomly generated set of fifteen asteroids for overall behavior. Another scenario included multiple closely placed asteroids that had no velocity and an asteroid far away moving directly toward the ship. This scenario was designed to promote the FIS to differentiate properly between the closure rate and the distance to asteroids. Another scenario was created with a wall of asteroids moving in the same direction to also differentiate between distance and the closure rate. Training over these scenarios and potentially others in the future aim to help the GA learn the differences between the two inputs in the solution space.

$$Fitness = AsteroidKills - 30 * Deaths \tag{6}$$

The GA parameters varied for each of the FIS setups trained on. As the solution space increases with the total number of rules, the number of generations and individuals also increases to accompany this. The training time also increases heavily to match this larger solution space. Crossover rates were selected to encourage a faster convergence which may come at the cost of finding a local minimum rather than the global minimum. The mutation rates were also selected to cover the global search. The fittest member of each population was recorded and used for plotting of the training over each generation.

Table 1. Genetic Algorithm Chromosome Crossover

Parent 1	0	1	2	0	1	2	1	2	1
Parent 2	2	1	1	1	2	1	1	0	2
				↓					
Child 1	0	1	2	0	2	1	1	0	2
Child 2	2	1	1	1	1	2	1	2	1

3 Results

The first FIS using 3 membership functions for each input or output has rules shown in Table 2. The GA used to train these rules was run with 100 generations and a population size of 20 for approximately 4 h. The crossover rate was set at 0.9, the mutation rate at 0.3, and the elitism rate at 0.1 to keep the top two members. The fitness function is also plotted over the generations in Fig. 3.

Table 2. FIS 3 × 3 Rule Base

Threat Level Rules		Distance to		
		Close	Medium	Far
Closure Rate	Negative	2	0	0
	Zero	2	1	0
	Positive	2	1	1

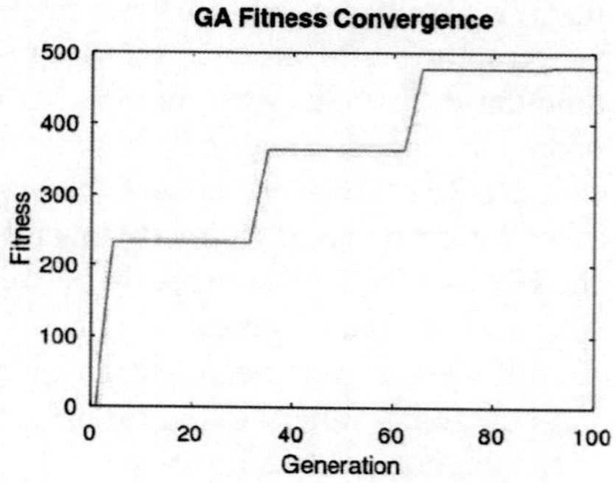

Fig. 3. GA Fitness Convergence

The next FIS using 5 membership functions for each input or output has rules shown in Table 3. The GA used to train these rules was run with 100 generations and a population size of 40 for approximately 7 h. The crossover rate was set at 0.9, the mutation rate at 0.3, and the elitism rate at 0.1 to keep the top four members. The fitness function is also plotted over the generations in Fig. 4.

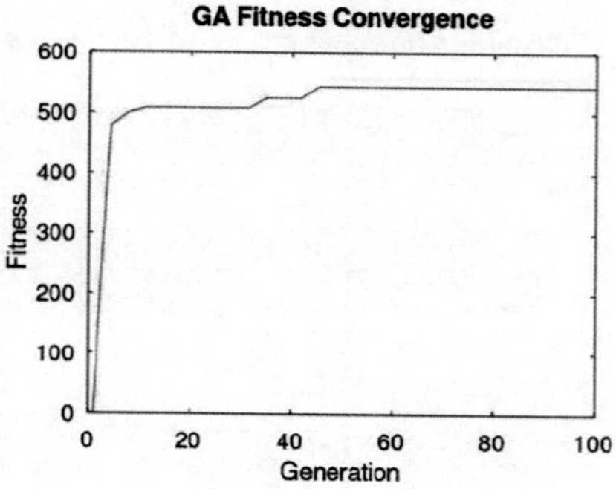

Fig. 4. GA Fitness Convergence

Table 3. FIS 5 × 5 Rule Base

Threat Level Rules		Distance to				
		Very Close	Close	Medium	Far	Very Far
Closure Rate	Very Negative	2	1	2	1	0
	Negative	1	0	2	2	2
	Zero	4	2	1	1	0
	Positive	4	1	1	0	0
	Very Positive	1	0	1	0	0

The final FIS using 7 membership functions for each input or output has rules shown in Table 4. The GA used to train these rules was run with 100 generations and a population size of 80 for approximately 15 h. The crossover rate was set at 0.9, the mutation rate at 0.3, and the elitism rate at 0.1 to keep the top eight members. The fitness function is also plotted over the generations in Fig. 5.

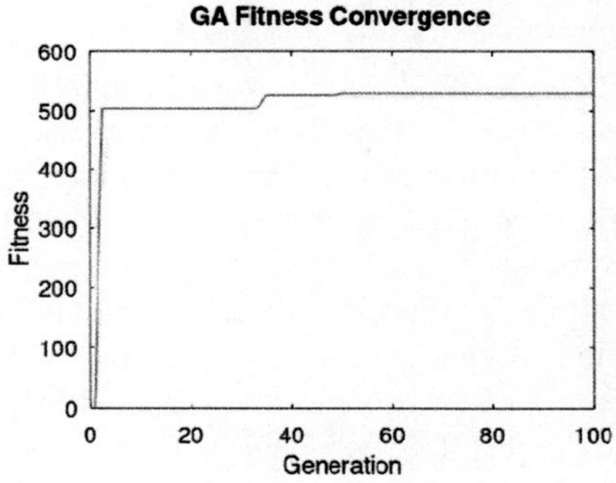

Fig. 5. GA Fitness Convergence

Table 4. FIS 7 × 7 Rule Base

Threat Level Rules		Distance to						
		Very Close	Close	Small Close	Medium	Small Far	Far	Very Far
Closure Rate	Very Negative	0	2	2	0	2	2	2
	Negative	2	6	3	3	4	1	2
	Small Negative	0	5	0	5	2	1	1
	Zero	1	3	1	0	2	0	1
	Small Positive	2	5	1	4	0	0	0
	Positive	2	3	3	1	0	0	1
	Very Positive	6	6	4	1	0	0	1

4 Discussion

The smallest FIS trained with a total of 9 rules finished training much quicker than the larger scenarios as each computation in each scenario is quicker and the solution space is greatly reduced. The final fitness from training was 490. The average evaluation time for each time step was 0.0021 s. It is expected that even the smallest fuzzy system is more capable than a human operator not only for the quick decision making between asteroids but the quick response and aiming as a result of the aiming and firing subsystem. With the smallest system, the rules are straightforward and distance seems to be prioritized most. The closure rate is also a factor and positive values are generally seen as high threat levels which matches initial judgment. Notably, close distances always result in a maximum threat level.

The next FIS with 25 total rules was trained in a significantly larger time than the smaller system. The training also resulted in a larger fitness score of 545 than the smaller FIS. The average evaluation time for each time step was 0.0024 s. The overall behavior seems to be in line with expectations but there are some rules that are not consistent with human judging. For example, the very close distance and very positive closure rate should result in a very large threat but is only a "1" after training. Since other nearby rules are correct, this behavior is likely drowned out and the threat output is still large as expected.

The seven by seven FIS with 49 rules took over 12 h to train and resulted in a fitness score between the two smaller FISs. This likely is due to the overall training process being more difficult considering the nearly doubled solution space and the increased computation time per time step. Each time step took 0.00469 s also doubling when compared to the smaller systems. The rule base also indicates a not fully trained system since there are heavy variations in adjacent rules that should be smoothed out by proper training. This larger system specifically may need to be trained on different scenarios or a wider variety of scenarios to achieve convergence of the rule base. The larger systems when fully trained may exhibit different behaviors than expected which makes them desirable. Specific rules for very far distances and very negative closure rate may be one of these emerging behaviors of interest. Since the map edges wrap at the borders, asteroids very far away without considering the map may actually be much closer. The closure rate would also be swapped from very negative to very positive in this case so we might expect a larger threat value. This behavior is somewhat elicited in both the five by five and seven by seven controllers but it is unclear if it is due to improper training.

5 Conclusion

Three total FISs have been created and trained for controlling the threat assessment of asteroids in the python Kessler Game. A subsystem for aiming and firing has been developed to predict the trajectory of asteroids and match the timing to shoot at the required angle. The produced controllers outperform human control and produce decisions considering all asteroids within milliseconds. Human control resulted in fitness values in the 200 to 300 range with extreme variance. The medium sized five by five FIS has shown the best performance in the scenarios tested on. The smaller three by three system was the fastest computationally and marginally produced the worst results. This smaller system also was the only one that managed a stable convergence to the final rule base where others struggled. The largest system took the longest time to train, had the longest computation time, and produced average results due to the poor training of the system. The trade off between training time and performance likely favors the smaller systems where they could be trained within thirty minutes. Other variations and improvements on the GA could be done to significantly speed up this process and open up the opportunity for more dynamic fitness functions.

Further modifying the fitness function and training time may be needed to optimize any of the final systems chosen for use. Despite the fuzzy systems converging according to the fitness function, some rules stand out as not in accordance with others. This is likely a result of the training scenarios being too easy for the FIS to perform. The randomized scenarios and some of the hand crafted scenarios place the ship in either too much or too little danger where the effects of the FIS do not matter. One of the training scenarios was removed specifically because of this. A scenario where all asteroids never come near the ship resulted in nearly any controller succeeding since the ship never came into danger during this. A larger variety of hand crafted scenarios and more randomized scenarios per fitness to obtain a better average of the controller's performance during training. This would increase training time but can be alleviated by introducing threading to the fitness function.

Since the FIS developed will be used in conjunction with other subsystems that also need to be trained and tested, the three by three system will be used in future work. In conjunction with an avoidance FIS the systems will be trained together and the rule base will likely change in accordance with the ship's new ability to maneuver. Evaluation and training time are also seen as very important for the performance since there is a limited time in the eXplainable AI challenge to complete the overall controller.

Acknowledgements. Without the help and guidance of Dr. Kelly Cohen, this work would not be possible. Additionally, thanks to the Thales group who is responsible for both the Kessler Game and their eXplainable AI challenge.

References

1. Badia, A.P., et al.: Agent57: outperforming the atari human benchmark. CoRR, abs/2003.13350 (2020). https://arxiv.org/abs/2003.13350
2. Bellemare, M.G., et al.: The arcade learning environment: an evaluation platform for general agents. J. Artif. Intell. Res. **47**, 253–279 (2013). https://doi.org/10.1613/jair.3912
3. Clare, M.C.A., et al.: Explainable artificial intelligence for bayesian neural networks: toward trustworthy predictions of ocean dynamics. J. Adv. Model. Earth Syst. **14**(11) (2022). https://doi.org/10.1029/2022ms003162
4. Ernest, N., et al.: Genetic fuzzy based artificial intelligence for unmanned combat aerial vehicle control in simulated air combat missions (2016)
5. Thales Group. Kessler-game (2023). https://github.com/ThalesGroup/kessler-game
6. Hafner, D., et al.: Mastering atari with discrete world models. In: International Conference on Learning Representations (2021). https://openreview.net/forum?id=0oabwyZbOu

7. Jiménez-Luna, J., Grisoni, F., Schneider, G.: Drug discovery with explainable artificial intelligence. Nat. Mach. Intell. **2**, 573–584 (2020). https://doi.org/10.1038/s42256-020-00236-4
8. Meima, N., Mallon, S.: Effectiveness of connectionist Q-learning strategies on agent performance in asteroids (2018)
9. Zadeh, L.A.: Fuzzy logic. Computer **21**(4), 83–93 (1988). https://doi.org/10.1109/2.53

Calibration Error Estimation Using Fuzzy Binning

Geetanjali Bihani(✉) and Julia Taylor Rayz

Computer and Information Technology, Purdue University,
West Lafayette, IN 47904, USA
{gbihani,jtaylor1}@purdue.edu

Abstract. Neural network-based decisions tend to be overconfident, where their raw outcome probabilities do not align with the true decision probabilities. Calibration of neural networks is an essential step towards more reliable deep learning frameworks. Prior metrics of calibration error primarily utilize crisp bin membership-based measures. This exacerbates skew in model probabilities and portrays an incomplete picture of calibration error. In this work, we propose a Fuzzy Calibration Error metric (FCE) that utilizes a fuzzy binning approach to calculate calibration error. This approach alleviates the impact of probability skew and provides a tighter estimate while measuring calibration error. We compare our metric with ECE across different data populations and class memberships. Our results show that FCE offers better calibration error estimation, especially in multi-class settings, alleviating the effects of skew in model confidence scores on calibration error estimation. We make our code and supplementary materials available at: https://github.com/bihani-g/fce.

Keywords: Language Models · Calibration · Fine-tuning · Fuzzy theory · Classification · Natural Language Processing

1 Introduction

Neural network-based decision-making systems have evolved rapidly in the recent decade. Within the domain of natural language processing, deep learning has shaped the current evolution in language modeling. These neural network-based language models are trained on large text corpora and can be fine-tuned across a wide range of NLP tasks and further improved using synthetic semantic enhancement schemes [1], yielding state-of-the-art performance [2–5]. Ideally, a neural model should output reliable and confident prediction probabilities. But recent works have shown that neural networks are unreliable and output highly overconfident predictions, resulting in over-estimation of the model's confidence in decisions [6–8]. This leads to model miscalibration, i.e. a lack of alignment between a

G. Bihani and J. T. Rayz—Contributing equally.

K. Cohen et al. (Eds.): NAFIPS 2023, LNNS 751, pp. 91–100, 2023.
https://doi.org/10.1007/978-3-031-46778-3_9

model's decision probabilities and its actual likelihood of correctness. This lack of calibration can severely impact the trustworthiness of a model's decisions.

A widely adopted measure of the degree of miscalibration is Expected Calibration Error (ECE) [9], used to measure neural network reliability [10–12]. The highly overconfident output prediction probabilities of neural networks result in a left-skewed probability distribution [13]. Since ECE utilizes a fixed-width crisp binning scheme, this skew results in higher probability bins largely contributing to the calibration error estimation, while lower probability bins are ignored [13–15]. To overcome these limitations, prior works have proposed alternative binning strategies such as equal-frequency binning [14], adaptive binning [15], replacing binning with smoothed kernel density estimation [16], and more. Most calibration error estimation techniques rely on crisp binning, which discards edge probabilities (probabilities that typically lie on the bin edge) that could have contributed to a more accurate calibration error estimation. Although some works have utilized fuzzification of prediction probabilities for downstream NLP tasks [17], the calibration impacts of such fuzzification are yet to be studied. We hypothesize that fuzzifying the binning scheme would allow edge probabilities to contribute toward more accurate calibration error estimation. Moreover, fuzzy binning would increase the visibility of lower probability scores by allowing them to have partial membership in higher probability bins, minimizing the skew problem in calibration error estimation.

Towards testing this hypothesis, we propose a new metric for estimating calibration error, i.e. Fuzzy Calibration Error (FCE), that utilizes fuzzy binning instead of crisp binning to allow edge probability contributions and minimize skew in calculating calibration error. We perform empirical evaluation across different classification settings, comparing FCE with the baseline calibration error estimation metric ECE.

Our results show that, unlike ECE, FCE better captures miscalibration in lower probability bins and provides a tighter and less skewed estimate of calibration error. These improvements are more visible in multi-class settings, where the skew in confidence scores exacerbates the calibration error estimation problem.

The contributions of this work are summarized as follows:

- We propose Fuzzy Calibration Error (FCE) metric which uses fuzzy binning to account for edge probabilities and minimize skew in calibration error estimation
- We perform empirical evaluation across a wide range of classification settings and show the benefits of using FCE over ECE in minimizing the impacts of probability skew on calibration error estimation

2 Background

2.1 Neural Network Calibration

Neural network calibration refers to the process of adjusting a neural network model's output probabilities to reflect the true probabilities of the events it is predicting. With the increased application of neural network architectures in high-risk real-world settings, their calibration has become an extensively studied topic in recent years [18–20]. Recent research has focused on improving the calibration of neural networks, particularly in the context of deep learning. Various methods have been proposed to achieve better calibration, including temperature scaling [6], isotonic regression [21], and histogram binning [22].

2.2 Expected Calibration Error

Expected calibration error (ECE) is a scalar measure of calibration error that calculates the weighted average of the difference between the accuracy of a model and its average confidence level over a set of bins defined by the predicted probabilities. Estimation of expected accuracy from finite samples is done by grouping predictions into M interval bins (each of size $\frac{1}{M}$), and the accuracy of each bin is calculated. Let B_m be a bin containing samples whose prediction confidence lies within the interval $I_m = \left(\frac{m-1}{M}, \frac{m}{M}\right]$. Then the accuracy of B_m, where y_i and $\hat{y}_i$ portray predicted and true class labels, is calculated as shown in Eq. 1.

$$\mathrm{acc}\left(B_m\right) = \frac{1}{|B_m|} \sum_{i \in B_m} \mathbf{1}\left(\hat{y}_i = y_i\right) \tag{1}$$

The average predicted confidence of B_m, is calculated as shown in Eq. 2, where $\hat{p}_i$ refers to the prediction probability of the ith instance in B_m.

$$\mathrm{conf}\left(B_m\right) = \frac{1}{|B_m|} \sum_{i \in B_m} \hat{p}_i \tag{2}$$

In an ideal scenario, for a perfectly calibrated model, $\mathrm{acc}\left(B_m\right) = \mathrm{conf}\left(B_m\right)$ for all m bins where $m \in \{1, \ldots, M\}$.

Finally, ECE is calculated as shown in Eq. 3, where n is total number of samples [9].

$$\mathrm{ECE} = \sum_{m=1}^{M} \frac{|B_m|}{n} \mid \mathrm{acc}\left(B_m\right) - \mathrm{conf}\left(B_m\right) \tag{3}$$

3 Fuzzy Calibration Error

In this work, we propose Fuzzy Calibration Error (FCE), a metric that transforms raw prediction probabilities into soft bin membership values for calibration error estimation. This transformation has two benefits:

1. Allows edge probability contributions when calculating calibration error
2. Minimize probability skew effects by increasing visibility of lower probability bins in calibration error estimation

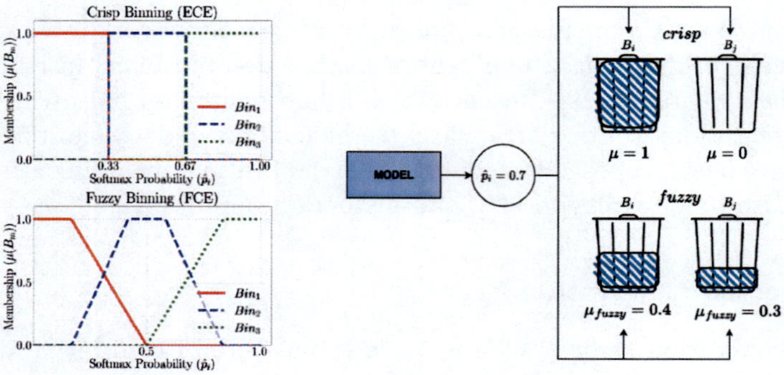

Fig. 1. Crisp binning (Top left) and fuzzy binning (Bottom left) of prediction probabilities, where the number of bins $M = 3$. An example of the difference in bin assignment based on $\hat{p_i}$ in crisp vs fuzzy binning (Right).

To perform fuzzification, we utilize trapezoidal membership functions to map raw softmax prediction probabilities to fuzzy bin membership values. The difference between crisp and fuzzy binning of model prediction probabilities is shown in Fig. 1, with $M = 3$ bins, and can be extended to any number of bins where $M > 3$. While ECE only allows for crisp membership within each bin, FCE offers a more flexible binning approach, with partial memberships allowed across multiple bins.

Fuzzy Calibration Error (FCE) calculates the weighted average of the difference between accuracy and average model confidence over a set of M fuzzy bins. Estimation of expected accuracy from finite samples is done by grouping predictions into M fuzzy bins, and the accuracy of each bin is calculated. Let B_m be a bin containing samples whose prediction confidence lies within the interval $I_m = \left(\frac{m-1}{M}, \frac{m}{M}\right]$. Then the accuracy for bin B_m, where y_i and $\hat{y}_i$ portray predicted and true class labels, is calculated as shown in Eq. 4.

$$\text{acc}_{fuzzy}(B_m) = \frac{1}{|\mu_{fuzzy}(B_m)|} \sum_{i \in B_m} \mu_{fuzzy}(B_m)(\hat{y}_i = y_i) \tag{4}$$

Then, the average fuzzy predicted confidence of B_m, is calculated as shown in Eq. 5.

$$\text{conf}_{fuzzy}(B_m) = \frac{1}{|\mu_{fuzzy}(B_m)|} \sum_{i \in B_m} \mu_{fuzzy}(B_m) \cdot \hat{p}_i \tag{5}$$

Finally, FCE is calculated as shown in Eq. 6. Unlike ECE where the average is taken over the number of samples in B_m i.e., n, we take the average over the total fuzzy membership in B_m i.e., $\sum_{m=1}^{M} \mu_{fuzzy}(B_m)$.

$$FCE = \frac{1}{\sum_{m=1}^{M} \mu_{fuzzy}(B_m)} \sum_{m=1}^{M} |\mu(B_m)| \cdot |\text{acc}_{fuzzy}(B_m) - \text{conf}_{fuzzy}(B_m)| \tag{6}$$

4 Experiments

To evaluate the impact of fuzzy binning on calibration error estimation, we perform empirical evaluations across different classification settings. We fine-tune language models for text classification and measure their calibration performance.

4.1 Experimental Setup

Datasets. We consider three text classification datasets to run our analyses, which vary in terms of class distributions, briefly described below.

- 20 NEWSGROUPS (20NG): The 20 NEWSGROUPS dataset [23] is a collection of newsgroup documents containing approximately 20,000 documents with an (almost) balanced class distribution across 20 newsgroups/topics.
- AGNEWS (AGN): The AG's news topic classification dataset [24] is a collection of approximately 128,000 news articles, from 4 sources. This dataset is widely used in clustering, classification and information retrieval.
- IMDB: The IMDB Movie reviews dataset [25] is a collection of 50,000 movie reviews from the Internet Movie Database (IMDb). Each review is assigned either a positive or negative label, and the data is widely used to train models for binary sentiment classification tasks.

We further simulate varying data resource settings to compare miscalibration across different fine-tuning regimes. This is achieved by using a limited portion of the training data to perform fine-tuning, and has been done in prior works [26].

Metrics. To evaluate calibration across different fine-tuning setups, we use ECE (refer to Eq. 3), FCE (refer to Eq. 6), and overconfidence (OF), described below.

- Overconfidence (OF): Overconfidence is the expectation of model prediction probabilities $\hat{p}_i$ (confidence scores) over incorrect predictions and is calculated as shown in Eq. 7.

$$\text{OF} = \frac{1}{|k|} \sum_{i \in incorrect} \hat{p}_i \tag{7}$$

Here k is the total number of incorrect predictions made by a given model.

Fine-Tuning Setup. We implement text classification using a fine-tuned BERT [27]. Since the focus of our work is not to create the most accurate fine-tuned model but to compare the efficacy of ECE and FCE across skewed prediction probabilities, we only fine-tune over one epoch and collect miscalibrated prediction probabilities.

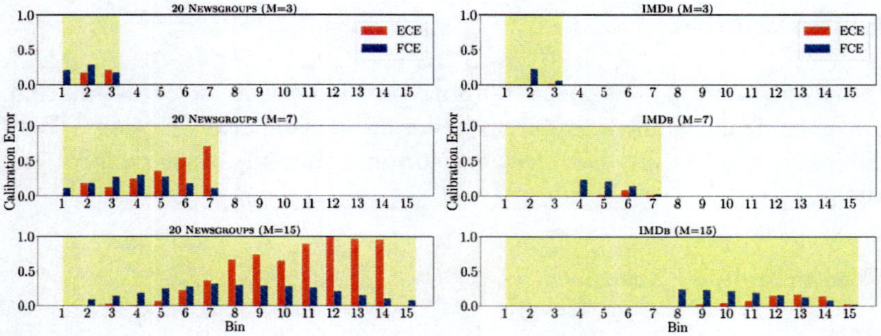

Fig. 2. Variation in calibration error estimated using ECE and FCE across different bin sizes (top to bottom) and class distributions (left vs right)

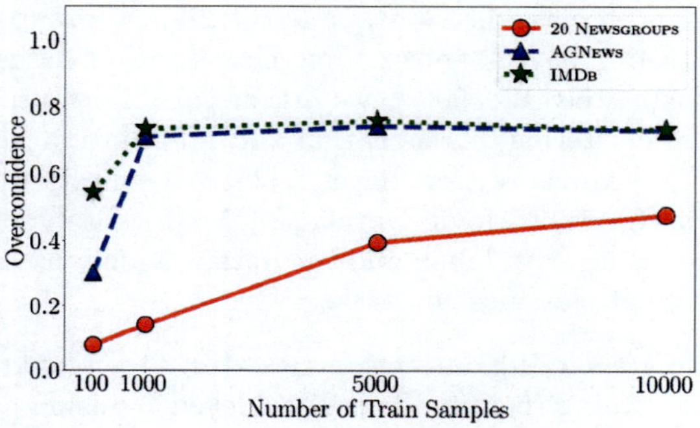

Fig. 3. Variation in model overconfidence (OF) across different sample sizes

4.2 Results

Fuzzy Binning in FCE Better Captures Lower Probability Bins and Edge Probabilities: While ECE bins are highly impacted by the leftward skew in prediction probabilities, FCE yields a more uniformly distributed binning scheme. This can be seen in Fig. 2, where the primary contributors of ECE calculations are the higher probability bins, barely including lower probability bins in calculations. On the other hand, FCE is more uniformly spread across the probability range, better capturing lower probability bins and offering immunity against highly skewed prediction probabilities.

Model Overconfidence in Multi-class Classification Settings is Low but Continuously Increasing: Refer to Fig. 3 to observe the changes in overconfidence in model predictions. Although, a multi-class classification dataset like 20 Newsgroups results in lower overconfidence in predictions in limited data regimes, as compared to datasets with fewer classes, this overconfidence increases as the number of samples during fine-tuning increases. On the other hand,

Table 1. Variations in ECE and FCE across different fine-tuning settings. Here, Δ calculates the average difference in estimated calibration error when binning is performed using fewer bins ($M \in [2..7]$) versus more bins ($M \in [8..15]$).

	ECE	Δ_{ECE}	FCE	Δ_{FCE}
Fine-tuning samples	AGNEWS			
100	15.41	**2.36**	32.50	0.00
1000	3.33	0.63	11.41	0.46
5000	0.71	0.41	7.77	0.71
10000	0.80	0.78	6.86	0.66
	IMDB			
100	5.00	**1.71**	22.50	0.00
1000	3.42	1.51	12.01	0.24
5000	1.49	0.23	7.41	0.82
10000	0.26	0.22	8.01	0.84
	20 NEWSGROUPS			
100	1.31	0.20	5.90	0.00
1000	29.21	**4.47**	38.83	0.27
5000	9.99	1.54	24.05	0.11
10000	2.28	1.30	16.18	0.39

[1] ECE, FCE, Δ_{ECE} and Δ_{FCE} values are scaled by a factor of 10.

datasets with fewer classes i.e., i.e., AGNEWS and IMDB output highly overconfident predictions in limited data regimes, but this overconfidence plateaus as one keeps adding more samples.

Unlike ECE, FCE is not Sensitive to the Binning Strategy and Underlying Data Used for Training: ECE is a highly sensitive calibration error estimation metric, and is easily influenced by slight changes in data and/or binning strategies. Table 1 shows variations in Δ, which calculates the average difference in estimated calibration error when binning is performed using fewer bins ($M \in [2..7]$) versus more bins ($M \in [8..15]$). While ECE displays larger variations in calibration error estimation due to binning choices, FCE is fairly immune to these choices and shows minimal Δ in most cases. Further, Fig. 4 shows that the distribution of ECE across probability bins is highly variable, and usually leftward skewed. On the other hand, FCE bins are more evenly distributed and as shown in Table 1, output more conservative calibration error estimates.

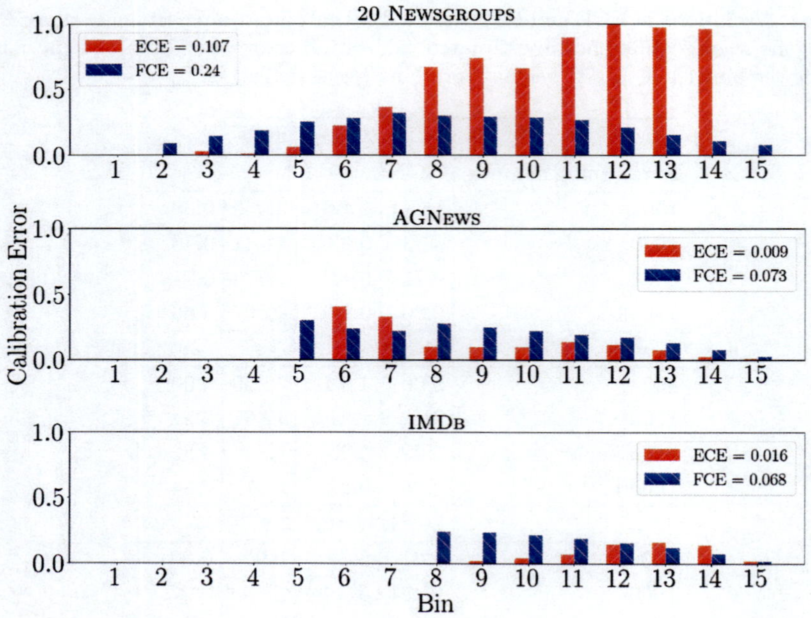

Fig. 4. Binning of prediction probabilities across $M = 15$ bins (model fine-tuned on $n = 5000$ samples)

5 Conclusion

Overconfidence in neural networks lends to the problem of erroneous estimation of calibration error. ECE, a widely adopted metric of measuring calibration error across model decisions has recently come under scrutiny for being biased towards high probability bins. To address this limitation, we propose a new calibration error metric, i.e. Fuzzy Calibration Error (FCE). This metric transforms raw model confidence scores into fuzzy bin memberships, allowing more visibility of lower probability bins within the calibration error calculations. Our results show that FCE offers a tighter estimate of calibration error and the benefits of this metric are more prominent in multi-class classification settings, where skew in model confidence largely affects calibration error estimation using ECE.

Acknowledgments. This work was partially supported by the Department of Justice grant #15PJDP-21-GK-03269-MECP.

References

1. Bihani, G., Rayz, J.T.: Low anisotropy sense retrofitting (LASeR): towards isotropic and sense enriched representations. In: NAACL-HLT 2021, p. 81 (2021)
2. Chen, Q., Zhuo, Z., Wang, W.: BERT for joint intent classification and slot filling. arXiv preprint arXiv:1902.10909 (2019)

3. Devlin, J., Chang, M.-W., Lee, K., Toutanova, K.: BERT: pre-training of deep bidirectional transformers for language understanding. arXiv preprint arXiv:1810.04805 (2018)
4. Radford, A., et al.: Language models are unsupervised multitask learners. OpenAI Blog **1**(8), 9 (2019)
5. Yang, Z., Dai, Z., Yang, Y., Carbonell, J., Salakhutdinov, R.R., Le, Q.V.: XLNet: generalized autoregressive pretraining for language understanding. In: Advances in Neural Information Processing Systems, vol. 32 (2019)
6. Guo, C., Pleiss, G., Sun, Y., Weinberger, K.: On calibration of modern neural networks. In: ICML 2017 (2017)
7. Kong, L., Jiang, H., Zhuang, Y., Lyu, J., Zhao, T., Zhang, C.: Calibrated language model fine-tuning for in- and out-of-distribution data. ArXiv (2020). https://doi.org/10.18653/v1/2020.emnlp-main.102
8. Jiang, Z., Araki, J., Ding, H., Neubig, G.: How can we know when language models know? On the calibration of language models for question answering. Trans. Assoc. Comput. Linguist. **9**, 962–977 (2021)
9. Naeini, M.P., Cooper, G., Hauskrecht, M.: Obtaining well calibrated probabilities using Bayesian binning. In: Proceedings of the AAAI Conference on Artificial Intelligence, vol. 29 (2015)
10. Ovadia, Y., et al.: Can you trust your model's uncertainty? Evaluating predictive uncertainty under dataset shift. In: Advances in Neural Information Processing Systems, vol. 32 (2019)
11. Huang, Y., Li, W., Macheret, F., Gabriel, R.A., Ohno-Machado, L.: A tutorial on calibration measurements and calibration models for clinical prediction models. J. Am. Med. Inform. Assoc. **27**(4), 621–633 (2020)
12. Tack, J., Mo, S., Jeong, J., Shin, J.: CSI: novelty detection via contrastive learning on distributionally shifted instances. In: Advances in Neural Information Processing Systems, vol. 33, pp. 11839–11852 (2020)
13. Nixon, J., Dusenberry, M.W., Zhang, L., Jerfel, G., Tran, D.: Measuring calibration in deep learning (2019)
14. Roelofs, R., Cain, N., Shlens, J., Mozer, M.C.: Mitigating bias in calibration error estimation. In: International Conference on Artificial Intelligence and Statistics, pp. 4036–4054. PMLR (2022)
15. Ding, Y., Liu, J., Xiong, J., Shi, Y.: Revisiting the evaluation of uncertainty estimation and its application to explore model complexity-uncertainty trade-off. In: Proceedings of the IEEE/CVF Conference on Computer Vision and Pattern Recognition Workshops, pp. 4–5 (2020)
16. Zhang, J., Kailkhura, B., Han, T.Y.-J.: Mix-n-match: ensemble and compositional methods for uncertainty calibration in deep learning. In: International Conference on Machine Learning, pp. 11117–11128. PMLR (2020)
17. Bihani, G., Rayz, J.T.: Fuzzy classification of multi-intent utterances. In: Rayz, J., Raskin, V., Dick, S., Kreinovich, V. (eds.) NAFIPS 2021. LNNS, vol. 258, pp. 37–51. Springer, Cham (2022). https://doi.org/10.1007/978-3-030-82099-2_4
18. Thulasidasan, S., Chennupati, G., Bilmes, J.A., Bhattacharya, T., Michalak, S.: On mixup training: improved calibration and predictive uncertainty for deep neural networks. In: Advances in Neural Information Processing Systems, vol. 32 (2019)
19. Malinin, A., Gales, M.: Predictive uncertainty estimation via prior networks. In: Advances in Neural Information Processing Systems, vol. 31 (2018)
20. Hendrycks, D., Liu, X., Wallace, E., Dziedzic, A., Krishnan, R., Song, D.: Pretrained transformers improve out-of-distribution robustness. In: Proceedings of the

58th Annual Meeting of the Association for Computational Linguistics, pp. 2744–2751. Association for Computational Linguistics, Online (2020). https://doi.org/10.18653/v1/2020.acl-main.244

21. Platt, J., et al.: Probabilistic outputs for support vector machines and comparisons to regularized likelihood methods. Adv. Large Margin Classif. **10**(3), 61–74 (1999)
22. Zadrozny, B., Elkan, C.: Obtaining calibrated probability estimates from decision trees and naive Bayesian classifiers. In: ICML, vol. 1, pp. 609–616 (2001)
23. Mitchell, T.: Twenty newsgroups data set. UCI Machine Learning Repository (1999)
24. Zhang, X., Zhao, J.J., LeCun, Y.: Character-level convolutional networks for text classification. In: NIPS (2015)
25. Maas, A.L., Daly, R.E., Pham, P.T., Huang, D., Ng, A.Y., Potts, C.: Learning word vectors for sentiment analysis. In: Proceedings of the 49th Annual Meeting of the Association for Computational Linguistics: Human Language Technologies, pp. 142–150. Association for Computational Linguistics, Portland (2011)
26. Kim, J., Na, D., Choi, S., Lim, S.: Bag of tricks for in-distribution calibration of pretrained transformers. arXiv. arXiv:2302.06690 [cs] (2023)
27. Devlin, J., Chang, M.-W., Lee, K., Toutanova, K.: BERT: pre-training of deep bidirectional transformers for language understanding. In: Proceedings of the 2019 Conference of the North American Chapter of the Association for Computational Linguistics: Human Language Technologies, Volume 1 (Long and Short Papers), pp. 4171–4186. Association for Computational Linguistics, Minneapolis (2019). https://doi.org/10.18653/v1/N19-1423

Developing System Requirements for Trustworthy AI Enabled Refueling Spacecraft

Elizabeth Rochford(✉) and Kelly Cohen

Aerospace Engineering, University of Cincinnati, Cincinnati, OH, USA
rochfoer@mail.uc.edu

Abstract. Refueling satellites is a relatively new concept that will only be first operational in the coming years. Aerial refueling of aircraft has taken place for decades. However, refueling satellites has one big challenge that refueling aircraft does not. Due to the stringent requirements, refueling satellites requires the use of intelligent autonomous systems which leads to the question of trust in the system. Trust needs to be built upon a comprehensive understanding of the system which in turn need to be integrated into the development of the system requirements. Model-based systems engineering (MBSE) is a proven solution to best capture the systems engineering approach increasing understandability. Furthermore, by providing a visual representation of a system, model-based systems engineering allows for better understanding, traceability, earlier detection of errors, and more.

1 Introduction

Refueling of aerospace vehicles is a concept that has been proven time and time again to be a force multiplier, operationally essential, safe, and cost effective. By not having to land, in air refueling extends the mission time of each aircraft that can be refueled in air. Transferring the concept to a satellite application, being able to refuel satellites in space will extend the operational lifetime of each satellite. In-orbit refueling of satellites is referred to as the "Holy Grail" of servicing satellites [1]. This will save research and development time and money, as well as save on launch costs and reduce associated launch risks, because now satellites won't need to be replaced after they run out of fuel. It is acknowledged that technology is constantly evolving, and new satellites are made to reflect that new technology but increasing the lifespan of satellites through refueling can also provide "flexibility in fleet planning [1]." Most satellites are able to continue operating well past the time they run out of fuel. According to Dr. Bryan L. Benedict, from the perspective of a client, they would save 28 million dollars each year per satellite that can be refueled rather than replaced [1].

When discussing refueling, we begin to think of aerial aircraft refueling, which is a similar concept to satellite refueling. The biggest difference between in-air refueling of military aircraft and satellites in space is the human aspect. When refueling military aircraft, the pilots of both vehicles have trust in the other which is based on elaborate

K. Cohen et al. (Eds.): NAFIPS 2023, LNNS 751, pp. 101–114, 2023.
https://doi.org/10.1007/978-3-031-46778-3_10

training, certification and a well-defined operational doctrine. When that human interaction is replaced with AI technology, that trust needs to be re-established. Before satellites will ever be refueled in space, there will need to be trust in the autonomous systems. To have trust in an AI enabled system, there needs to be an understanding of the AI model based on explainability, transparency coupled with verification and validation.

2 Literature Review

2.1 Space Trusted Autonomy Readiness Levels

Space trusted autonomy readiness levels are a new concept. A ground-breaking article was published in October 2022 [2]. In the past, technology readiness levels (TRLs) have been used to evaluate the state of maturity of space technologies. There are many issues with TRLs when it comes to space autonomy. First, TRLs were initially designed for non-autonomous systems which may have included certain aspects which involved temporal automation (e.g. landing sequence and deployment of a Mars rover). Traditional TRLs don't necessarily focus on information processing but rather size and weight of the system. The quintessential issue is trust [2]. Humans are assumed to be trustworthy. As a result, traditional TRLs don't discuss what it means to be trustworthy. When responsibilities are passed from a human to an autonomous system, there is a lack of trust. TRLs assume that humans are in control, which is not always the case, leading to an autonomous system not being appropriately assessed. Another issue is that an autonomous system will rarely be able to be tested in the exact operating conditions. Humans cannot perfectly replicate the space environment on Earth, and it is seldom feasible to send a prototype into space for testing [3].

Space Trusted Autonomy Readiness (STAR) levels were produced to account for "autonomous capability and trust" [2]. It is important to note that STAR levels are not meant to replace TRLs but compliment them. There are 9 STAR levels that all have a defined maturity in regard to assurance, context, implementation, and operation [2]. The authors state "the need to balance readiness levels becomes acute in this two-dimensional representation of readiness" shown in Fig. 2 [2]. From this figure, we can see that high trust and low technology readiness leads to overconfidence, while low trust and high technology readiness leads to under confidence. Both are undesirable outcomes (Fig. 1).

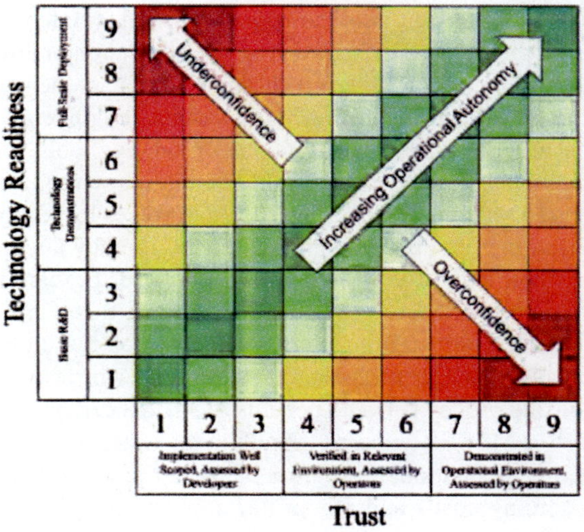

Fig. 1. Technology Readiness versus Trust Levels [2]

2.2 Sparse Requirements Systems Engineering

Systems engineers must decide how to handle these situations, and frequently they move forward using an SRSE approach [3]. Systems engineering is all about managing risk. SRSE is the "conscious decision to proceed toward system definition, design and procurement before all requirements are fully defined" [3]. This approach moves forward while accepting the risk of potential additional time and costs. In addition to potential time and cost risks, the risk of the fact that the product might not fully meet the design requirements is also accepted. The design is still expected to be sufficient for the project.

When it comes to using SRSE for assured autonomy, there are implications that must be considered. For assured autonomy, in order to train and test the system, you must have a complete set of input data. However, it is impossible to have a fully defined set of input data without having a fully defined system. In addition, there cannot be a set of outputs without a set of inputs. Therefore, using the SRSE approach, there will not be a fully defined set of inputs and outputs. For assured autonomy in learning systems there are even more challenges. The 2015–2018 Technology Investment Strategy report discusses four challenges concerning assured autonomy in learning systems [4]. The state-space explosion, unpredictable environments, emergent behavior, and human-machine interfacing and communication are these four challenges [3].

2.3 Model-Based Systems Engineering

Model-Based Systems Engineering (MBSE) is the "formalized application of modeling to support system requirements, design, analysis, verification and validation activities beginning in the conceptual design phase and continuing throughout development and later life cycles" [5]. The model will typically analyze the life cycle of the system by looking at behavior and performance analysis, simulations and testing, and requirements traceability. The goal of MBSE is to provide traceability, full system understanding, and to automate the design process. MBSE allows for early detection of errors in the system as well as an understanding of how a design change will impact the system.

Compared to the traditional document-based systems engineering, MBSE can be easily understood by all stakeholders. As with other modeling tools, there is a standard way of doing this within the model which eliminates multiple voices that are commonly seen in document-based systems engineering. Multiple voices leaves room for miscommunication and misinterpretation. Documents can be written by anyone, and everyone has their own voice, their own way of saying things. In MBSE, there is one way to go about modeling a system. MBSE is a language of its own that can be understood by anyone without misinterpretation. Human error is a common reason projects lose time and money. MBSE can help to decrease the risk of human error in systems engineering which will save time and money [6]. The benefits of MBSE can be visualized in Fig. 3 [7]. The upfront cost of MBSE compared to traditional systems engineering is greater. This greater cost is due to additional resources being put towards defining the process, infrastructure and training costs, and model development, verification, and curation. However, over time and the course of a full project, MBSE proves to be lower in cost than traditional systems engineering. MBSE allows for early error detections, reuse of the model, overall reduction of risks, improved communication among the team, and more. These are all benefits of MBSE that help to save costs over the course of a project when compared to traditional systems engineering [8].

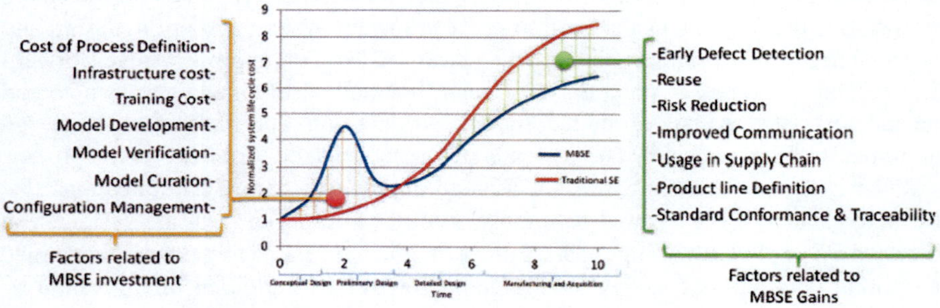

Fig. 2. MBSE gain compared to traditional SE [7]

2.4 Concept of Refueling Aerospace Vehicles

Refueling vehicles has been a concept for a long time with the earliest in-air refueling experiment occurring in the 1920s [9]. The concept of in-air refueling has been well known among the aerospace industry, especially with the military applications. With in-air refueling of aircraft, both the aircraft providing the fuel and the aircraft receiving the fuel are flown by humans. As discussed above, having that human element allows for an automatic trust in the system. When discussing refueling satellites, both the refueling vehicle and the satellite are not directly controlled by humans which leads to a lack of trust. When humans lack trust, they tend to choose alternative methods that are trustworthy. The goal is to prove trust so that AI can be used for refueling vehicles.

Currently, Lockheed Martin, Northrop Grumman, and Orbit Fab have teamed up to design a refueling vehicle [10, 11]. Together, they are focused on sending a refueling vehicle into Geosynchronous orbit (GEO) [10]. This will be a commercially available design and would cost customers $20 million to refuel their satellites just once [10]. This project is still in the process of being designed but is expected to be launched in 2025 [10].

The Space Force is also designing a system to test on-orbit refueling and servicing [12]. Tetra-5 is a prototype system consisting of up to three spacecraft that will work together to carry out the system missions [12]. The Space Force is using Tetra-5 as a proof of concept [12]. The system will be launched in GEO with the intention to show rendezvous and docking capabilities [12]. The lifespan of the system will only be two years [12].

2.5 Trustworthy AI

The definition of trustworthy is worthy of confidence [13]. Synonyms for trustworthy are reliable, calculable, responsible, and safe. All words that will describe trustworthy artificial intelligence. There are many definitions for artificial intelligence, but the U.S. Department of State defines AI as "a machine-based system that can, for a given set of human defined objects, make predictions, recommendations, or decisions influencing real or virtual environments" [14]. Trustworthy AI is an important part of refueling satellites. AI will be used to monitor system surroundings to ensure safety of the system, to self-diagnose and heal if anything goes wrong with the system, and so much more. In addition to this, trustworthy AI will allow for more operational flexibility, superior performance, and to be more fault tolerant and continue to operate if the system can guarantee safe and effective operations.

For all AI systems, having a large, good quality data set to train the AI is crucial [15]. The training data sets the foundation for every decision the system will make for its whole life cycle. Trustworthy AI is accurate, reliable, understandable, and ethical [14]. Understandability is a key component to having trustworthy AI. The algorithm needs to be transparent or, in other words, inspectable [16]. This will allow the user to have confidence in the system, adding to the trust in the AI.

Bias is one of the biggest challenges in creating trustworthy AI [16]. Bias is present in every human made thing, including AI algorithms. Typically, bias is introduced in the training data [15]. As mentioned above, AI needs a large data set to train from. In order to get a large data set, values are typically created by humans which can unconsciously add bias to the system. While bias is still an ongoing issue, one way to help mitigate harmful biases is to have a very diverse team working with the system [16]. A team of all different educational, cultural, and political backgrounds can help to minimize bias. Trustworthy AI will also allow for more people to be able to modify and work with the algorithms, opening the door for a diverse team.

AI is the future of so many industries, including aerospace. AI will help cure cancer, allow for better protection of our country, allow for safer travel, and so much more.

However, until the challenges of creating understandable AI are addressed, there will be hesitation to using AI [17].

3 Methodology

3.1 Systems Engineering

Refueling vehicles and satellites are systems that are in the process of or are already designed. The subsections below will discuss how requirements were developed from initial design to disposal, and how requirements and life cycle research was done to determine safety constraints as well as typical timelines for projects.

3.1.1 Requirements

Before developing a set of requirements, the stakeholders must be defined. For this system the stakeholders are listed below.

- **Customer:** In this case, it is assumed that the customer will be a private company (not the government). This customer will want its project to be successful.
- **Government:** While the government isn't the direct customer, the success of this project is important for future government projects and designs. Refueling satellites will have numerous government/defense/military applications.
- **Manufacturers:** The companies that manufacture and assemble the system will want the system to be successful and perform as expected. They will want to ensure the parts and assembly meet expectations and do not fail.
- **Control Center:** The engineers and scientists on Earth that are responsible for communicating to the system. The control center will also be responsible for intermediate level maintenance.
- **International Space Station:** The engineers and scientists on the ISS are responsible for depot-level maintenance.

The first step was determining expected cost, timeline, lifespan, and weight. These values were estimated through research on average costs and timelines for developing satellites. The system design cost, launch cost, and system weight were estimated based on the OSAM 1 on-orbit servicing vehicle concept [18]. OSAM 1 is a refueling vehicle that would launch to low earth orbit (LEO) [18].

In order to create system requirements, it was important to consider what type of things could threaten the system. The design must account for the environment that the system will operate in. The main threats to the system in its operational environment are radiation, charged particles, collision with other objects in space, and gravity. These threats along with customer defined needs are what drive the capabilities of the system. However, not all requirements are made to account for potential threats. A lot of requirements are written to simply meet customer needs. The customer determined that this system needs to be able to safely connect with satellites, to be able to travel in any direction, to transfer fuel from its own fuel tank to the fuel tank of the satellite, etc.

After determining capabilities, subsystems were identified to meet necessary capabilities. These subsystems include a tether to connect to satellites, a propulsive system to control speed and direction, an AI control system to monitor proximity to other objects, radiation levels, the ability to communicate with the control center, to self-diagnose any issues based on an elaborate prognostics and heath management system, and a fuel tank to store fuel as well as transfer the fuel from the tank to the satellite. Note that transfer of fuel in a zero-gravity environment has its own unique set of challenges. The goal of the requirements for this utilization phase was ensuring safety, to make sure that if the system senses danger, it can protect itself.

3.1.2 MBSE Tool Introduction

For this project, Eclipse Capella™ was used to create models [19]. Capella was created by Thales in 2007 and is an open-source program that provides a user-friendly way to utilize MBSE [19]. The organization and architecture of Capella allows for the user to work on multiple diagrams at once while still maintaining that connectivity of a single component. For this project, the system description and requirements were very high level. In the Capella models, the organizational level diagrams were utilized.

There are several benefits to using Eclipse Capella™. While it was mentioned above, it is a huge benefit that Capella is open source. Many MBSE tools are expensive to use, making it hard for companies to implement on a large scale. Open-source programs, such as Capella, also have forums and lots of tools to help people learn the software [20]. Eclipse Capella™ has numerous training materials to help users learn and feel confident in using the software [20]. The training material courses that Capella offers are free and very informative. These courses, along with the general layout of Capella, make this program a user-friendly software that even beginners can use. Figure 8 shows the general layout of Capella [21]. The layout is organized and well thought out, making it very user friendly and easy to learn.

4 Results and Discussion

4.1 System Requirements

Below are top level requirements for a refueling vehicle. The requirements are organized into phases. The utilization phase is further broken down into system capabilities. These requirements would need adaptation for each refueling vehicle as every project has different needs as well as timelines. For this application, it was assumed that the system would be launched into low-earth orbit (LEO) and is to be used for nonfederal applications. The system needs to utilize trustworthy AI, maintain communication with the control center located on Earth, maintain safety in its operating environment, transfer fuel from itself to a satellite, travel through space and rendezvous with other systems, have self-diagnostic and healing capabilities, and follow proper disposal instructions. Various requirements were made based on standard design processes in the aerospace industry and other companies also creating refueling vehicles (Table 1).

Table 1. System Requirements for Refueling Vehicles

Phase	Number	Requirement	Validation
Design	1.1	The design team shall have a PDR and SRR presentation and meeting before starting any testing or construction	Demonstration
	1.2	The team shall have a CDR presentation and meeting before final design is decided upon	Demonstration
	1.3	The system shall be designed and built in 7 years	Demonstration
	1.4	The system shall have a digital twin for testing and simulation purposes	Demonstration
	1.5	The design team shall consist of experts in the AI field	Demonstration
Production and Testing	2.1	The system shall cost no more than $230 million to manufacture	Demonstration
	2.2	The system shall weigh no more than 6500 kg [15]	Ground test
	2.3	The system shall cost no more than $50 million to launch to LEO [15]	Demonstration
	2.4	The system shall be designed and tested to a STAR Level 7 [2]	Simulation
	2.5	The system shall be designed and tested to TrRL Level 8 [2]	Simulation
Utilization	3.1	The system shall be able to provide service to all satellites as requested by the mission control center	Simulation
	3.2	The system shall be able to locate given satellites and navigate to the satellite in a fuel-efficient manner	Simulation
	3.3	The system shall provide continuous feedback to control center during and after fueling	Ground Test

(continued)

Table 1. (*continued*)

Phase	Number	Requirement	Validation
AI	3.4	The system will have a human centered AI system to enable it locate or refuel satellite and perform all tasks until intermediate level maintenance and astronauts in ISS for depot maintenance	Simulation
	3.4.1	The system shall utilize trustworthy AI to enable fault tolerance	Ground Test
	3.5	System software updates shall be applied to the system immediately after verification and validation testing is deemed complete	Ground Test
Safe connection	3.6	The system shall be able to safely connect with given satellite and conduct refueling operation	Simulation
	3.6.1	The system shall have 8 thrusters to provide 6 DOF [22]	Flight Test
	3.6.2	The system shall provide continuous feedback to control center during and after fueling	Ground Test
Proximity Awareness	3.7	The system shall continuously use proximity sensors and a run-time assurance system to determine if objects are less than 1 km away [23]	Flight Test
	3.7.1	The control center shall monitor space object locations and give coordinates to system	Simulation
Space Weather	3.8	Solar weather probes shall constantly monitor space weather conditions and radiation levels	Flight Test
Space Weather	3.9	The system shall put electronics into "safe mode" when there is increased radiation in the environment	Ground Test

(*continued*)

Table 1. *(continued)*

Phase	Number	Requirement	Validation
Space Weather	3.10	The system electronics shall be radiation hardened to protect against radiation levels	Ground Test
Space Weather	3.11	The system shall perform a reboost maneuver if there are increased drag forces	Flight Test
Cyber Attacks	3.12	System shall be compliant to NIST 800–171: Protecting Controlled Unclassified Information in Nonfederal Systems and Organizations [24]	Ground Test
Self-Diagnostic Capabilities	3.13	The system shall continuously self-monitor and diagnose any failures and decide on operating at a fault tolerant mode if it can guarantee safe and effective operations	Ground Test
	3.13.1	The control center shall be available to respond to any failure in less than an hour from detection	Demonstration
	3.14	The system shall travel at a velocity of no more than 6 cm/second while docking with satellites [25]	Flight Test
	3.15	The system shall be able to transfer fuel to a satellite	Simulation
	3.15.1	The system shall utilize a nozzle tool to transfer fuel to satellite [26]	Simulation
	3.15.2	The system shall calculate the amount of fuel and fueling time for each satellite given fuel tank size given by the control center	Ground Test
Disposal	4.1	The system lifespan shall be 10 years [27]	Simulation
	4.2	At completion of the mission, the system shall return to the International Space Station	Simulation

(continued)

Table 1. (*continued*)

Phase	Number	Requirement	Validation
	4.3	Scientists on the ISS shall recycle parts from the system	Demonstration
	4.3.1	Recycled system parts shall be upcycled to create a new refueling vehicle	Simulation
	4.3.2	Any parts that cannot be upcycled shall be replaced with new parts sent from Earth	Simulation

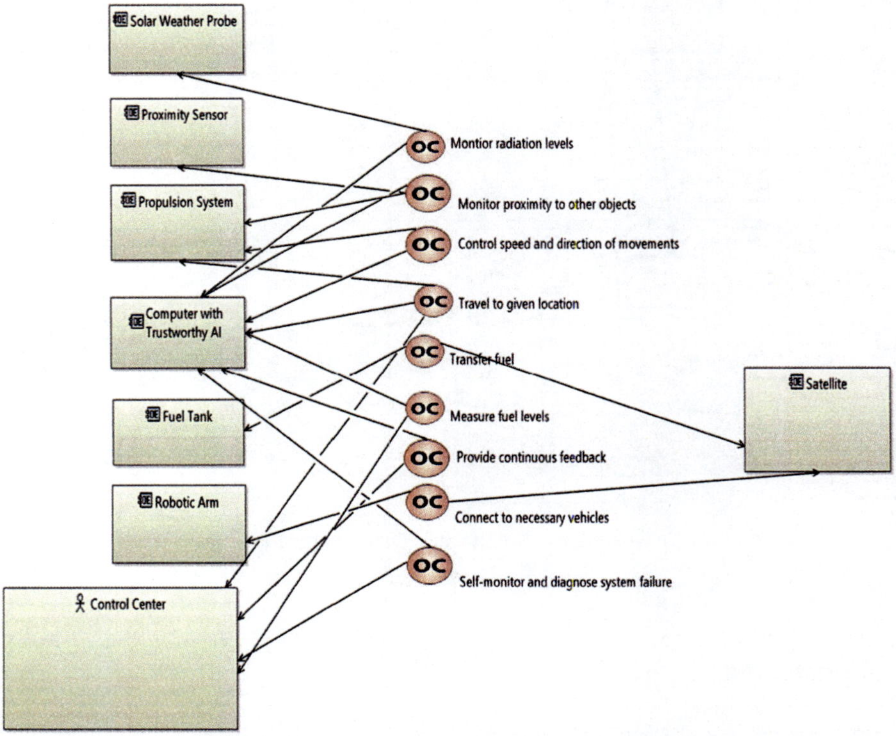

Fig. 3. Capella Operational Capabilities Diagram for Refueling Vehicles

4.2 MBSE Models

Eclipse Capella was used to create MBSE models. These models were kept high-level and focused on requirements. Organizational and Systems models were created. This section will explain and discuss the meaning of each model as well as how the models were created. To start, the system was broken down into its major subsystems: solar weather probe, proximity sensor, propulsion system, computer with trustworthy AI, fuel tank, and a robotic arm. Next, the control center and external systems that the refueling vehicle would frequently interact with were identified. Then, the customer needs were broken down into major capabilities. The models below show how the subsystems, external systems, and major capabilities are connected and the involvement that each have with a variety of functions (Fig. 4).

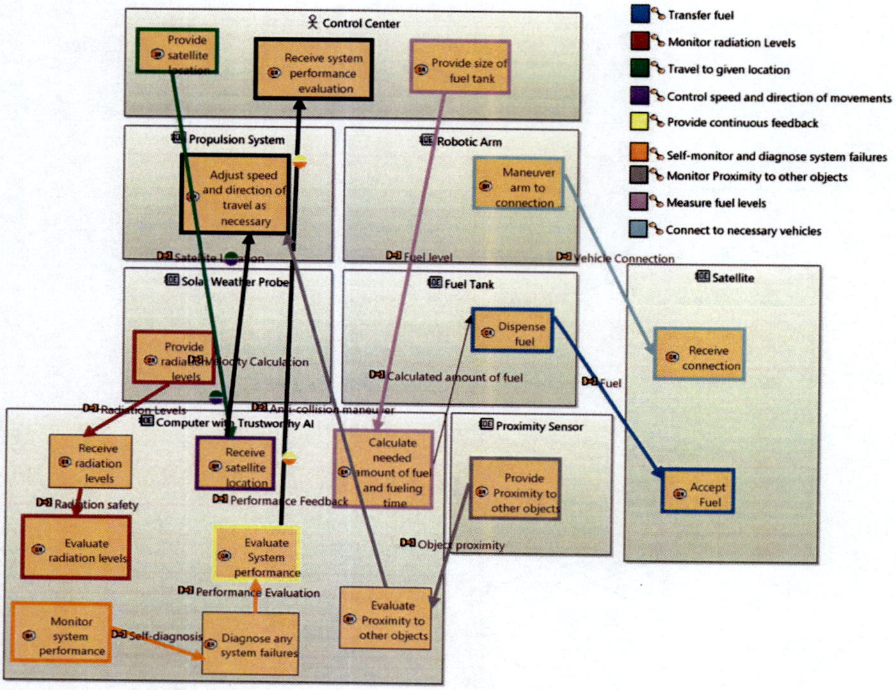

Fig. 4. Capella Operational Architecture Diagram for Refueling Vehicles

5 Conclusion

The models presented were formed based on requirements written. The requirements were based on various research of satellite safety, behavior, and design processes. Trustworthy AI is a major challenge with proving feasibility of refueling satellites. Space trusted autonomy readiness levels as well as trust readiness levels were evaluated and chosen for the project. These safety and trust levels will help provide a standard to design the system to.

The MBSE models that were created show that a complicated system can be broken down into less complicated parts that are easier to understand. The models also help to show the connection between subsystems and how the system will interact with its surroundings. The models help to provide a greater understanding of the system and its processes. By providing a way to understand the system, it will be more trustworthy. In addition, the MBSE models will provide traceability and early detection of errors. The also allow for all stakeholders to be on the same page.

References

1. Benedict, B.L.: Investing in satellite life extension - fleet planning options for spacecraft owner/operators. In: AIAA SPACE 2014 Conference and Exposition (2014). https://doi.org/10.2514/6.2014-4445
2. Hobbs, K.L., et al.: Space trusted autonomy readiness levels (2022)
3. Gosian, G., Cohen, K.: Sparse requirements systems engineering and implications for assured autonomy. J. Defense Manag. (2021)
4. United States, Congress, Research and Engineering. *Autonomy Community of Interest (COI) Test and Evaluation, Verification and Validation (TEVV) Working Group: Technology Investment Strategy 2015–2018*, Office of the Assistant Secretary of Defense (2015)
5. Hart, L.E.: Introduction To Model-Based System Engineering (MBSE) and SysML. INCOSE. Delaware Valley INCOSE Chapter Meeting. https://www.incose.org/docs/default-source/delaware-valley/mbse-overview-incose-30-july-2015.pdf. Accessed 15 Feb 2023
6. Shevchenko, N.: An Introduction to Model-Based Systems Engineering (MBSE)." Software Engineering Institute Blog, Carnegie Mellon University (2020). https://insights.sei.cmu.edu/blog/introduction-model-based-systems-engineering-mbse/. Accessed 15 Feb 2023
7. Els, Peter. "Model-Based Systems Save Development Time and Money." *Automotive IQ* (2022). https://www.automotive-iq.com/autonomous-drive/columns/model-based-systems-save-development-time-and-money
8. Madni, A., Purohit, S.: Economic analysis of model-based systems engineering. Systems **7**(1), 12 (2019). https://doi.org/10.3390/systems7010012
9. "Aerial Refueling." *Wikipedia*, Wikimedia Foundation (2023). https://en.wikipedia.org/wiki/Aerial_refueling
10. Foust, J.: Orbit Fab Secures New Investor to Support Satellite Refueling Efforts. *SpaceNews* (2023). https://spacenews.com/orbit-fab-secures-new-investor-to-support-satellite-refueling-efforts/
11. "Refueling Satellites in Space." *Lockheed Martin* (2021). https://www.lockheedmartin.com/en-us/news/features/2021/refueling-satellites-in-space.html
12. United States, Congress, Office of Public Affairs. *Space Systems Command Selects Orion Space Solutions for Tetra-5 Other Transaction Agreement* (2022). https://www.ssc.spaceforce.mil/Portals/3/Documents/PRESS%20RELEASES/Tetra-5.pdf?ver=otxbrmw6mqtW5c3d4v084w%3D%3D. Accessed 15 Feb 2023
13. "Trustworthy Definition & Meaning." *Merriam-Webster*, Merriam-Webster. https://www.merriam-webster.com/dictionary/trustworthy
14. "Artificial Intelligence (AI) - United States Department of State." *U.S. Department of State*, U.S. Department of State (2021). https://www.state.gov/artificial-intelligence/
15. Matthews, Sky. "Challenges in Systems Engineering of Intelligent and Autonomous Systems." IBM Engineering (2019)
16. Felderer, M., Rudolf, R.: An Overview of Systems Engineering Challenges for Designing AI-Enabled Aerospace Systems (2020). https://doi.org/10.2514/6.2021-0564.vid
17. U.S. Department of Defense Responsible Artificial Intelligence Strategy and Implementation Pathway (2022)

18. Hatty, I.: Viability of on-orbit servicing spacecraft to prolong the operational life of satellites. J. Space Safety Eng. **9**(2), 263–268 (2022). https://doi.org/10.1016/j.jsse.2022.02.011
19. "Discover Capella in 2 Minutes." *YouTube*, Eclipse Capella, 21 Sept. 2020, https://www.youtube.com/watch?v=WSzlN4YT3gM. Accessed 15 Feb. 2023
20. Guindon, C.: Why Capella? Taking on the Challenges of Complex Systems Engineering. *The Eclipse Foundation.* https://www.eclipse.org/community/eclipse_newsletter/2017/december/article1.php
21. "Features and Benefits." *Capella MBSE Tool - Features*, Eclipse Capella. https://www.eclipse.org/capella/features.html
22. Space Technology 7." *NASA*, NASA Jet Propulsion Laboratory. https://www.jpl.nasa.gov/nmp/st7/TECHNOLOGY/thrusters.html
23. Lea, R.: What Is a Geosynchronous Orbit? *Space.com*, Space (2015). https://www.space.com/29222-geosynchronous-orbit.html
24. United States, Congress, Ross, Ron, et al. *Protecting Controlled Unclassified Information in Nonfederal Systems and Organizations*, National Institute of Standards and Technology (2016)
25. "TRACK AND CAPTURE OF THE ORBITER WITH THE SPACE STATION REMOTE MANIPULATOR SYSTEM" (PDF). NASA. Archived (PDF) from the original on August 7, 2020. Retrieved July 7, 2017
26. "NASA's Exploration & in-Space Services." NASA, NASA. https://nexis.gsfc.nasa.gov/rrm_refueling_task.html
27. Borthomieu, Y., Gianfranco P.: 14 - Satellite Lithium-Ion Batteries. Lithium-Ion Batteries: Advances and Applications, pp. 311–344. Elsevier, Amsterdam (2014)

Theoretical Explanation of Bernstein Polynomials' Efficiency

Vladik Kreinovich(✉)

Department of Computer Science, University of Texas at El Paso,
500 W. University, El Paso, TX 79968, USA
vladik@utep.edu

Abstract. Many fuzzy data processing problems can be reduced to problems of interval computations. In many applications of interval computations, it turned out to be beneficial to represent polynomials on a given interval $[\underline{x}, \overline{x}]$ as linear combinations of *Bernstein* polynomials $(x - \underline{x})^k \cdot (\overline{x} - x)^{n-k}$. In this paper, we provide a theoretical explanation for this empirical success: namely, we show that under reasonable optimality criteria, Bernstein polynomials can be uniquely determined from the requirement that they are optimal combinations of optimal polynomials corresponding to the interval's endpoints.

1 Formulation of the Problem

Polynomials are Often Helpful. If we know the quantities $x_1, \ldots, x_n$ with fuzzy uncertainty, how can we estimate the uncertainty of the result $y = f(x_1, \ldots, x_n)$ of applying a data processing algorithm f to these values? It is known (see, e.g., [1,11,18,19]) that under reasonable conditions, for each α, the α-cut $\mathbf{y}(\alpha)$ of y is equal to the range

$$f(\mathbf{x}_1(\alpha), \ldots, \mathbf{x}_n(\alpha)) = \{f(x_1, \ldots, x_n) : x_1 \in \mathbf{x}_1(\alpha), \ldots, x_n \in \mathbf{x}_n(\alpha)\}.$$

For fuzzy numbers, α-cuts are intervals. In this case, the problem of computing the corresponding range is known as the problem of *interval computations*; see, e.g., [10,13,14].

In many areas of numerical analysis, in particular, in interval computations, it turns out to be helpful to approximate a dependence by a polynomial. For example, in interval computations, Taylor methods – in which the dependence is approximated by a polynomial – turned out to be very successful; see, e.g., [2–4,9,12,16].

The Efficiency of Polynomials Can be Theoretically Explained. The efficiency of polynomials is not only an empirical fact, this efficiency can also be theoretically justified. Namely, in [17], it was shown that under reasonable assumptions on the optimality criterion – like invariance with respect to selection a starting point and a measuring unit for describing a quantity – every function from the optimal class of approximating functions is a polynomial.

K. Cohen et al. (Eds.): NAFIPS 2023, LNNS 751, pp. 115–126, 2023.
https://doi.org/10.1007/978-3-031-46778-3_11

How to Represent Polynomials in a Computer: Straightforward Way and Bernstein Polynomials. In view of the fact that polynomials are efficient, we need to represent them inside the computer. A straightforward way to represent a polynomial is to store the coefficients at its monomials. For example, a natural way to represent a quadratic polynomial $f(x) = c_0 + c_1 \cdot x + c_2 \cdot x^2$ is to store the coefficients c_0, c_1, and c_2.

It turns out that in many applications in which we are interested in functions defined on a given interval $[\underline{x}, \overline{x}]$, we get better results if instead, we represent a general polynomial as a linear combination of *Bernstein polynomials*, i.e., functions proportional to $(x - \underline{x})^k \cdot (\overline{x} - x)^{n-k}$, and store coefficients of this linear combination. For example, in this representation, it is faster to estimate the maximum or minimum on a given polynomial on a given 1-D or multi-D interval; see, e.g., [5–8, 15, 20].

For example, a general quadratic polynomial on the interval $[0, 1]$ can be represented as

$$f(x) = a_0 \cdot x^0 \cdot (1-x)^2 + a_1 \cdot x^1 \cdot (1-x)^1 + a_2 \cdot x^2 \cdot (1-x)^0 = a_0 \cdot x^2 + a_1 \cdot x \cdot (1-x) + a_2 \cdot (1-x)^2;$$

to represent a generic quadratic polynomial in a computer, we store the values a_0, a_1, and a_2. (To be more precise, we store values proportional to a_i.)

For polynomials of several variables defined on a box $[\underline{x}_1, \overline{x}_1] \times \ldots \times [\underline{x}_n, \overline{x}_n]$, we can use similar multi-dimensional Bernstein polynomials which are proportional to $\prod_{i=1}^{n} (x_i - \underline{x}_i)^{k_i} \cdot (\overline{x}_i - x_i)^{n-k_i}$.

Natural Questions. Natural questions are:

- why is the use of these basic functions more efficient than the use of standard monomials $\prod_{i=1}^{n} x_i^{k_i}$?
- are Bernstein polynomials the best or there are even better expressions?

Towards Possible Answers to These Questions. To answer these questions, we take into account that in the 1-D case, an interval $[\underline{x}, \overline{x}]$ is uniquely determined by its endpoints $\underline{x}$ and $\overline{x}$. Similarly, in the multi-D case, a general box $[\underline{x}_1, \overline{x}_1] \times \ldots \times [\underline{x}_n, \overline{x}_n]$ is uniquely determined by two multi-D "endpoints" $\underline{x} = (\underline{x}_1, \ldots \underline{x}_n)$ and $\overline{x} = (\overline{x}_1, \ldots, \overline{x}_n)$. It is therefore reasonable to design the basic polynomials as follows:

- first, we find two polynomial functions $\underline{f}(x)$ and $\overline{f}(x)$, where $x = (x_1, \ldots, x_n)$, related to each of the endpoints;
- then, we use some combination operation $F(a, b)$ to combine the functions $\underline{f}(x)$ and $\overline{f}(x)$ into a single function $f(x) = G(\underline{f}(x), \overline{f}(x))$.

In this paper, we use the approach from [17] to prove that if we select the optimal polynomials $\underline{f}(x)$ and $\overline{f}(x)$ on the first stage and the optimal combination operation on the second stage, then the resulting function $f(x)$ is proportional to a Bernstein polynomial.

In other words, we prove that under reasonable optimality criteria, Bernstein polynomials can be uniquely determined from the requirement that they are optimal combinations of optimal polynomials corresponding to the interval's endpoints.

2 Optimal Functions Corresponding to Endpoints: Towards a Precise Description of the Problem

Formulation of the Problem: Reminder. Let us first find optimal polynomials corresponding to endpoints $\underline{x}$ and $\overline{x}$.

We consider applications in which the dependence of a quantity y on the input values $x_1, \ldots, x_n$ is approximated by a polynomial $y = f(x) = f(x_1, \ldots, x_n)$. For each of the two endpoints $\underline{x}$ and $\overline{x}$, out of all polynomials which are "related" to this point, we want to find the one which is, in some reasonable sense, optimal.

How to Describe this Problem in Precise Terms. To describe this problem in precise terms, we need to describe:

- what it means for a polynomial to be "related" to the point, and
- what it means for one polynomial to be "better" than the other.

Physical Meaning. To formalize the two above notions, we take into account that in many practical applications, the inputs numbers x_i are values of some physical quantities, and the output y also represent the value of some physical quantity.

Scaling and Shift Transformations. The numerical value of each quantity depends on the choice of a measuring unit and on the choice of the starting point. If we replace the original measuring unit by a unit which is $\lambda > 0$ times smaller (e.g., use centimeters instead of meters), then instead of the original numerical value y, we get a new value $y' = \lambda \cdot y$.

Similarly, if we replace the original starting point with a new point which corresponds to y_0 on the original scale (e.g., as the French Revolution did, select 1789 as the new Year 0), then, instead as the original numerical value y, we get a new numerical value $y' = y - y_0$.

In general, if we change both the measuring unit and the starting point, then instead of the original numerical value y, we get the new value $\lambda \cdot y - y_0$.

We Should Select a Family of Polynomials. Because of scaling and shift, for each polynomial $f(x)$, the polynomials $\lambda \cdot f(x) - y_0$ represent the same dependence, but expressed in different units. Because of this fact, we should not select a *single* polynomial, we should select the entire *family* $\{\lambda \cdot f(x) - y_0\}_{\lambda, y_0}$ of polynomials representing the original dependence for different selections of the measuring unit and the starting point.

Scaling and Shift for Input Variables. In many practical applications, the inputs numbers x_i are values of some physical quantities. The numerical value of each such quantity also depends on the choice of a measuring unit and on the choice of the starting point. By using different choices, we get new values $x'_i = \lambda_i \cdot x_i - x_{i0}$, for some values λ_i and x_{i0}.

Transformations Corresponding to a Given Endpoint $x^{(0)} = \left(x_1^{(0)}, \ldots, x_n^{(0)}\right)$. Once the endpoint is given, we no longer have the freedom of changing the starting point, but we still have re-scalings: $x_i - x_i^{(0)} \to \lambda_i \cdot \left(x_i - x_i^{(0)}\right)$, i.e., equivalently, $x_i \to x'_i = x_i^{(0)} + \lambda_i \cdot \left(x_i - x_i^{(0)}\right)$.

What is Meant by "the Best" Family? When we say "the best" family, we mean that on the set of all the families, there is a relation $\geq$ describing which family is better or equal in quality. This relation must be transitive (if $\mathcal{F}$ is better than $\mathcal{G}$, and $\mathcal{G}$ is better than $\mathcal{H}$, then $\mathcal{F}$ is better than $\mathcal{H}$).

Final Optimality Criteria. The preference relation $\geq$ is not necessarily asymmetric, because we can have two families of the same quality. However, we would like to require that this relation be *final* in the sense that it should define a unique *best* family $\mathcal{F}_{\text{opt}}$, for which $\forall \mathcal{G}\, (\mathcal{F}_{\text{opt}} \geq \mathcal{G})$.

Indeed, if none of the families is the best, then this criterion is of no use, so there should be *at least one* optimal family.

If *several* different families are equally best, then we can use this ambiguity to optimize something else: e.g., if we have two families with the same approximating quality, then we choose the one which is easier to compute. As a result, the original criterion was not final: we obtain a new criterion: $\mathcal{F} \geq_{\text{new}} \mathcal{G}$, if either $\mathcal{F}$ gives a better approximation, or if $\mathcal{F} \sim_{\text{old}} \mathcal{G}$ and $\mathcal{G}$ is easier to compute. For the new optimality criterion, the class of optimal families is narrower.

We can repeat this procedure until we obtain a final criterion for which there is only one optimal family.

Optimality Criteria Should be Invariant. Which of the two families is better should not depend on the choice of measuring units for measuring the inputs x_i. Thus, if $\mathcal{F}$ was better than $\mathcal{G}$, then after re-scaling, the re-scaled family $\mathcal{F}$ should still be better than the re-scaled family $\mathcal{G}$.

Thus, we arrive at the following definitions.

3 Optimal Functions Corresponding to Endpoints: Definitions and the Main Result

Definition 1. *By a* family, *we mean a set of functions from* $\mathbb{R}^n \to \mathbb{R}$ *which has the form* $\{C \cdot f(x) - y_0 : C, y_0 \in \mathbb{R}, C > 0\}$ *for some polynomial* $f(x)$. *Let* $\mathcal{F}$ *denote the class of all possible families.*

Definition 2. *By a* optimality criterion $\preceq$ *on the class* $\mathcal{F}$, *we mean a pre-ordering relation on the set* $\mathcal{F}$, *i.e., a transitive relation for which* $F \preceq F$ *for every* F. *We say that a family* F *is optimal with respect to the optimality criterion* $\preceq$ *if* $G \preceq F$ *for all* $G \in \mathcal{F}$.

Definition 3. *We say that the optimality criterion is* final *if there exists one and only one optimal family.*

Definition 4. *Let $x^{(0)}$ be a vector. By a $x^{(0)}$-*rescaling *corresponding to the values $\lambda = (\lambda_1, \ldots, \lambda_n)$, $\lambda_i > 0$, we mean a transformation $x \to x' = T_{x^{(0)},\lambda}(x)$ for which*

$$x'_i = x_i^{(0)} + \lambda_i \cdot \left(x_i - x_i^{(0)}\right).$$

*By a $x^{(0)}$-*rescaling *of a family $F = \{C \cdot f(x) - y_0\}_{C,y_0}$, we mean a family $T_{x^{(0)},\lambda}(F) = \{C \cdot f(T_{x^{(0)},\lambda}(x))\}_{C,y_0}$. We say that an optimality criterion is $x^{(0)}$-*scaling-invariant *if for every F, G, and λ, $F \preceq G$ implies $T_{x^{(0)},\lambda}(F) \preceq T_{x^{(0)},\lambda}(G)$.*

Proposition 1.

- *Let $\preceq$ be a final $x^{(0)}$-scaling-invariant optimality criterion. Then every polynomial from the optimal family has the form*

$$f(x) = A + B \cdot \prod_{i=1}^{n} \left(x_i - x_i^{(0)}\right)^{k_i}. \tag{1}$$

  for some natural numbers $k_1, \ldots, k_n$.
- *For every sequence of natural numbers k_i, there exists a final $x^{(0)}$-scaling-invariant optimality criterion for which the family (1) is optimal.*

Comment. For readers' convenience, all the proofs are placed in the special (last) Proofs section.

Discussion. As we have mentioned, the value of each quantity is defined modulo a starting point. It is therefore reasonable, for y, to select a starting point so that $A = 0$. Thus, we get the dependence

$$f(x) = B \cdot \prod_{i=1}^{n} \left(x_i - x_i^{(0)}\right)^{k_i}.$$

Once the starting point for y is fixed, the only remaining y-transformations are scalings $y \to \lambda \cdot y$.

4 Optimal Combination Operations

In the previous section, we described the optimal functions corresponding to the endpoints $\underline{x}$ and $\overline{x}$. What is the optimal way of combining these functions? Since we are dealing only with polynomial functions, it is reasonable to require that a combination operation transform polynomials into polynomials.

Definition 5. *By a* combination operation, *we mean a function $K : \mathbb{R}^2 \to \mathbb{R}$ for which, if $\underline{f}(x)$ and $\overline{f}(x)$ are polynomials, then the composition $K\left(\underline{f}(x), \overline{f}(x)\right)$ is also a polynomial.*

Lemma 1. *A function $K(a, b)$ is a combination operation if and only if it is a polynomial.*

Discussion. Similarly to the case of optimal functions corresponding to individual endpoint, the numerical value of the function $K\left(\underline{a}, \overline{a}\right)$ depends on the choice of the measuring unit and the starting point: an operation that has the form $K\left(\underline{a}, \overline{a}\right)$ under one choice of the measuring unit and starting point has the form $C \cdot K\left(\underline{a}, \overline{a}\right) - y_0$ under a different choice. Thus, we arrived at the following definition.

Definition 6. *By a* C-family, *we mean a set of functions from* $\mathbb{R}^2 \to \mathbb{R}$ *which has the form* $\{C \cdot K(a, b) - y_0 : C, y_0 \in \mathbb{R}, C > 0\}$ *for some combination operation* $K(a, b)$. *Let* $\mathcal{K}$ *denote the class of all possible C-families.*

Definition 7. *By an* optimality criterion $\preceq$ *on the class* $\mathcal{K}$ *of all C-families, we mean a pre-ordering relation on the set* $\mathcal{K}$, *i.e., a transitive relation for which* $F \preceq F$ *for every C-family* F. *We say that a C-family* F *is optimal with respect to the optimality criterion* $\preceq$ *if* $G \preceq F$ *for all* $G \in \mathcal{K}$.

Definition 8. *We say that the optimality criterion is* final *if there exists one and only one optimal C-family.*

Discussion. From the previous section, we know that both functions $\underline{f}(x)$ and $\overline{f}(x)$ are determined modulo scaling $\underline{f}(x) \to \underline{\lambda} \cdot \underline{f}(x)$ and $\overline{f}(x) \to \overline{\lambda} \cdot \overline{f}(x)$. Thus, it is reasonable to require that the optimality relation not change under such re-scalings.

Definition 9. *By a* C-rescaling *corresponding to the values* $\lambda = \left(\underline{\lambda}, \overline{\lambda}\right)$, *we mean a transformation* $T_\lambda\left(\underline{a}, \overline{a}\right) = \left(\underline{\lambda} \cdot \underline{a}, \overline{\lambda} \cdot \overline{a}\right)$. *By a* C-rescaling *of a family*

$$F = \{C \cdot K(\underline{a}, \overline{a}) - y_0\}_{C, y_0},$$

we mean a family $T_\lambda(F) = \{C \cdot K(T_\lambda(a))\}_{C, y_0}$. *We say that an optimality criterion is* C-scaling-invariant *if for every* F, G, *and* λ, $F \preceq G$ *implies* $T_\lambda(F) \preceq T_\lambda(G)$.

Proposition 2. *Let* $\preceq$ *be a final C-scaling-invariant optimality criterion. Then every combination operation from the optimal family has the form*

$$K(\underline{a}, \overline{a}) = A + B \cdot \underline{a}^{\underline{k}} \cdot \overline{a}^{\overline{k}}.$$

5 Conclusions

By applying this optimal combination operation from Sect. 4 to the optimal functions corresponding to $x^{(0)} = \underline{x}$ and $x^{(0)} = \overline{x}$ (described in Sect. 3), we conclude that the resulting function has the form

$$f(x_1, \ldots, x_n) = K\left(\underline{f}(x_1, \ldots, x_n), \overline{f}(x_1, \ldots, x_n)\right)$$

$$= A + B \cdot \left(\prod_{i=1}^{n} (x_i - \underline{x}_i)^{\underline{k}_i} \right)^{\underline{k}} \cdot \left(\prod_{i=1}^{n} (\overline{x}_i - x_i)^{\overline{k}_i} \right)^{\overline{k}}.$$

Modulo an additive constant, this function has the form

$$f(x_1, \ldots, x_n) = B \cdot \prod_{i=1}^{n} (x_i - \underline{x}_i)^{\underline{k}'_i} \cdot \prod_{i=1}^{n} (\overline{x}_i - x_i)^{\overline{k}'_i},$$

where $\underline{k}'_i = \underline{k}_i \cdot \underline{k}$ and $\overline{k}'_i = \overline{k}_i \cdot \overline{k}$.

These are Bernstein polynomials. Thus, Bernstein polynomials can indeed by uniquely determined as the result of applying an optimal combination operation to optimal functions corresponding to $\underline{x}$ and $\overline{x}$.

6 Proofs

Proof of Proposition 1

1°. Let us first prove that the optimal family F_{opt} is $x^{(0)}$-scaling-invariant, i.e., $T_{x^{(0)},\lambda}(F_{\text{opt}}) = F_{\text{opt}}$.

Since F_{opt} is an optimal family, we have $G \preceq F_{\text{opt}}$ for all families G. In particular, for every family G and for every λ, we have $T_{x^{(0)},\lambda^{-1}}(G) \preceq F_{\text{opt}}$. Since the optimal criterion is $x^{(0)}$-scaling-invariant, we conclude that

$$T_{x^{(0)},\lambda}\left(T_{x^{(0)},\lambda^{-1}}(G)\right) \preceq T_{x^{(0)},\lambda}(F_{\text{opt}}).$$

One can easily check that if we first re-scale the family with the coefficient λ^{-1}, and then with λ, then we get the original family G back. Thus, the above conclusion takes the form $G \preceq T_{x^{(0)},\lambda}(F_{\text{opt}})$. This is true for all families G, hence the family $T_{x^{(0)},\lambda}(F_{\text{opt}})$ is optimal. Since the optimality criterion is final, there is only one optimal family, so $T_{x^{(0)},\lambda}(F_{\text{opt}}) = F_{\text{opt}}$. The statement is proven.

2°. For simplicity, instead of the original variables x_i, let us consider auxiliary variables $z_i = x_i - x_i^{(0)}$. In terms of these variables, re-scaling takes a simpler form $z_i \to \lambda_i \cdot z_i$. Since $x_i = z_i + x_i^{(0)}$, the dependence $f(x_1, \ldots, x_n)$ take the form

$$g(z_1, \ldots, z_n) = f\left(z_1 + x_1^{(0)}, \ldots, z_n + x_n^{(0)}\right).$$

Since the function $f(x_1, \ldots, x_n)$ is a polynomial, the new function $g(z_1, \ldots, z_n)$ is a polynomial too.

3°. Let us now use the invariance that we have proved in Part 1 of this proof to find the dependence of the function $f(z)$ on each variable z_i. For that, we will use invariance under transformations that change z_i to $\lambda_i \cdot z_i$ and leave all other coordinates z_j $(j \neq i)$ intact.

Let us fix the values z_j of all the variables except for z_i. Under the above transformation, invariance implies that if $g(z_1, \ldots, z_{i-1}, z_i, z_{i+1}, \ldots, z_n)$ is a function from the optimal family, then the re-scaled function $g(z_1, \ldots, z_{i-1}, \lambda_i \cdot z_i, z_{i+1}, \ldots, z_n)$ belongs to the same family, i.e.,

$$g(z_1, \ldots, z_{i-1}, \lambda_i \cdot z_i, z_{i+1}, \ldots, z_n) = C(\lambda_i) \cdot g(z_1, \ldots, z_{i-1}, z_i, z_{i+1}, \ldots, z_n) - y_0(\lambda_i)$$

for some values C and y_0 depending on λ_i. Let us denote

$$g_i(z_i) = g(z_1, \ldots, z_{i-1}, z_i, z_{i+1}, \ldots, z_n).$$

Then, the above condition takes the form

$$g_i(\lambda \cdot z_i) = C(\lambda_i) \cdot g_i(z_i) - y_0(\lambda_i).$$

It is possible that the function $g_i(z_i)$ is a constant. If it is not a constant, this means that there exist values $z_i \neq z_i'$ for which $g_i(z_i) \neq g_i(z_i')$. For these two values, we get

$$g_i(\lambda_i \cdot z_i) = C(\lambda_i) \cdot g_i(z_i) - y_0(\lambda_i);$$
$$g_i(\lambda_i \cdot z_i') = C(\lambda_i) \cdot g_i(z_i') - y_0(\lambda_i).$$

By subtracting these equations, we conclude that

$$g_i(\lambda_i \cdot z_i) - g_i(\lambda_i \cdot z_i') = C(\lambda_i) \cdot (g_i(z_i) - g_i(z_i')),$$

hence

$$C(\lambda_i) = \frac{g_i(\lambda_i \cdot z_i) - g_i(\lambda_i \cdot z_i')}{g_i(z_i) - g_i(z_i')}.$$

Since the function $g_i(z_i)$ is a polynomial, the right-hand side is a smooth function of λ. Thus, the dependence of $C(\lambda_i)$ on λ_i is differentiable (smooth). Since $y_0(\lambda_i) = C(\lambda_i) \cdot g_i(z_i) - g_i(\lambda_i \cdot z_i)$, and both C and g_i are smooth functions, the dependence $y_0(\lambda_i)$ is also smooth.

Since all three functions C, y_0, and g_i are differentiable, we can differentiate both sides of the equality $g_i(\lambda_i \cdot z_i) = C(\lambda_i) \cdot g_i(z_i) - y_0(\lambda_i)$ by λ_i and take $\lambda_i = 1$. This leads to the formula

$$z_i \cdot \frac{dg_i}{dz_i} = C_1 \cdot g_i(z_i) - y_1,$$

where we denoted $C_1 \stackrel{\text{def}}{=} \frac{dC}{d\lambda_i}_{|\lambda_i=1}$ and $y_1 \stackrel{\text{def}}{=} \frac{dy_0}{d\lambda_i}_{|\lambda_i=1}$.

By moving all the terms related to g_i to one side and all the terms related to z_i to the other side, we get

$$\frac{dg_i}{C_1 \cdot g_i - y_1} = \frac{dz_i}{z_i}.$$

We will consider two possibilities: $C_1 = 0$ and $C_1 \neq 0$.

3.1°. If $C_1 = 0$, then the above equation takes the form

$$-\frac{1}{y_1} \cdot dg_i = \frac{dz_i}{z_i}.$$

Integrating both sides, we get

$$-\frac{1}{y_1} \cdot g_i = \ln(z_i) + \text{const},$$

thus $g_i = -y_1 \cdot \ln(z_i) + \text{const}$. This contradicts to the fact that the dependence $g_i(z_i)$ is polynomial. Thus, $C_1 \neq 0$.

3.2°. Since $C_1 \neq 0$, we can introduce a new variable $h_i = g_i - \frac{y_1}{C_1}$. For this new variable, we have $dh_i = dg_i$. Hence the above differential equation takes the simplified form

$$\frac{1}{C_1} \cdot \frac{dh_i}{h_i} = \frac{dz_i}{z_i}.$$

Integrating both sides, we get

$$\frac{1}{C_1} \cdot \ln(h_i) = \ln(z_i) + \text{const},$$

hence

$$\ln(h_i) = C_1 \cdot \ln(z_i) + \text{const},$$

and

$$h_i = \text{const} \cdot z_i^{C_1}.$$

Thus,

$$g_i(z_i) = h_i(z_i) + \frac{y_1}{C_1} = \text{const} \cdot z_i^{C_1} + \frac{y_1}{C_1}.$$

Since we know that $g_i(z_i)$ is a polynomial, the power C_1 should be a non-negative integer, so we conclude that

$$g_i(z_i) = A \cdot z_i^{k_i} + B$$

for some values A_i, B_i, and k_i which, on general, depend on all the other values z_j.

4°. Since the function $g(z_1, \ldots, z_n)$ is a polynomial, it is continuous and thus, the value k_i continuously depends on z_j. Since the value k_i is always an integer, it must therefore be constant – otherwise we would have a discontinuous jump from one integer to another. Thus, the integer k_i is the same for all the values z_j.

5°. Let us now use the above dependence on each variable z_i to find the dependence on two variables. Without losing generality, let us consider dependence on the variables z_1 and z_2.

Let us fix the values of all the other variables except for z_1 and z_2, and let us define

$$g_{12}(z_1, z_2) = g(z_1, z_2, z_3, \ldots, z_n).$$

Our general result can be applied both to the dependence on z_1 and to the dependence on z_2. The z_1-dependence means that $g_{12}(z_1, z_2) = A_1(z_2) \cdot z_1^{k_1} + B_1(z_2)$, and the z_1-dependence means that $g_{12}(z_1, z_2) = A_2(z_1) \cdot z_2^{k_2} + B_2(z_1)$. Let us consider two possible cases: $k_1 = 0$ and $k_1 \neq 0$.

5.1°. If $k_1 = 0$, this means that $g_{12}(z_1, z_2)$ does not depend on z_1 at all, so both A_2 and B_2 do not depend on z_1, hence we have $g_{12}(z_1, z_1) = A_2 \cdot z_2^{k_2} + B_2$.

5.2°. Let us now consider the case when $k_1 \neq 0$. For $z_1 = 0$, the z_1-dependence means that $g_{12}(0, z_2) = B_1(z_2)$, and the z_2-dependence implies that $B_1(z_2) = g_{12}(0, z_2) = A_2(0) \cdot z_2^{k_2} + B_2(0)$.

For $z_1 = 1$, the z_1-dependence means that $g_{12}(1, z_2) = A_1(z_2) + B_1(z_2)$. On the other hand, from the z_2-dependence, we conclude that $A_1(z_2) + B_1(z_2) = g_{12}(1, z_2) = A_2(1) \cdot z_2^{k_2} + B_2(1)$. We already know the expression for $B_1(z_2)$, so we conclude that

$$A_1(z_2) = g_{12}(1, z_2) - B_1(z_2) = (A_2(1) - A_2(0)) \cdot z_2^{k_2} + (B_2(1) - B_2(0)).$$

Thus, both $A_1(z_2)$ and $B_1(z_2)$ have the form $a + b \cdot z^{k_2}$, hence we conclude that

$$g_{12}(z_1, z_2) = (a + b \cdot z_2^{k_2}) \cdot z_1^{k_1} + (c + d \cdot z_2^{k_2}) = c + a \cdot z_1^{k_1} + d \cdot z_2^{k_2} + b \cdot z_1^{k_1} \cdot z_2^{k_2}.$$

Previously, we only considered transformations of a single variable, let us now consider a joint transformation $z_1 \to \lambda_1 \cdot z_1$, $z_2 \to \lambda_2 \cdot z_2$. In this case, we get

$$g(\lambda_1 \cdot z_1, \lambda_2 \cdot z_2) = c + a \cdot \lambda_1^{k_1} \cdot z_1^{k_1} + d \cdot \lambda_2^{k_2} \cdot z_2^{k_2} + b \cdot \lambda_1^{k_1} \cdot \lambda_2^{k_2} \cdot z_1^{k_1} \cdot z_2^{k_2}.$$

We want to make sure that

$$g(\lambda_1 \cdot z_1, \lambda_2 \cdot z_2) = C(\lambda_1, \lambda_2) \cdot g(z_1, z_2) - y_0(\lambda_1, \lambda_2),$$

i.e., that

$$\begin{aligned} & c + a \cdot \lambda_1^{k_1} \cdot z_1^{k_1} + d \cdot \lambda_2^{k_2} \cdot z_2^{k_2} + b \cdot \lambda_1^{k_1} \cdot \lambda_2^{k_2} \cdot z_1^{k_1} \cdot z_2^{k_2} \\ &= C(\lambda_1, \lambda_2) \cdot (c + a \cdot z_1^{k_1} + d \cdot z_2^{k_2} + b \cdot z_1^{k_1} \cdot z_2^{k_2}) - y_0(\lambda_1, \lambda_2). \end{aligned}$$

Both sides are polynomials in z_1 and z_2; the polynomials coincide for all possible values z_1 and z_2 if and only if all their coefficients coincide. Thus, we conclude that

$$a \cdot \lambda_1^{k_1} = a \cdot C(\lambda_1, \lambda_2);$$
$$d \cdot \lambda_2^{k_2} = d \cdot C(\lambda_1, \lambda_2);$$
$$c \cdot \lambda_1^{k_1} \cdot \lambda_2^{k_2} = c \cdot C(\lambda_1, \lambda_2).$$

If $a \neq 0$, then by dividing both sides of the a-containing equality by a, we get $C(\lambda_1, \lambda_2) = \lambda_1^{k_1}$. If $d \neq 0$, then by dividing both sides of the d-containing equality by d, we get $C(\lambda_1, \lambda_2) = \lambda_2^{k_2}$. If $c \neq 0$, then by dividing both sides of the c-containing equality by c, we get $C(\lambda_1, \lambda_2) = \lambda_1^{k_1} \cdot \lambda_2^{k_2}$. These three formulas are incompatible, so only one of three coefficients a, d, and c is different from 0 and two other coefficients are equal to 0. In all three cases, the dependence has the form

$$g_{12}(z_1, z_2) = a + \text{const} \cdot z_1^{\ell_1} \cdot z_2^{\ell_2}.$$

6°. Similarly, by considering more variables, we conclude that

$$g(z_1, \ldots, z_n) = a + \text{const} \cdot z_1^{\ell_1} \cdot \ldots \cdot z_n^{\ell_n}.$$

By plugging in the values z_i in terms of x_i, we get the conclusion of the first part of the proposition. The first part of the proposition is proven.

7°. For each sequence of natural numbers k_i, we can form an optimality criterion in which:

- the family $\mathcal{F}$ described by the formula (1) is better than any other family $\mathcal{G}$, i.e., $\mathcal{G} \preceq \mathcal{F}$ and $\mathcal{F} \not\preceq \mathcal{G}$, and
- all other families $\mathcal{G}$ and $\mathcal{H}$ are equivalent to each other: $\mathcal{G} \preceq \mathcal{H}$ and $\mathcal{H} \preceq \mathcal{G}$.

The proposition is proven.

Proof of Lemma 1. Let us first show that if the function $K(a,b)$ is a combination operation, then $K(a,b)$ is a polynomial. Indeed, by definition of a combination operation, if we take $\underline{f}(x) = x_1$ and $\overline{f}(x) = x_2$, then the function $f(x) = K\left(\underline{f}(x), \overline{f}(x)\right) = K(x_1, x_2)$ is a polynomial.

Vice versa, if $K(x_1, x_2)$ is a polynomial, then for every two polynomials $\underline{f}(x)$ and $\overline{f}(x)$, the composition $f(x) = K\left(\underline{f}(x), \overline{f}(x)\right)$ is also a polynomial. The lemma is proven.

Proof of Proposition 2. Due to Lemma, Proposition 2 follows from Proposition 1 – for the case of two variables.

Acknowledgments. This work was supported in part by the National Science Foundation grants 1623190 (A Model of Change for Preparing a New Generation for Professional Practice in Computer Science), and HRD-1834620 and HRD-2034030 (CAHSI Includes), and by the AT&T Fellowship in Information Technology. It was also supported by the program of the development of the Scientific-Educational Mathematical Center of Volga Federal District No. 075-02-2020-1478, and by a grant from the Hungarian National Research, Development and Innovation Office (NRDI).

The author is thankful to the anonymous referees for valuable suggestions.

References

1. Belohlavek, R., Dauben, J.W., Klir, G.J.: Fuzzy Logic and Mathematics: A Historical Perspective. Oxford University Press, New York (2017)
2. Berz, M., Hoffstätter, G.: Computation and application of Taylor polynomials with interval remainder bounds. Reliable Comput. **4**, 83–97 (1998)
3. Berz, M., Makino, K.: Verified integration of ODEs and flows using differential algebraic methods on high-order Taylor models. Reliable Comput. **4**, 361–369 (1998)
4. Berz, M., Makino, K., Hoefkens, J.: Verified integration of dynamics in the solar system. Nonlinear Anal. Theory Methods Appl. **47**, 179–190 (2001)
5. Garloff, J.: The Bernstein algorithm. Interval Comput. **2**, 154–168 (1993)
6. Garloff, J.: The Bernstein expansion and its applications. J. Am. Rom. Acad. **25–27**, 80–85 (2003)
7. Garloff, J., Graf, B.: Solving strict polynomial inequalities by Bernstein expansion. In: Munro, N. (ed.) The Use of Symbolic Methods in Control System Analysis and Design. IEE Control Engineering, London, vol. 56, pp. 339–352 (1999)

8. Garloff, J., Smith, A.P.: Solution of systems of polynomial equations by using Bernstein polynomials. In: Alefeld, G., Rohn, J., Rump, S., Yamamoto, T. (eds.) Symbolic Algebraic Methods and Verification Methods - Theory and Application, pp. 87–97. Springer, Vienna (2001). https://doi.org/10.1007/978-3-7091-6280-4_9
9. Hoefkens, J., Berz, M.: Verification of invertibility of complicated functions over large domains. Reliable Comput. **8**(1), 1–16 (2002)
10. Jaulin, L., Kiefer, M., Didrit, O., Walter, E.: Applied Interval Analysis, with Examples in Parameter and State Estimation, Robust Control, and Robotics. Springer, London (2001). https://doi.org/10.1007/978-1-4471-0249-6
11. Klir, G., Yuan, B.: Fuzzy Sets and Fuzzy Logic. Prentice Hall, Upper Saddle River (1995)
12. Lohner, R.: Einschliessung der Lösung gewöhnlicher Anfangs- und Randwertaufgaben und Anwendungen. Ph.D. thesis, Universität Karlsruhe, Karlsruhe, Germany (1988)
13. Mayer, G.: Interval Analysis and Automatic Result Verification. de Gruyter, Berlin (2017)
14. Moore, R.E., Kearfott, R.B., Cloud, M.J.: Introduction to Interval Analysis. SIAM, Philadelphia (2009)
15. Nataraj, P.S.V., Arounassalame, M.: A new subdivision algorithm for the Bernstein polynomial approach to global optimization. Int. J. Autom. Comput. **4**, 342–352 (2007)
16. Neumaier, A.: Taylor forms - use and limits. Reliable Comput. **9**, 43–79 (2002)
17. Nedialkov, N.S., Kreinovich, V., Starks, S.A.: Interval arithmetic, affine arithmetic, Taylor series methods: why, what next? Numer. Algorithms **37**, 325–336 (2004)
18. Nguyen, H.T., Walker, C.L., Walker, E.A.: A First Course in Fuzzy Logic. Chapman and Hall/CRC, Boca Raton (2019)
19. Novák, V., Perfilieva, I., Močkoř, J.: Mathematical Principles of Fuzzy Logic. Kluwer, Boston, Dordrecht (1999)
20. Ray, S., Nataraj, P.S.V.: A new strategy for selecting subdivision point in the Bernstein approach to polynomial optimization. Reliable Comput. **14**, 117–137 (2010)

Forcing the Network to Use Human Explanations in Its Inference Process

Javier Viaña(✉) and Andrew Vanderburg

MIT Kavli Institute for Astrophysics and Space Research, Massachusetts Institute of Technology, Cambridge 02139, USA
vianajr@mit.edu

Abstract. We introduce the concept of ForcedNet, a neural network that has been trained to generate a simplified version of human-like explanations in its hidden layers. The main difference with a regular network is that the ForcedNet has been educated such that its inner reasoning reproduces certain patterns that could be somewhat considered as human-understandable explanations. If designed appropriately, a ForcedNet can increase the model's transparency and explainability. We also propose the use of support features, hidden variables that complement the explanations and contain additional information to achieve high performance while the explanation contains the most important features of the layer. We define the optimal value of support features and what analysis can be performed to select this parameter. We demonstrate a simple ForcedNet case for image reconstruction using as explanation the composite image of the saliency map that is intended to mimic the focus of the human eye. The primary objective of this work is to promote the use of intermediate explanations in neural networks and encourage deep learning development modules to integrate the possibility of creating networks like the proposed ForcedNets.

Keywords: Artificial intelligence · deep learning · neural networks · explainable AI · explainability · AI ethics · image processing · machine learning

1 Introduction

We are in a time when the exponential use of artificial intelligence is prioritizing performance over the ability to explain and justify the outcomes from a human perspective. The field of eXplainable AI (XAI), which started becoming popular in 2016 [1], has been slowly developing over the last decade, but there is not yet a generalized method to understand the internal inference process of the widely used deep neural networks [2–5]. Nevertheless, some essential ideas such as extracting meaning from neural networks or relying on existing prior expert knowledge were already studied in the 90's [6,7], which years later evolved into the field of XAI.

K. Cohen et al. (Eds.): NAFIPS 2023, LNNS 751, pp. 127–140, 2023.
https://doi.org/10.1007/978-3-031-46778-3_12

This need for understanding becomes more acute with the entry of legislation such as the General Data Protection Regulation in the European Union on algorithmic decision-making and the "right to explanation" when it comes to human data [8–10]. However, many of the XAI technologies used for tasks that leverage human data, such as user experience enhancement or travel demand analysis, do not meet yet these requirements [11, 12].

In healthcare, the integration of AI largely depends on the trustworthiness of the algorithm chosen. Explainability is playing a critical role in order to achieve the validation and verification capabilities desired [13]. In fact, this need for trustworthiness has fostered the development of novel XAI for specific medical applications that can later be extended to other areas [14–16].

In engineering, the opposite happened, where human supervision became obsolete in those processes that were automated with black box AI and it was not until the lack of transparency was evident that the awareness of XAI began to be raised. Over time, several methods have appeared that claim to be explainable in the field, e.g., from the generative design of motors [17], to their prognosis, health monitoring and fault diagnosis [18,19].

Many of the techniques used to generate explanations in neural networks focus on what are known as post-hoc explanations. This means that once the system is trained, we add different algorithms that can extract posterior reasoning [20–22]. Some examples include, semantic web technologies [23], contrastive sample generation (GRACE) [24], or extracting global symbolic rules [25]. Nonetheless, opting for methods that extract explanations after the training implies that the network is not "aware" of our intention to explain its reasoning. In other words, the opportunity to add this information during training is lost, which is not only attractive for performance reasons, but also to educate the pipeline so that its inner process is more human understandable.

Some techniques such as DeConvNet [26], Guided BackProp [27], Layer-wise Relevance Propagation [28] and the Deep Taylor Decomposition [29] seek to explain the classifier decisions by propagating the output back through the network to map the most relevant features of the encoding process. However, more recent work [30] has shown that these methods do not produce valid explanations for simple linear models. An alternative approach that has demonstrated to be very successful is generating more interpretable latent spaces in autoencoders [31–34]. These often encode the information taking into account expert knowledge in order to improve the understanding of the latent variables. Nevertheless, their training does not rely on any ground truth latent explanation. Symmetric autoencoders have also been utilized to represent coherent information via reordering of the data [35], which can certainly help to understand the underlying decisions of the machine.

Another option is to develop new algorithms that are more transparent by nature and also high performing [36]. The main drawback is that such development implies a bottom-up reformulation of the neural networks and the learning formulas of the backpropagation. CEFYDRA is an example of these novel net-based algorithms that, in its case, replaces the neural unit with a fuzzy inference system in order to reason its outputs [37–39]. Combining case-based reasoning with deep learning has made significant advances in XAI as well, where the networks leverage already seen scenarios and adapt their solutions to solve new problems [40–42]. Other researchers have considered adding constrains to the learning process [43–45] but it has not been done with the intention of forcing the machine to think or replicate human reasoning. On the other hand, there have also been attempts to improve explainability by simulating human reasoning, but not within the neural network itself [46].

2 Architecture

2.1 The ForcedNet

The growth of XAI has raised concerns about the quality of the explanations used to explain the algorithms [47–50]. Logically, an explanation is valid as long as it's understandable for the person who digests it. Therefore, its validity depends on the recipient, which makes that assessment a highly subjective task. Regardless, if we consider the explanations as one more feature of the dataset, we could even integrate them into the training process, to educate the machine in a human manner. Such human-like reasoning could possibly help opening the black box [51–53].

In this work we introduce the concept of ForcedNet, a neural network that has been "forced" to produce a simplified version of a human-like explanation in one of its intermediate layers and then leverages fully or partially this explanation to generate the desired output. We define these simplified versions human-like explanations as any type of information that can help understand the reasoning process of the algorithm from the human perspective. The choice of the best explanation format is problem specific, and even for one same task there might be several types of useful explanations, such as visual or textual. As it was mentioned in the introduction, the validity of an explanation is subjective since it involves the perception and assessment of a human, which might vary.

Figure 1 is the depiction of an example ForcedNet architecture. In addition to the usual inputs and outputs, we also have the desired explanations, and the support features (whose purpose will be described later) in what we identify as the explanation layer of the network. This layer divides the left and right sections which are the reasoning and inference sections, respectively.

The reasoning section is responsible for producing the explanation of the input. This explanation should be a description that not only encodes the information of the input but at the same time is understandable from the human perspective. The inference section is responsible for generating the desired output from the human-understandable explanation.

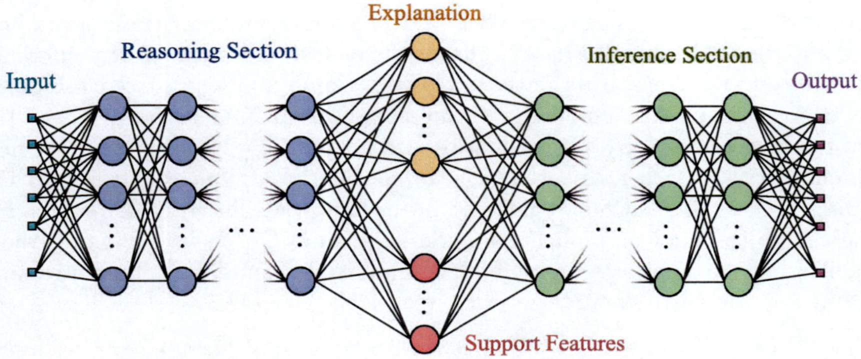

Fig. 1. Schematic representation of the proposed network, ForcedNet, composed of a central explanation layer that contains the human-understandable explanation and the support features.

Generating the final output using only the information encoded in the explanation might be a difficult task. Indeed, the performance of the inference section depends on the quality of the explanation chosen for the problem. For that matter, we introduce the concept of support features, which precisely support the explanation features encapsulating additional information of the input that is not present in the explanation. These features are not understandable from the human perspective as the explanation features, but in some problems might be necessary. To grasp their importance, let us consider the following image reconstruction task depicted in Fig. 2, where $\boldsymbol{x}$, and $\boldsymbol{y}$ denote the input and output respectively. The explanation chosen, $\boldsymbol{t}$, is a short linguistic description of the image. The support vector (support features arranged in a vector format), $\boldsymbol{s}$, has S features that have no apparent meaning, but together with the embedded explanation, the inference section is able to retrieve most of the image. In Fig. 2, $\hat{\boldsymbol{v}}$ denotes a prediction of the variable $\boldsymbol{v}$.

The training of such a system can leverage a triple backpropation method, where in each step we train the reasoning section, the inference section, and the full system separately as shown in Fig. 3. Note that the backpropagation in the reasoning and the inference sections excludes the weights associated to the neurons of the support vector because we have no prior data for these features.

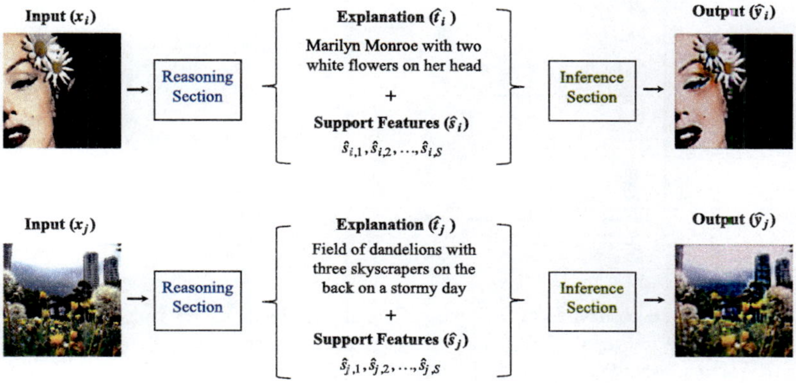

Fig. 2. Two examples of a ForcedNet that uses the image embedding as the explanation and performs the image reconstruction task based on this textual definition of the image. Images publicly available at Flickr [54]. This example is schematic and only serves to understand better the idea behind a ForcedNet.

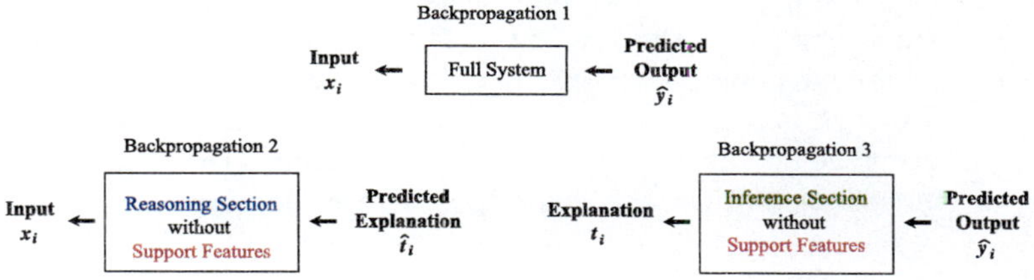

Fig. 3. Schematic representation of the triple backpropagation method for the training of a ForcedNet, given a single explanation layer.

2.2 Design Considerations

Choosing the best number of support features should be evaluated carefully and is problem specific. There is an optimal trade-off between explainability (low S) and performance (high S). Alternatively, one can study the evolution in performance of the following two pipelines as we vary the value of S to make a better decision:

- Network A, which uses the explanation of T fixed number of features and S support features, i.e., a total of $T + S$ hidden features.
- Network B, which only uses S hidden features, and it does not produce any explanation.

Fixing the same training hyperparameters, we increase the value of S in both networks and keep track of their performance, which we denote by ρ_A and ρ_B respectively (same figure of merit must be the chosen, e.g., for regression tasks: Root Mean Squared Error, or Mean Absolute Error, etc.). When the $\rho_A \approx \rho_B$ for a given value S, it means that network A is not using the explanation

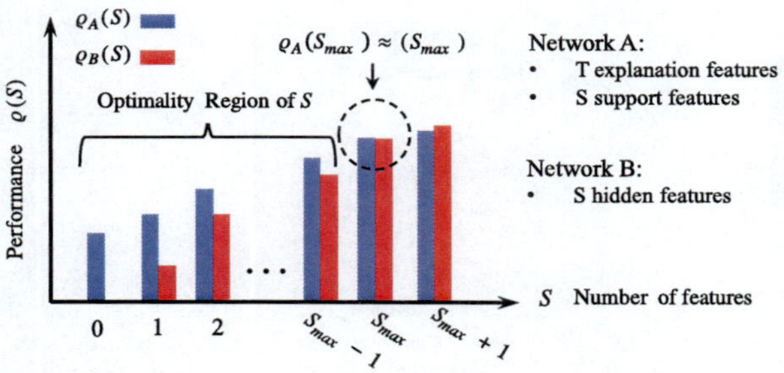

Fig. 4. Performance comparison for two different networks as we increase the value of S while we fix the value of T. Network A is a ForcedNet that utilizes T explanation features together with S support features. Network B is a standard autoencoder that uses S hidden features.

features to produce the outputs. We denote this limiting value of S with S_{max} (Fig. 4). In other words, the network A has sufficient information with the S_{max} support features to fulfill its task, since network B has been able to obtain similar performance without the use of the F features. Thus, the explanations generated by A are meaningless because pipeline A may not be using them at all. The optimal value S, is $S_{opt} \in \mathbb{Z}^+ : 0 \leqslant S_{opt} < S_{max}$, which marks the trade-off between performance and explainability, because it ensures that the explanation features are indeed used in the reasoning to generate the output.

3 Case Study

We chose the image reconstruction task to exemplify a simple proof of concept of a ForcedNet. To generate the intermediate explanations, we decided to leverage the saliency map ($\boldsymbol{m}$) of the image, which defines the regions of the image on which the human eye focuses first. Our explanation is composed of both the original image and the saliency map. We injected Gaussian noise and increased the transparency proportionally to those pixels that were less important according to the saliency map, while we kept resolved those areas that were more relevant. In other words, in the explanation we tried to capture the way in which the human eye focuses on an image, distorting the surrounding information while locating the machine's center of attention on the vital pixels.

For the reasoning section of the network, we utilized DeepGaze II [55], a pre-trained model that has demonstrated the best performance to date predicting saliency maps on datasets like MIT300 saliency benchmark [56], where it reported the following metrics: AUC = 88%, sAUC = 77%, NSS = 2.34. DeepGaze II was trained in two phases. In its pre-training phase it used the SALICON dataset [57], which consists of 10000 images with pseudofixations from a mouse-contingent task, and was fine-tunned using the MIT1003 dataset [58].

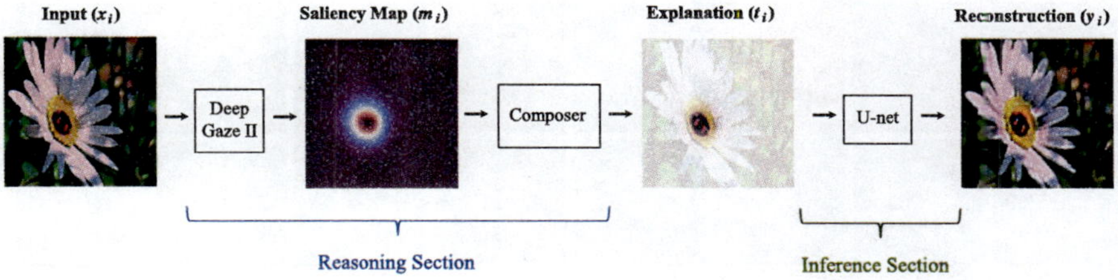

Fig. 5. Processing pipeline of the image reconstruction task fully based on the human-like attention map (the explanation).

Because we are using a pre-trained model for the reasoning section we cannot consider the use of support features. However, the scope of this research was simply to introduce the idea of ForcedNets and to illustrate a simple example with the reconstruction task chosen. In future work, we will study the effect of different support features in a ForcedNet trained back-to-back.

For the inference section, we chose a shallow convolutional autoencoder. The specifications of the architecture are shown in Table 1.

Table 1. Architecture of the convolutional autoencoder chosen.

Layer	Filter	Kernel Size	Activation Function	Padding	Strides
Convolution 2D	16	3 × 3	ReLu	Same	2
Convolution 2D	8	3 × 3	ReLu	Same	2
Deconvolution 2D	8	3 × 3	ReLu	Same	2
Deconvolution 2D	16	3 × 3	ReLu	Same	2
Convolution 2D	3	3 × 3	Sigmoid	Same	2

We chose a dataset of 1,000 RGB flower images of 128 × 128 pixels obtained from Flickr [54]. This selection of images is publicly available in [59].

First, we predicted the feature map of all the images using DeepGaze II, and then we obtained the composed image that served as the explanation matrix. We then trained the convolutional autoencoder with the generated explanations and the desired outputs. Figure 5 shows the different steps of the architecture.

For the training of the inference section, we chose a learning rate of 10^{-4} for 3,000 epochs with 20 steps each. We used 0.7, 0.2, 0.1, training, validation and testing splits.

Since the reasoning section of the ForcedNet was already trained we did not require the triple backpropagation method. However, for future reference, the reasoning and the inference sections should be trained simultaneously so that both can learn from all the three types of data (inputs, outputs, and explanations).

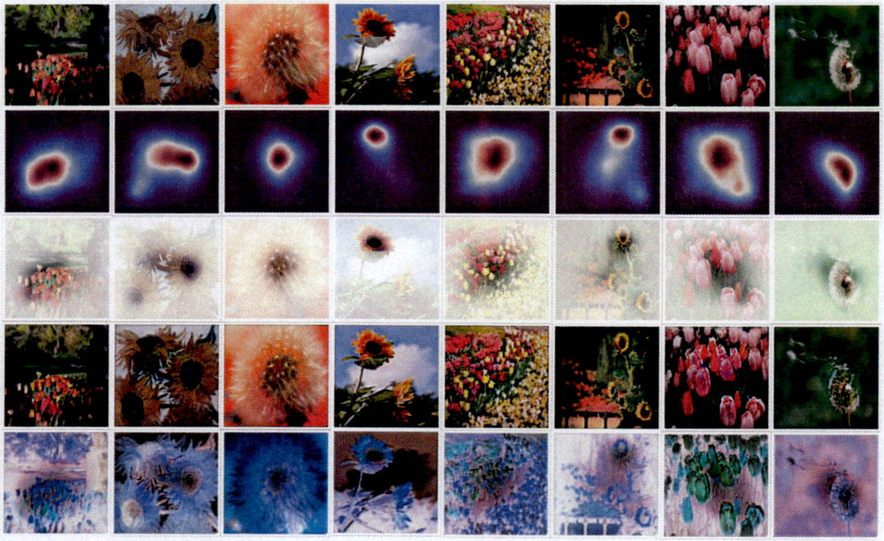

Fig. 6. Eight training instances of the dataset. First row represents the original image, second row the saliency map predicted by DeepGaze II, third row the composed attention map (the human-like explanation), row four the output of the convolutional autoencoder (reconstructed image), row five the difference image between the input and the output of the system in normalized space to appreciate better the errors.

Additionally, DeepGaze II required an additional reference image map that measured the human bias to look at pixels on the central region of the image. This map was generated accordingly as described in [55] for the chosen dataset.

4 Results

After training, the Mean Absolute Error (MAE) for 10-fold cross validation was 0.0107, 0.0109, 0.0108 for each set, respectively. To calculate the MAE we used the average error for all the pixels of the image.

Figure 6 and Fig. 7 show some prediction examples for both the training and the testing datasets of the ForcedNet considered. It can be appreciated that the convolutional autoencoder is able to reconstruct most of the image using only the information available on the explanation data. At the same time, the explanation layer gives us that additional knowledge in the prediction of the areas in the image where the machine focuses more.

Fig. 7. Eight testing instances of the dataset. First row represents the original image, second row the saliency map predicted by DeepGaze II, third row the composed attention map (the human-like explanation), row four the output of the convolutional autoencoder (reconstructed image), row five the difference image between the input and the output of the system in normalized space to appreciate better the errors.

5 Discussion

With this study, we are trying to demonstrate that is viable to force the network to have intermediate layers where we have explanations, and that there are different but simple ways in which we can conceive simplified behaviors of human reasoning that might even help the machine learning process.

Having the explainability requirement in the middle of the pipeline might increase the difficulty of the learning process, as it imposes a constrain in the machine's internal inference process. However, there might be scenarios where we see the opposite, e.g., if the human description contains useful information for the task chosen, it could even help the training guiding the weights to more optimal combinations. On the other hand, it does require an extra effort from the human to generate the dataset of explanation training instances.

The architecture explained in this study is scalable to designs with several explainable layers. In other words, one could stack different reasoning sections together to form a chain of explanations as shown in Fig. 8. That kind of pipeline would guide the thought process of the network even more than the case studied. The motivation behind such a system lies in validation and verification purposes or simply because we decide to constrain the internal inference process of the network in a smarter way than letting the optimization to brute force input-to-output backpropagation. The number of backpropagation methods that one

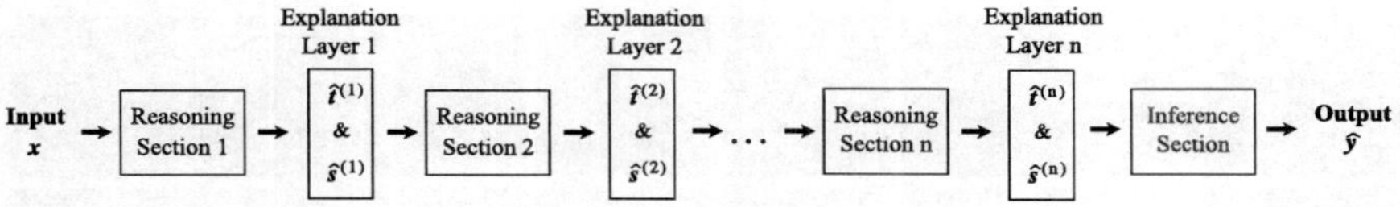

Fig. 8. Schematic representation of a ForcedNet composed of several explanation layers. Each explanation layer is generated by the corresponding reasoning section, but there is only one inference section.

can perform simultaneously in each step given n explanation layers is $\frac{n^2+3n+2}{2}$, which comes from the combinations of $n+2$ elements (n explanation layers and both the extremes of the pipeline) grouped in pairs. Note that n^2+3n will always generate a positive even number if $n \in \mathbb{Z}^+$.

Future work could focus on demonstrating the use of ForcedNets in image reconstruction leveraging the image embedding as the explanation. Further experiments should also be conducted to test the benefit of support features and the optimal choice of S.

The authors believe that the community would be benefited from an open-source module for neural network development that could automate plugging explanations in the middle of the network and choosing for the right number of support features.

6 Conclusions

We have presented the concept of ForcedNets as a tool to guide the learning of a neural network to generate intermediate human-understandable explanations. This technique has been demonstrated with a simple image reconstruction example where the explanation is a composition image of the saliency map. The use of support features has been discussed as a method to ensure high performance even when the explanations are too simple or do not capture all the necessary information of the previous layers. The authors believe that this technique could significantly help to obtain the desired explainability in neural networks, as long as there are explanations for the chosen problem and that these can be easily incorporated into the learning process.

Acknowledgments. For their insightful discussions on the topic and comments on the paper, we thank Faraz Faruqi and Mariona Badenas from Massachusetts Institute of Technology, Jagoba Zuluaga and Eric del Sastre from Harvard Medical School. The authors would also like to extend special thanks to the anonymous reviewers that peer-reviewed this work.

Authors' contributions. Made the conception and design of the study and performed data analysis and interpretation: Javier Viaña. Supervised the work: Andrew Vanderburg.

Availability of data and materials. All images in this archive are licensed under the Creative Commons By-Attribution License. Data publicly available in Flickr [54], selection of training data stored in GitHub [59].

Code developed for the project publicly available in GitHub[59]. Pre-trained DeepGaze II model available in GitHub[60].

Financial Support and Sponsorship. This work was supported by two NASA Grants, the NASA Extremely Precise Radial Velocity Foundation Science Program (No. 80NSSC22K0848) and the NASA Astrophysical Data Analysis Program (No. 80NSSC22K1408).

References

1. Gunning, D.: Explainable artificial intelligence (XAI). Defense Advanced Research Projects Agency (DARPA), nd Web, vol. 2, no. 2 (2017)
2. Xu, F., Uszkoreit, H., Du, Y., Fan, W., Zhao, D., Zhu, J.: Explainable AI: a brief survey on history, research areas, approaches and challenges. In: Tang, J., Kan, M.-Y., Zhao, D., Li, S., Zan, H. (eds.) NLPCC 2019. LNCS (LNAI), vol. 11839, pp. 563–574. Springer, Cham (2019). https://doi.org/10.1007/978-3-030-32236-6_51
3. Adadi, A., Berrada, M.: Peeking inside the black-box: a survey on explainable artificial intelligence (XAI). IEEE Access **6**, 52138–52160 (2018)
4. Guidotti, R., Monreale, A., Ruggieri, S., Turini, F., Giannotti, F., Pedreschi, D.: A survey of methods for explaining black box models. ACM Comput. Surv. (CSUR) **51**(5), 1–42 (2018)
5. Angelov, P., Soares, E.: Towards explainable deep neural networks (xDNN). Neural Netw. **130**, 185–194 (2020). ID: 271125
6. Turner, H., Gedeon, T.D.: Extracting meaning from neural networks. In: Proceedings 13th International Conference on AI, vol. 1, pp. 243–252 (1993)
7. Thrun, S.: Explanation-based neural network learning. In: Thrun, S. (ed.) Explanation-Based Neural Network Learning: A Lifelong Learning Approach. The Kluwer International Series in Engineering and Computer Science, vol. 357, pp. 19–48. Springer, Boston (1996). https://doi.org/10.1007/978-1-4613-1381-6_2. ID: Thrun 1996
8. Goodman, B., Flaxman, S.: European union regulations on algorithmic decision-making and a "right to explanation". AI Mag. **38**(3), 50–57 (2017)
9. Wachter, S., Mittelstadt, B., Floridi, L.: Why a right to explanation of automated decision-making does not exist in the general data protection regulation. Int. Data Priv. Law **7**(2), 76–99 (2017)
10. Hamon, R., Junklewitz, H., Sanchez, I., Malgieri, G., De Hert, P.: Bridging the gap between AI and explainability in the GDPR: towards trustworthiness-by-design in automated decision-making. IEEE Comput. Intell. Mag. **17**(1), 72–85 (2022)
11. Ferreira, J.J., Monteiro, M.S.: What are people doing about XAI user experience? A survey on AI explainability research and practice. In: Marcus, A., Rosenzweig, E. (eds.) HCII 2020. LNCS, vol. 12201, pp. 56–73. Springer, Cham (2020). https://doi.org/10.1007/978-3-030-49760-6_4
12. Alwosheel, A., van Cranenburgh, S., Chorus, C.G.: Why did you predict that? Towards explainable artificial neural networks for travel demand analysis. Transp. Res. Part C: Emerg. Technol. **128**, 103143 (2021). ID: 271729

13. Markus, A.F., Kors, J.A., Rijnbeek, P.R.: The role of explainability in creating trustworthy artificial intelligence for health care: A comprehensive survey of the terminology, design choices, and evaluation strategies. J. Biomed. Inform. **113** (2021)
14. Ran, G., et al.: CA-net: comprehensive attention convolutional neural networks for explainable medical image segmentation. IEEE Trans. Med. Imaging **40**(2), 699–711 (2021)
15. Dasari, C.M., Bhukya, R.: Explainable deep neural networks for novel viral genome prediction. Appl. Intell. **52**(3), 3002–3017 (2022)
16. Biffi, C., et al.: Learning interpretable anatomical features through deep generative models: application to cardiac remodeling. In: Frangi, A.F., Schnabel, J.A., Davatzikos, C., Alberola-López, C., Fichtinger, G. (eds.) MICCAI 2018. LNCS, vol. 11071, pp. 464–471. Springer, Cham (2018). https://doi.org/10.1007/978-3-030-00934-2_52
17. Sasaki, H., Hidaka, Y., Igarashi, H.: Explainable deep neural network for design of electric motors. IEEE Trans. Magn. **57**(6), 1–4 (2021)
18. Grezmak, J., Wang, P., Sun, C., Gao, R.X.: Explainable convolutional neural network for gearbox fault diagnosis. Procedia CIRP **80**, 476–481 (2019)
19. Kim, M.S., Yun, J.P., Park, P.: An explainable convolutional neural network for fault diagnosis in linear motion guide. IEEE Trans. Ind. Inform. **17**(6), 4036–4045 (2020)
20. Jeyakumar, J.V., Noor, J., Cheng, Y.-H., Garcia, L., Srivastava, M.: How can I explain this to you? an empirical study of deep neural network explanation methods. In: Advances in Neural Information Processing Systems, vol. 33, pp. 4211–4222 (2020)
21. Keane, M.T., Kenny, E.M.: How case-based reasoning explains neural networks: a theoretical analysis of XAI using *Post-Hoc* explanation-by-example from a survey of ANN-CBR twin-systems. In: Bach, K., Marling, C. (eds.) ICCBR 2019. LNCS (LNAI), vol. 11680, pp. 155–171. Springer, Cham (2019). https://doi.org/10.1007/978-3-030-29249-2_11
22. Zhang, Q., Yang, Y., Liu, Y., Wu, Y.N., Zhu, S.-C.: Unsupervised learning of neural networks to explain neural networks. arXiv preprint arXiv:1805.07468 (2018)
23. Sarker, M.K., Xie, N., Doran, D., Raymer, M., Hitzler, P.: Explaining trained neural networks with semantic web technologies: First steps. arXiv preprint arXiv:1710.04324 (2017)
24. Le, T., Wang, S., Lee, D.: GRACE: generating concise and informative contrastive sample to explain neural network model's prediction. In: Proceedings of the 26th ACM SIGKDD International Conference on Knowledge Discovery & Data Mining, pp. 238–248 (2020)
25. Zhou, Z.-H., Jiang, Y., Chen, S.-F.: Extracting symbolic rules from trained neural network ensembles. AI Commun. **16**(1), 3–15 (2003)
26. Zeiler, M.D., Fergus, R.: Visualizing and understanding convolutional networks. In: Fleet, D., Pajdla, T., Schiele, B., Tuytelaars, T. (eds.) ECCV 2014. LNCS, vol. 8689, pp. 818–833. Springer, Cham (2014). https://doi.org/10.1007/978-3-319-10590-1_53
27. Springenberg, J.T., Dosovitskiy, A., Brox, T., Riedmiller, M.A.: Striving for simplicity: the all convolutional net. In: Bengio, Y., LeCun, Y. (eds.) 3rd International Conference on Learning Representations, ICLR, San Diego, CA, USA, 7–9 May 2015, Workshop Track Proceedings (2015)

28. Binder, A., Montavon, G., Lapuschkin, S., Müller, K.-R., Samek, W.: Layer-wise relevance propagation for neural networks with local renormalization layers. In: Villa, A.E.P., Masulli, P., Pons Rivero, A.J. (eds.) ICANN 2016. LNCS, vol. 9887, pp. 63–71. Springer, Cham (2016). https://doi.org/10.1007/978-3-319-44781-0_8
29. Montavon, G., Lapuschkin, S., Binder, A., Samek, W., Müller, K.-R.: Explaining nonlinear classification decisions with deep Taylor decomposition. Pattern Recogn. **65**, 211–222 (2017)
30. Kindermans, P.-J., et al.: Learning how to explain neural networks: PatternNet and PatternAttributionn. arXiv preprint arXiv:1705.05598 (2017)
31. Neumeier, M., Botsch, M., Tollkühn, A., Berberich, T.: Variational autoencoder-based vehicle trajectory prediction with an interpretable latent space, pp. 820–827 (2021)
32. Kim, J.-Y., Cho, S.-B.: Explainable prediction of electric energy demand using a deep autoencoder with interpretable latent space. Expert Syst. Appl. **186**, 115842 (2021). ID: 271506
33. Bodria, F., Guidotti, R., Giannotti, F., Pedreschi, D.: Interpretable latent space to enable counterfactual explanations. In: Pascal, P., Ienco, D. (eds.) DS 2022. LNCS, vol. 13601, pp. 525–540. Springer, Cham (2022). https://doi.org/10.1007/978-3-031-18840-4_37
34. Bölat, K., Kumbasar, T.: Interpreting variational autoencoders with fuzzy logic: a step towards interpretable deep learning based fuzzy classifiers. In: IEEE International Conference on Fuzzy Systems (FUZZ-IEEE), pp. 1–7 (2020)
35. Bharadwaj, P., Li, M., Demanet, L.: Redatuming physical systems using symmetric autoencoders. Phys. Rev. Res. **4**(2) (2022)
36. Rudin, C.: Stop explaining black box machine learning models for high stakes decisions and use interpretable models instead. Nat. Mach. Intell. **1**(5), 206–215 (2019)
37. Viaña, J., Ralescu, S., Kreinovich, V., Ralescu, A., Cohen, K.: Single hidden layer CEFYDRA: cluster-first explainable FuzzY-based deep self-reorganizing algorithm. In: Dick, S., Kreinovich, V., Lingras, P. (eds.) NAFIPS 2022. LNNS, vol. 500, pp. 298–307. Springer, Cham (2023). https://doi.org/10.1007/978-3-031-16038-7_29
38. Viaña, J., Ralescu, S., Kreinovich, V., Ralescu, A., Cohen, K.: Multiple hidden layered CEFYDRA: cluster-first explainable FuzzY-based deep self-reorganizing algorithm. In: Dick, S., Kreinovich, V., Lingras, P. (eds.) NAFIPS 2022. LNNS, vol. 500, pp. 308–322. Springer, Cham (2023). https://doi.org/10.1007/978-3-031-16038-7_30
39. Viaña, J., Ralescu, S., Kreinovich, V., Ralescu, A., Cohen, K.: Initialization and plasticity of CEFYDRA: cluster-first explainable FuzzY-based deep self-reorganizing algorithm. In: Dick, S., Kreinovich, V., Lingras, P. (eds.) NAFIPS 2022. LNNS, vol. 500, pp. 323–335. Springer, Cham (2023). https://doi.org/10.1007/978-3-031-16038-7_31
40. Park, J.H., Shin, C.-K., Im, K.H., Park, S.C.: A local weighting method to the integration of neural network and case based reasoning. In: Neural Networks for Signal Processing XI: Proceedings of the 2001 IEEE Signal Processing Society Workshop (IEEE Cat. No. 01TH8584), pp. 33–42. IEEE (2001)
41. Amin, K., Kapetanakis, S., Althoff, K.-D., Dengel, A., Petridis, M.: Answering with cases: a CBR approach to deep learning. In: Cox, M.T., Funk, P., Begum, S. (eds.) ICCBR 2018. LNCS (LNAI), vol. 11156, pp. 15–27. Springer, Cham (2018). https://doi.org/10.1007/978-3-030-01081-2_2

42. Corbat, L., Nauval, M., Henriet, J., Lapayre, J.-C.: A fusion method based on deep learning and case-based reasoning which improves the resulting medical image segmentations. Expert Syst. Appl. **147**, 113200 (2020)
43. Yang, Z., Zhang, A., Sudjianto, A.: Enhancing explainability of neural networks through architecture constraints. IEEE Trans. Neural Netw. Learn. Syst. **32**(6), 2610–2621 (2021)
44. Rieger, L., Singh, C., Murdoch, W., Yu, B.: Interpretations are useful: penalizing explanations to align neural networks with prior knowledge. In: International Conference on Machine Learning, pp. 8116–8126. PMLR (2020)
45. Shavlik, J.W., Towell, G.G.: Combining explanation-based learning and artificial neural networks. In: Proceedings of the Sixth International Workshop on Machine Learning, pp. 90–92. Elsevier (1989)
46. Blazek, P.J., Lin, M.M.: Explainable neural networks that simulate reasoning. Nature Comput. Sci. **1**(9), 607–618 (2021). ID: Blazek 2021
47. Fel, T., Vigouroux, D., Cadène, R., Serre, T.: How good is your explanation? Algorithmic stability measures to assess the quality of explanations for deep neural networks. In: Proceedings of the IEEE/CVF Winter Conference on Applications of Computer Vision, pp. 720–730 (2022)
48. Johs, A.J., Lutts, M., Weber, R.O.: Measuring explanation quality in XCBR. In: Proceedings of the 26th International Conference on Case-Based Reasoning, p. 75. Springer, Heidelberg (2018)
49. Pedreschi, D., Giannotti, F., Guidotti, R., Monreale, A., Ruggieri, S., Turini, F.: Meaningful explanations of black box AI decision systems. In: Proceedings of the AAAI Conference on Artificial Intelligence, vol. 33, pp. 9780–9784 (2019)
50. Dombrowski, A.-K., Anders, C.J., Müller, K.-R., Kessel, P.: Towards robust explanations for deep neural networks. Pattern Recogn. **121**, 108194 (2022)
51. Castelvecchi, D.: Can we open the black box of AI? Nat. News **538**(7623), 20 (2016)
52. Miller, T.: Explanation in artificial intelligence: insights from the social sciences. Artif. Intell. **267**, 1–38 (2019)
53. Kenny, E.M., Ford, C., Quinn, M., Keane, M.T.: Explaining black-box classifiers using post-hoc explanations-by-example: the effect of explanations and error-rates in xai user studies. Artif. Intell. **294**, 103459 (2021)
54. Flickr. www.flicker.com. Accessed 16 Nov 2022
55. Kümmerer, M., Wallis, T.S.A., Gatys, L.A., Bethge, M.: Understanding low- and high-level contributions to fixation prediction. In: 2017 IEEE International Conference on Computer Vision (ICCV), pp. 4799–4808 (2017)
56. Bylinskii, Z., et al.: MIT saliency benchmark
57. Jiang, M., Huang, S., Duan, J., Zhao, Q.: SALICON: Saliency in context. In: Proceedings of the IEEE Conference on Computer Vision and Pattern Recognition, pp. 1072–1080 (2015)
58. Judd, T., Ehinger, K., Durand, F., Torralba, A.: Learning to predict where humans look. In: IEEE International Conference on Computer Vision (ICCV) (2009)
59. Viaña, J.: ForcedNet for image reconstruction (2022). https://github.com/JavierVianaAi/forcednets-image-reconstruction
60. Matthias Kummerer. DeepGaze (2022). https://github.com/matthias-k/DeepGaze

Deep Learning ANFIS Architectures

Ben van Oostendorp(✉), Eric Zander, and Barnabas Bede

DigiPen Institute of Technology, Redmond, WA 98052, USA
{ben.vanoostendorp,eric.zander,b.bede}@digipen.edu

Abstract. To explore the capabilities and significance of neuro-fuzzy systems, we study deep learning with Adaptive Network-based Fuzzy Inference Systems (ANFIS). By incorporating ANFIS layers into neural networks and treating the fuzzy sets in antecedents as custom activation functions, we test the combination of a neural network's flexibility with the capabilities of ANFIS architectures in applications such as control and function approximation. This approach is well-supported by existing tools for automatic differentiation.

Keywords: Fuzzy systems · Neural networks · Deep learning · Neuro-fuzzy systems

1 Introduction

Fuzzy logic finds successful employment in applications such as control systems and function approximation [7,11]. Takagi-Sugeno-Kang (TSK) fuzzy systems offer one approach to developing performant fuzzy systems based on rules with fuzzy antecedents and a constant, linear, or higher order consequent [2,10]. Notably, one can successfully optimize such systems with genetic learning algorithms. However, Adaptive Network-based Fuzzy Inference Systems (ANFIS) [5] additionally provide a framework for combining neural networks with fuzzy systems and have found success in various applications [12]. Deep learning ANFIS architectures have been considered as potentially explainable alternatives to existing machine learning algorithms in medical imaging applications [1], and existing deep learning frameworks such as PyTorch provide automatic differentiation capabilities to support further experimentation [8]. In this paper, we discuss optimization of TSK systems in ANFIS architectures with deep learning.

2 Deep Learning ANFIS Architectures

The central ideas presented in this paper concerns the combination of ANFIS layers with other neural networks components. For example, joining ANFIS layers with dense or convolutional layers has been used in various applications [14,15]. We will explore two different types of ANFIS layers. These include Takagi-Sugeno-Kang fuzzy systems with either singleton or linear consequents.

K. Cohen et al. (Eds.): NAFIPS 2023, LNNS 751, pp. 141–148, 2023.
https://doi.org/10.1007/978-3-031-46778-3_13

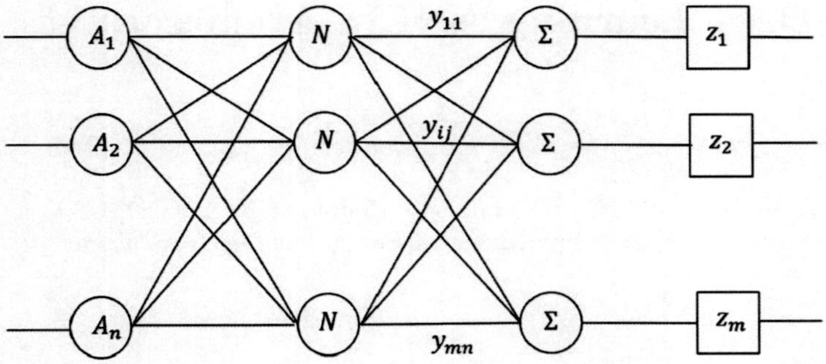

Fig. 1. Example of ANFIS layer with singleton consequences.

Let us consider fuzzy sets on a universe $X \subseteq \mathbb{R}$, $A_i : X \to [0, 1]$. In this paper we will consider Gaussian and triangular membership functions. The parameters of both the Gaussian and triangular fuzzy sets will be trainable parameters of the ANFIS layer. The fuzzy rules for a TSK fuzzy system with singleton consequents are of the form

$$\text{If } x \text{ is } A_i \text{ then } y = y_i, i = 1, ..., n.$$

The output of this type of system can be calculated as

$$TSK(x) = \frac{\sum_{i=1}^{n} A_i(x) \cdot y_i}{\sum_{i=1}^{n} A_i(x)}.$$

The fuzzy rules for a TSK fuzzy system with linear consequents are

$$\text{If } x \text{ is } A_i \text{ then } y = a_i \cdot x + b_i, i = 1, ..., n.$$

The output of this type of system can be calculated as

$$TSK(x) = \frac{\sum_{i=1}^{n} A_i(x) \cdot (a_i \cdot x + b_i)}{\sum_{i=1}^{n} A_i(x)}.$$

The individual rule outputs z_i in the singleton case and a_i, b_i in the linear case are also trainable parameters, $i = 1, ..., n$.

The above systems have one output, but may be easily generalized to the multi-output case by considering multiple outputs combined in parallel with their own parameter sets [3]. Also, we may rewrite the TSK system in a form using normalized firing levels.

The structure of an ANFIS layer with singleton outputs is illustrated in Fig 1. The previous layers of the network connect to the initial layer generating the inputs of the antecedents. The normalization layer outputs normalized firing levels. The outputs of the normalization layer are connected to the consequents that generate the final outputs of the ANFIS layer. We connect the outputs in a parallel structure to obtain a vector of the required dimension that can serve as input of other neural or ANFIS layers. In the case of linear consequents as

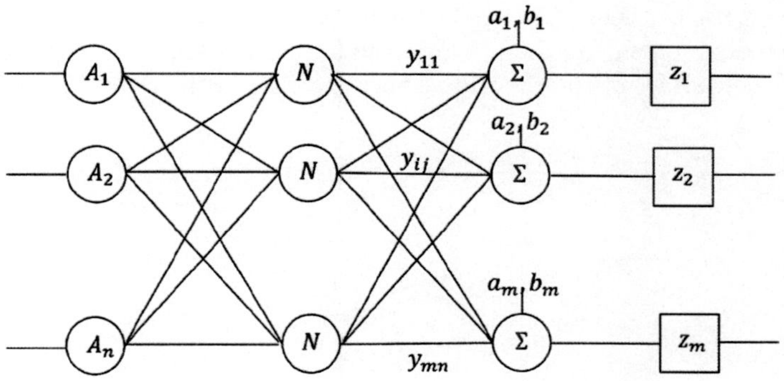

Fig. 2. Example of ANFIS layer with linear consequences.

in Fig. 2 we will connect the inputs of the ANFIS layer to the output through trainable weights.

Training algorithms for ANFIS architectures primarily rely on gradient-based methods, least squares methods, or a hybrid of these two [5]. In the present paper we use gradient descent algorithm together with ADAM optimization [6]. The training of the proposed ANFIS architectures is achieved using PyTorch's automatic differentiation module [8]. These tools allow us to combine ANFIS layers and neural layers for a rich structure with high flexibility.

3 Function Approximation

3.1 ANFIS Setup and Training

To test the approximation capabilities of all combinations of ANFIS learning, we train four different models with the same number of parameters. An ANFIS system can be 0th order with a singleton consequent or 1st order with a linear consequent. An ANFIS system can apply rule normalization or not given that the network itself can account for normalization in its learning. If we do not have normalization for a rule base in the case of Gaussian layers, the system becomes a Radial Basis Function Network (RBFN) [4]. If we use triangular membership, we obtain a different architecture where neurons have a triangular fuzzy membership activation function. In certain applications, rule normalization can create instability, especially when firing levels of all the rules are very small. Normalization is removed in such cases to improve stability. Each of these models are tested on the same functions.

Each model is comprised of 2 layers with 16 neurons each and a fuzzy rule base with 32 rules for a total of 449 learnable parameters. In the figures below, Gaussian membership functions were used and initialized from the normal distribution for both the mean and standard deviation.

In all figures, the expected output of the functions is shown in purple with each other model referenced by a different color. If only purple is shown, that is because the error between the approximated function and the actual function is too small to distinguish. At the bottom, the final Mean Squared Error (MSE) is shown for each model of the function approximated in the domain shown.

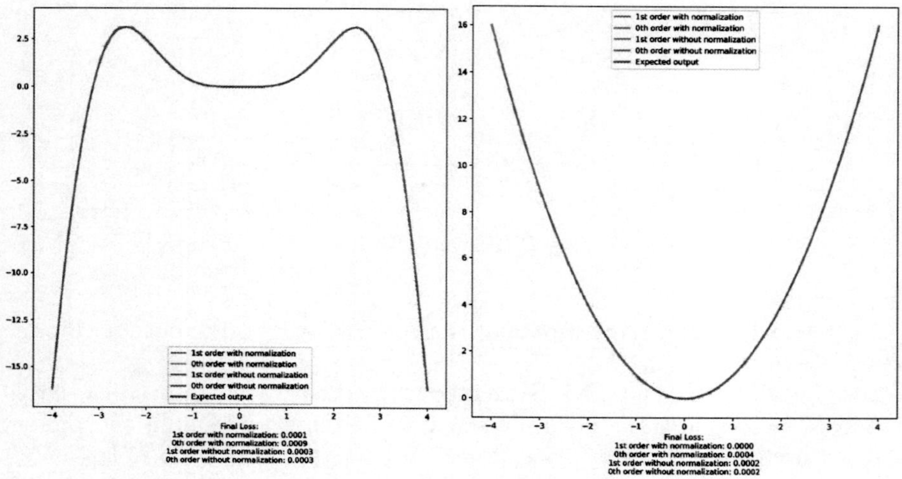

(a) $y = \frac{\sin(x) \cdot x^3}{3}$

(b) $y = x^2$

Fig. 3. Two simple functions that are differentiable

3.2 Experimental Results

In Fig. 3 we approximate two relatively simple functions with the expectation that all versions of the ANFIS can learn them well and easily. This is shown to be true; all of the losses for each model become very small with almost no visible deviations from the function. In Fig. 4 we approximate two more challenging functions: an approximation of a square wave and a function involving the absolute value and logarithm. Both of these functions have more complicated derivatives, and we can see some of the ANFIS models struggle to learn them with high accuracy. However, the 1st order ANFIS model with rule normalization (or standard ANFIS) learns the first function very well. The rest of the models appear to learn the second function better, with a 0th order ANFIS with rule normalization learning it the best.

In Fig. 5 we attempt to have the ANFIS models learn two non-differentiable functions that fall into the fractal family. Figure 5a is the Weierstrass function [13] is defined as

$$f(x) = \sum_{n=0}^{\infty} a^n \cos(b^n \pi x)$$

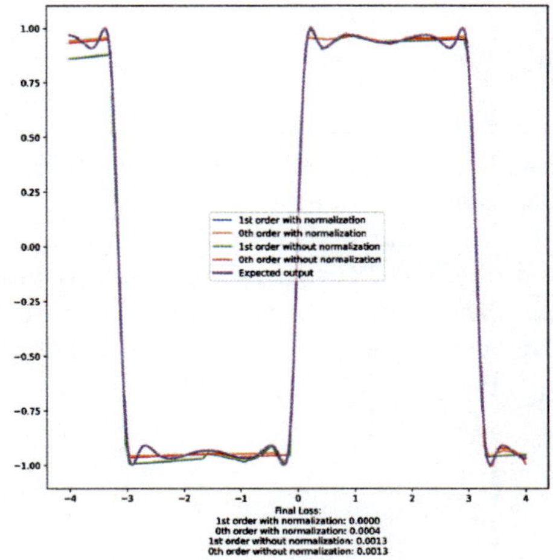

(a) $y = \sin(2 \cdot \sin(2 \cdot \sin(x)))$

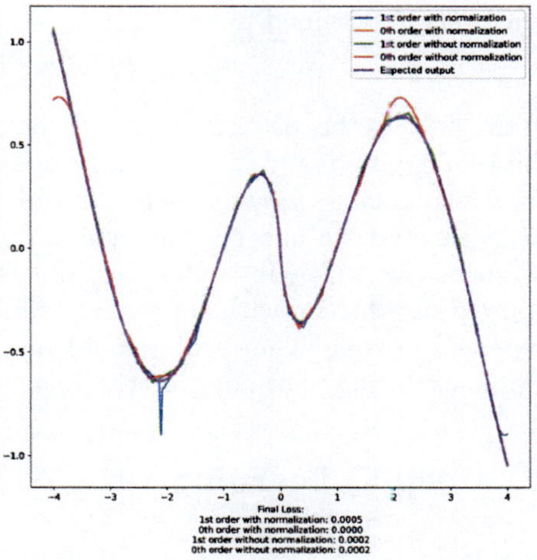

(b) $y = \log|x| \cdot \sin(x)$

Fig. 4. Two more complicated functions that are differentiable

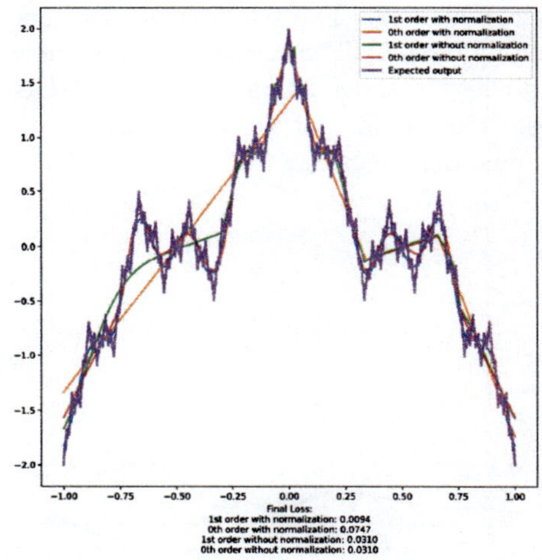

(a) Weierstrass function with a=.5, b=3, and sum from n=0 to n=20

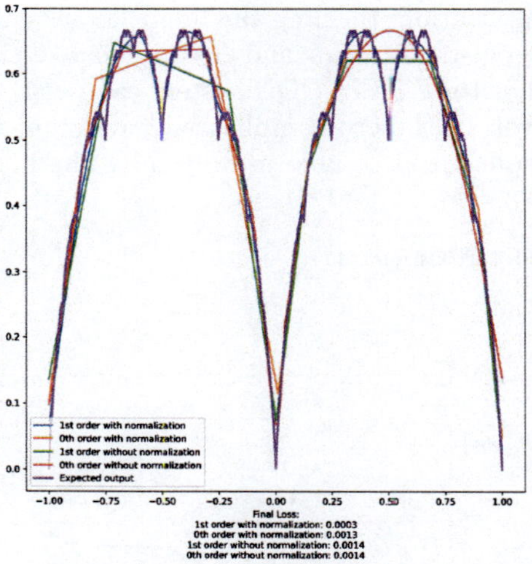

(b) Blancmange curve with sum from n=0 to n=20

Fig. 5. Two approximated fractals that are non-differentiable

where in our implementation, we use $a = .5$, $b = 3$, and sum from $n = 0$ to $n = 20$. Figure 5b is the Blancmange curve [17] is defined as

$$f(x) = \sum_{n=0}^{\infty} \frac{s(2^n x)}{2^n}$$

where $s(x)$ is defined by

$$s(x) = \min_{n \in \mathbb{Z}} |x - n|$$

so that $s(x)$ is the distance from x to the nearest integer. In our implementation of the Blancmange function, we sum from $n = 0$ to $n = 20$. Both of these functions are continuous everywhere but non-differentiable. This may make it challenging to learn, and we can see that some models struggle to learn these with high precision. However, in both cases, the 1st order ANFIS with rule normalization approximates them with the lowest error and the least visible deviation from the expected output. The results could improve with a larger network; the current architecture may not support the relative complexity of these functions.

4 Deep Q-Learning with ANFIS Layers

In this section we expand on the results discussed in [16] where we proposed a Deep Q-learning Neural Network with ANFIS layers (ANFIS DQN). We refer to the above for details on the proposed architecture. In [16], we considered a cart-pole application and showed how a deep Q-learning algorithm combined with ANFIS layers can solve the given environment. In this paper we consider another application: tackling the lunar lander environment [9]. The training parameters, hyper-parameters, and environment testing follow [16]. The results in Fig. 6 show that the ANFIS DQN system can perform roughly as well as a traditional DQN even in a more complicated environment. Given sufficient training time, both models will be able to fully solve the lunar lander problem.

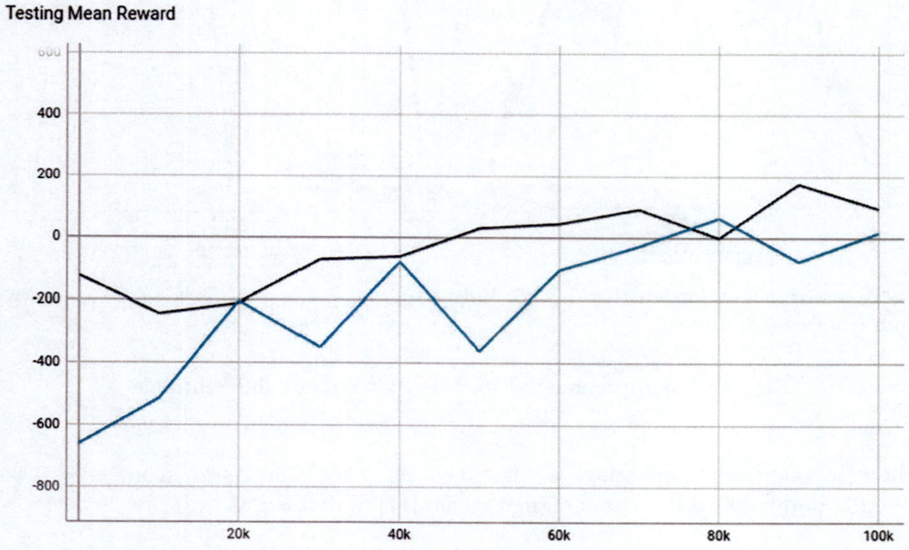

Fig. 6. ANFIS DQN (light blue) vs DQN (dark gray) in LunarLander-v2

5 Conclusion

In the present paper we illustrate the combination of ANFIS architectures with other neural layers in a flexible neuro-fuzzy architecture. The training of the proposed networks employs backpropagation and automatic differentiation via easy integration with PyTorch. We exemplify the proposed architectures in function approximation experiments and a reinforcement learning-based lunar lander task.

In future research we plan to test more complex architectures with other datasets and applications. Additionally, further experimentation with explainability mechanisms such as conversions from an ANFIS rule-base to a Mamdani fuzzy inference system or other novel methods stands to clarify the unique value of neuro-fuzzy systems.

References

1. Al-Ali, A., Elharrouss, O., Qidwai, U., Al-Maaddeed, S.: ANFIS-net for automatic detection of COVID-19. Sci. Rep. **11**(1), 17318 (2021)
2. Bede, B., Rudas, I.J.: Approximation properties of higher order Takagi-Sugeno fuzzy systems. In: 2013 Joint IFSA World Congress and NAFIPS Annual Meeting (IFSA/NAFIPS), pp. 368–373. IEEE (2013)
3. Benmiloud, T.: Multioutput adaptive neuro-fuzzy inference system. In: Proceedings of the 11th WSEAS International Conference on Nural Networks and 11th WSEAS International Conference on Evolutionary Computing and 11th WSEAS International Conference on Fuzzy Systems, NN 2010/EC 2010/FS 2010, , pp. 94–98. World Scientific and Engineering Academy and Society (WSEAS), Stevens Point (2010)
4. Huang, D.S.: Radial basis probabilistic neural networks: model and application. Int. J. Pattern Recogn. Artif. Intell. **13**(07), 1083–1101 (1999)
5. Jang, J.S.: ANFIS: adaptive-network-based fuzzy inference system. IEEE Trans. Syst. Man Cybern. **23**(3), 665–685 (1993). https://doi.org/10.1109/21.256541
6. Kingma, D.P., Ba, J.: Adam: a method for stochastic optimization. arXiv preprint arXiv:1412.6980 (2014)
7. Mamdani, E.H., Assilian, S.: An experiment in linguistic synthesis with a fuzzy logic controller. Int. J. Man Mach. Stud. **7**(1), 1–13 (1975)
8. Paszke, A., et al.: Automatic differentiation in pytorch. In: NIPS 2017 Workshop on Autodiff (2017). https://openreview.net/forum?id=BJJsrmfCZ
9. Ravichandiran, S.: Hands-On Reinforcement Learning with Python: Master Reinforcement and Deep Reinforcement Learning Using OpenAI Gym and TensorFlow. Packt Publishing Ltd. (2018)
10. Sugeno, M., Kang, G.: Structure identification of fuzzy model. Fuzzy Sets Syst. **28**(1), 15–33 (1988)
11. Takagi, T., Sugeno, M.: Fuzzy identification of systems and its applications to modeling and control. IEEE Trans. Syst. Man Cybern. **8**(1), 116–132 (1985)
12. Walia, N., Singh, H., Sharma, A.: ANFIS: adaptive neuro-fuzzy inference system-a survey. Int. J. Comput. Appl. **123**(13) (2015)

13. Weierstrass, K.: Über continuirliche Functionen eines reellen Arguments, die für keinen Werth des letzteren einen bestimmten Differentialquotienten besitzen. Cambridge Library Collection - Mathematics, vol. 2, pp. 71–74. Cambridge University Press (2013). https://doi.org/10.1017/CBO9781139567817.006
14. Yang, Y., Chen, Y., Wang, Y., Li, C., Li, L.: Modelling a combined method based on ANFIS and neural network improved by de algorithm: a case study for short-term electricity demand forecasting. Appl. Soft Comput. **49**, 663–675 (2016)
15. Yazdanbakhsh, O., Dick, S.: A deep neuro-fuzzy network for image classification. arXiv preprint arXiv:2001.01686 (2019)
16. Zander, E., van Oostendorp, B., Bede, B.: Reinforcement learning with Takagi-Sugeno-Kang fuzzy systems (2023, Submitted)
17. 貞治, 誘導函數ヲ有セザル連續函數ノ簡單ナル例. Tokyo Sugaku-Butsurigakkwai Hokoku **1**, F176–F177 (1901). https://doi.org/10.11429/subutsuhokoku1901.1.F176

Growth Kinetics of Gold Nanoparticles via P-Fuzzy Systems

Vinícius F. Wasques[1,2(✉)], Valéria Spolon Marangoni[1], and Shaian José Anghinoni[1]

[1] Ilum School of Science - CNPEM, Campinas, Brazil
vwasques@outlook.com, valeria.marangoni@ilum.cnpem.br, shaian220071@ilum.cnpem.br

[2] São Paulo State University - UNESP, Rio Claro, Brazil

Abstract. This article presents an application of *p*-fuzzy systems to estimate the maximum absorption wavelength of gold nanopartcles, as well as the stability point during their growth kinects. The classification of the wavelength and absorption values are modeled by fuzzy sets. Two fuzzy rule-based systems are provided in this study. The first aims to estimate the curves of wavelength by absorption. The second has the purpose of estimating the maximum absorbance with respect to time. The numerical method incorporated in the *p*-fuzzy system is the Euler's method, where the direction field is replaced by the output produced by the fuzzy rule-based system. For these fuzzy rule-based systems the Mamdani method is considered as the fuzzy inference. A simulation of the proposed method, considering a dataset of gold nanoparticles formation, is presented in order to validate the response provided by the *p*-fuzzy system.

Keywords: P-fuzzy systems · Fuzzy Rule-Based Systems · Chemical Reaction · Nanomaterials

1 Introduction

Gold nanomaterials present unique optical and electronic properties arising from the surface plasmons oscillation of their conduction electrons when excited by light at specific wavelengths. This oscillation, known as localized surface plasmon resonance (LSPR), depends on their size and shape and results in high absorption and scattering of light, making them useful for many application from medicine and sensors to catalysis and energy [1,2].

Since size, shape and properties are closely connected, the synthesis play an important role in nanostructures' properties. Gold nanostructures can be synthetized by a large variety of methods. The in-situ bottom-up synthesis, which is the most common chemical route, involves the reduction of gold ions ($HAuCl_4$)

V. F. Wasques—FAPESP n° 2023/03927-0 and S. J. Anghinoni - FAPESP n° 2023/02558-0.

K. Cohen et al. (Eds.): NAFIPS 2023, LNNS 751, pp. 149–159, 2023.
https://doi.org/10.1007/978-3-031-46778-3_14

to atomic gold in aqueous solution by a reducing agent such as sodium citrate, ascorbic acid and sodium boron hydride. A further stabilizing agent is often used to prevent agglomeration or further growth of the particles and it can be a charged (electrostatic stabilization) or a polymeric molecule (steric stabilization). Trisodium citrate, for example, is often used due to its ability to act as both reducing agent and stabilizer [3].

Understanding how the reaction proceeds is essential to predict the mechanism of formation of the particles, choose the stabilizer and reducing molecules and, therefore, develop more efficient synthesis approaches with high control of morphology. Despite the availability of a large amount of experimental and simulation data on size and size distribution of nanoparticles, it is still a very challenging task to design and control the size of nanoparticles systematically. In this proof-of-concept study, we apply *p*-fuzzy systems to predict the growth of gold nanoparticles using polyvinylpyrrolidone (PVP) and ascorbic acid as stabilizer and reducing agent, respectively.

The motivation to model the dynamic of this synthesis via fuzzy sets theory is based on the description of the phenomenon behavior through linguistic variables. For example, for *large* values of wavelength, it is expected to have a *small* absorbance. The words *large* and *small* describe subjective statements, since the dynamics of the phenomenon depends on several parameters such as the shape and size of the particles, for instance.

This paper studies two dynamics from the perspective of fuzzy sets theory. The first is to reproduce the behaviour of the wavelength having only the qualitative knowledge of the phenomenon, and for this modeling, a Fuzzy Rule-Based System (FRBS) is proposed. The second problem is to simulate the maximum absorbance curve. Since this problem depends on time, a *p*-fuzzy system is proposed. The information provided by the first FRBS is considered for the construction of a second FRBS, which feeds the *p*-fuzzy system. The construction of the fuzzy rules, as well as the mechanism of a *p*-fuzzy system, were motivated by some works found in the literature, such as [4,5]. The results provided here are compared with experimental data in order to validate the proposed modeling. These preliminary results reveal the potential of this strategy to predict the reaction kinetics of plasmonic nanostructures.

2 Preliminaries

2.1 Fuzzy Sets Theory

A fuzzy set is a generalization of a classical set, in the following sense, a classical set is defined by its characteristic function, that is, the characteristic function of a classical set A is given by χ_A, where $\chi_A(x) = 1$, if $x \in A$ or $\chi_A(x) = 0$, if $x \notin A$. This implies that a certain element must or must not satisfy the properties of the set A. This characterization lacks in the information about "how much" the element fulfills the requirement imposed by the set A. For example, let A be as the set of all particles that have a *small* size. Here, the expression small can be associated with nanoparticles, which has a range of size $[1, 100]$ nanometers (nm). So, depending on the size of a nanoparticle x, it may fulfill the property *small* with a greater degree than another nanoparticle y.

Motivated by the generalization of the characteristic function, Zadeh proposed the membership function, which is defined by $\varphi_A : X \to [0, 1]$, where the value $\varphi_A(x)$ means the degree of association of the element x in the set A, such that the closer the value $\varphi_A(x)$ is to 1, the greater its association with set A [6]. For example, a nanoparticle x with size less or equal than another nanoparticle y satisfies $\varphi_A(x) \geq \varphi_A(y)$, where A is the set of all particles that have a *small* size.

In the classical logic two connectors can be considered *and* and *or*, which are modeled by the minimum and maximum operators, respectively. These connectors can be also extended to the fuzzy logic by t-norms and s-norms. A t-norm is a binary operator $t : [0, 1] \times [0, 1] \to [0, 1]$ that satisfies the properties of commutative, associative, monotonicity and has 1 as an identity element. The minimum operator $\wedge : [0, 1] \times [0, 1] \to [0, 1]$ given by $\wedge(u, v) = \min\{u, v\}$ is an example of a t-norm. On the other hand, an s-norm is a binary operator $s : [0, 1] \times [0, 1] \to [0, 1]$ that also satisfies the properties of commutative, associative and monotonicity, but in contrast to a t-norm, it has 0 as an identity element. The maximum operator $\vee : [0, 1] \times [0, 1] \to [0, 1]$ given by $\vee(u, v) = \max\{u, v\}$ is an example of an s-norm [7].

The Cartesian product of fuzzy sets can be defined in terms of the minimum operator, that is, let A and B be fuzzy sets. The Cartesian product $A \times B$ is a fuzzy set, whose membership function is given by $\varphi_{A\times B}(a, b) = \varphi_A(a) \wedge \varphi_B(b) = \min\{\varphi_A(a), \varphi_B(b)\}$.

A fuzzy binary relation $\mathcal{R}$ of the universes U_1 and U_2 is defined by any subset of $U_1 \times U_2$, whose membership function is given by $\varphi_{\mathcal{R}} : U_1 \times U_2 \to [0, 1]$, where $\varphi_{\mathcal{R}}(x, y)$ represents the membership degree of association between x and y with respect to relation $\mathcal{R}$ [8].

This paper is dedicated to model sentences in the form of "If *input*, then *output*", where each fuzzy set in the input is called by *antecedent* and each fuzzy set in the output is called by *consequent*. For example,

If the wavelength of the wave is *verylong*, then the absorbance is *small*. (1)

Here the fuzzy set that describe the linguistic variable *very long* is the antecedent and the fuzzy set that describe *small* is the consequent. This sentence is also called by a fuzzy rule.

The FRBS studied here will be modeled by the Mamdani method. Recall that, given a fuzzy rule "If x is A_i, then y is B_i", it is possible to define a fuzzy relation $\mathcal{M}$ from the antecedents A_i and the consequents B_i, for all $i = 1, \ldots, n$. So, considering n rules with one antecedent and one consequent, which is the problem considered in this paper, the fuzzy relation proposed by Mamdani is given by

$$\varphi_{\mathcal{M}}(x, y) = \varphi_{\mathcal{R}_1}(x, y) \vee \varphi_{\mathcal{R}_2}(x, y) \vee \cdots \vee \varphi_{\mathcal{R}_n}(x, y),$$

where

$$\varphi_{\mathcal{R}_i}(x, y) = \varphi_{A_i}(x) \wedge \varphi_{B_i}(y), \quad \forall i = 1, \ldots, n.$$

In other words, the fuzzy relation $\mathcal{M}$ is nothing more than the union of the fuzzy Cartesian products between the antecedents and consequents of each rule. Hence, the fuzzy set of the output has membership function given by

$$\varphi_B(u) = \sup_x \{\varphi_A(x) \wedge \varphi_{\mathcal{M}}(x,u)\},$$

where sup stands for the supremum operator.

Figure 1 illustrates how the Mamdani method works, considering a *crisp* input value.

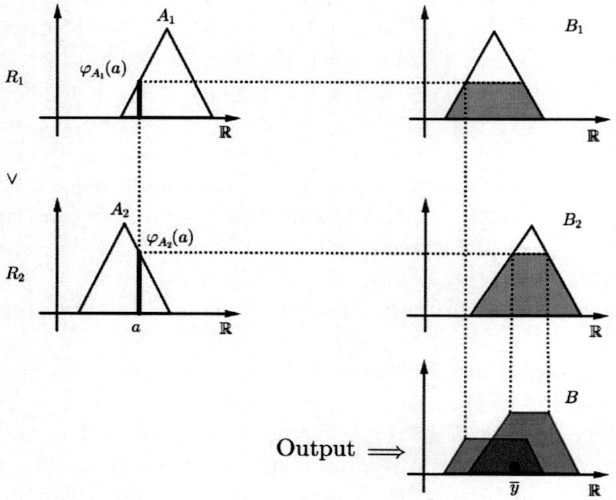

Fig. 1. Mamdani method from a crisp input value a. The first and second rules ($\mathcal{R}_1$ and $\mathcal{R}_2$) are represented by the first and second line, respectively. The fuzzy set B is the fuzzy output produced from the Mamdani method. The value $\overline{y}$, represented by the black dot, stands for the center of mass of the fuzzy set B.

A FRBS, specifically fuzzy controllers, is characterized by four essential components: an input module (fuzzyfication), where the input variables are modeled by fuzzy sets; a rule base module; a fuzzy inference module, which consists with methods that allow to manipulate each input in order to obtain an output, here the Mamdani method will be considered; and an output module (defuzzyfication), that here will be given by the center of mass (CM)

$$CM(B) = \frac{\sum_{i=1}^{n} x_i \varphi_B(x_i)}{\sum_{i=1}^{n} \varphi_B(x_i)}.$$

Figure 2 illustrates the mechanism of a Fuzzy Rule-Based System.

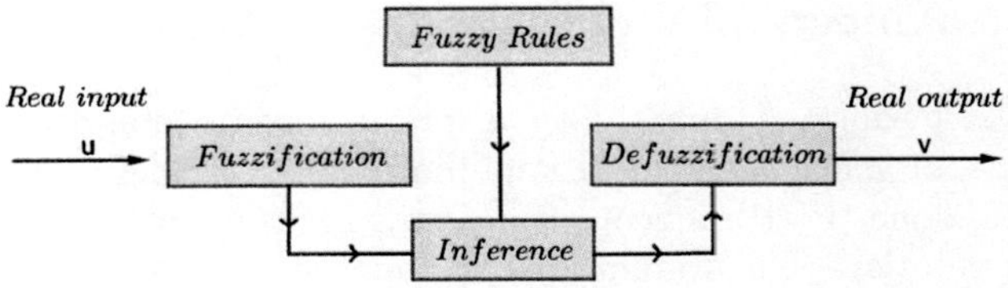

Fig. 2. Diagram of a Fuzzy Rule-Based System, where the input (u) and the output (v) are real values (crisps).

2.2 P-Fuzzy Systems

A *p*-fuzzy systems mean partially fuzzy systems, where the term partially refers to qualitative knowledge of the phenomenon that is not necessarily described in a deterministic way. In the context of differential equations, this kind of approach studies problems where the direction field of a differential equation, associated with an Initial Value Problem (IVP) or a Boundary Problem (BVP), is known only qualitatively. For example, an IVP is given by

$$\begin{cases} \frac{dx}{dt} = f(x(t)) \\ x(t_0) = x_0 \in \mathbb{R} \end{cases}.$$

In theoretical models the direction field $f(x(t))$ is given. However, in practical problems it is not always possible to provide this information. So, a *p*-fuzzy system, through a FRBS that incorporates prior knowledge of the phenomenon by an expert or dataset, replaces the direction field by the Euler's numerical method

$$x_{n+1} = x_n + h(FRBS_f(x_n)),$$

where h is the size of the step and $FRBS_f(x_n)$ is the output obtained by the FRBS via the input x_n.

Figure 3 represents how the *p*-fuzzy system works.

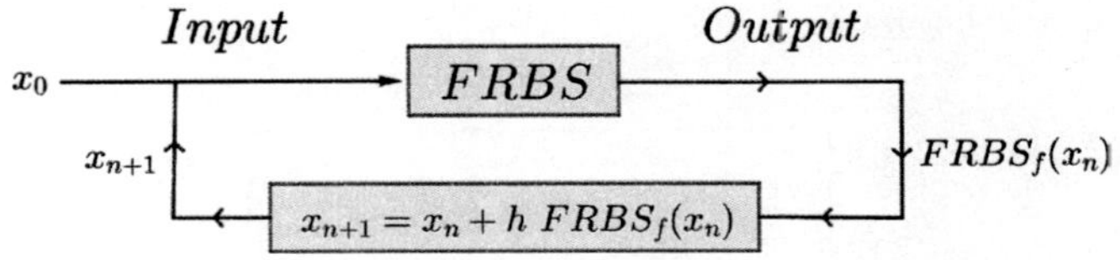

Fig. 3. Diagram of a *p*-fuzzy system, where the numerical method considered is the Euler's method.

This paper proposes a FRBS to describe the maximum absorbance × wavelength of gold nanoparticles. From this system, the *p*-fuzzy method will be applied to simulate the chemical reaction. Next, a brief contextualization of the problem of gold nanoparticles formation and the results obtained through laboratory experiments will be presented.

3 Growth Kinects of Gold Nanoparticle Formation

The kinetics is performed directly on a quartz cuvette standard rectangular with a volume of 3.5 mL and at room temperature. Briefly, 3 mL of 2%wt polyvinylpyrrolidone (PVP) aqueous solution is mixture with 40 µL of 0.01 mol L^{-1} $HAuCl_4$ solution. The system is kept under stirring for 5 min followed by addition of 100 µL 0.1 mol L^{-1} ascorbic acid. Immediately, the color changes from yellow to colorless and the cuvette is quickly placed on an ultraviolet-visible (UV-Vis) spectrophotometer. Measurements are performed constantly and in high speed until stabilization. At the end of the reaction, the suspension presents a dark red color, typical of spherical gold nanoparticles.

3.1 Results

The spectra show the evolution of the localized surface plasmon resonance (LSPR) peak with time (Fig. 4). As one can see, the absorption increases in intensity as the reaction proceeds, suggesting an increase in the particle size and number. The reaction proceeds fast in the first 2 min and then slower until the stabilization (Fig. 5).

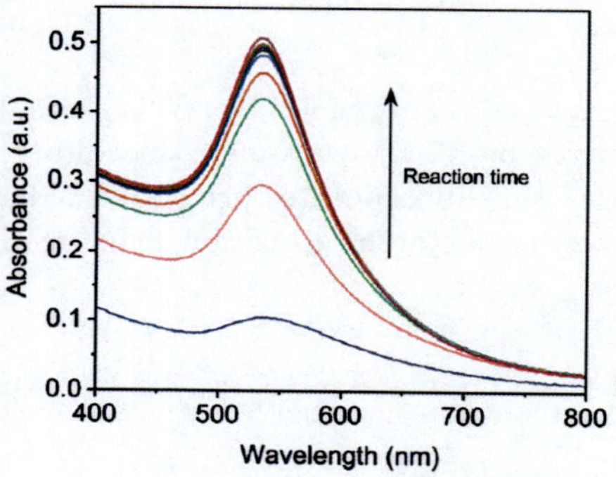

Fig. 4. Absorption kinetics of gold nanoparticles formation. UV-Vis spectra recorded during nanoparticle formation.

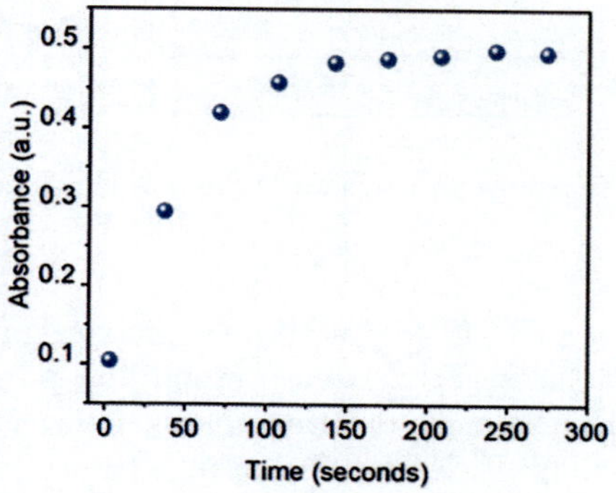

Fig. 5. Absorption kinetics of gold nanoparticles formation. Intensity of plasmon resonance band vs time obtained from the UV-Vis spectra.

4 Chemical Reaction of Gold Nanoparticles via P-Fuzzy System

First, the modeling will start describing the maximum absorbance of gold nanoparticles from its wavelength. So the domain of wavelength considered here is $[400, 800]$ nanometers (nm) and the domain of maximum absorbance is $[0, 0.58]$, which are the same considered in the laboratory experiments.

For the classification of the wavelength will be considered six fuzzy sets:

- *Low* (L);
- *Medium Low* (ML);
- *Medium* (M);
- *Medium High* (MH);
- *High* (H);
- *Very High* (VH) (see Fig. 6).

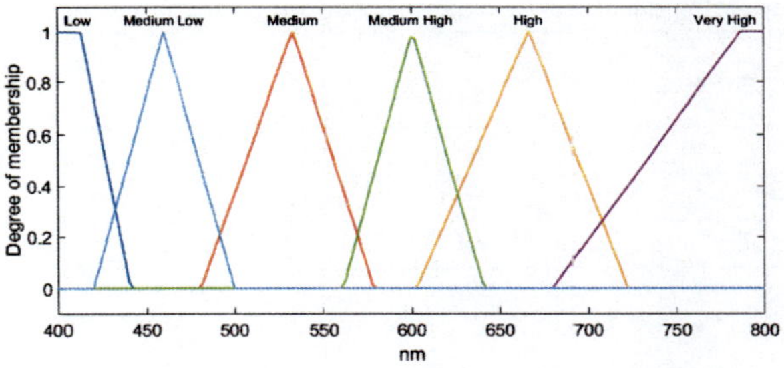

Fig. 6. Fuzzy sets *Low* (L), *Medium Low* (ML), *Medium* (M), *Medium High* (MH), *High* (H) and *Very High* (VH) for the classification of the wavelength of gold nanoparticles.

On the other hand, for the maximum absorbance will be considered four fuzzy sets:

- *Low* (L);
- *Medium* (M);
- *High* (H);
- *Very High* (VH) (see Fig. 7).

From the FRBS given by Table 1 and the Mamdani method, the curve generated by the fuzzy inference system, where the domain is the input wavelength and the codomain is the output maximum absorbance, simulates the curve depicted in Fig. 4.

Note that the curve depicted in Fig. 8 is qualitatively similar to the ones presented in Fig. 4. The purpose of this fuzzy modeling stage is to calibrate the fuzzy sets in the fuzzyfication step, so they can be reused in the simulation of the *p*-fuzzy system. Recall that the curves provided in Fig. 4 were obtained experimentally. So, the fuzzy sets of the antecedents and consequents were determined by a dataset [7,8].

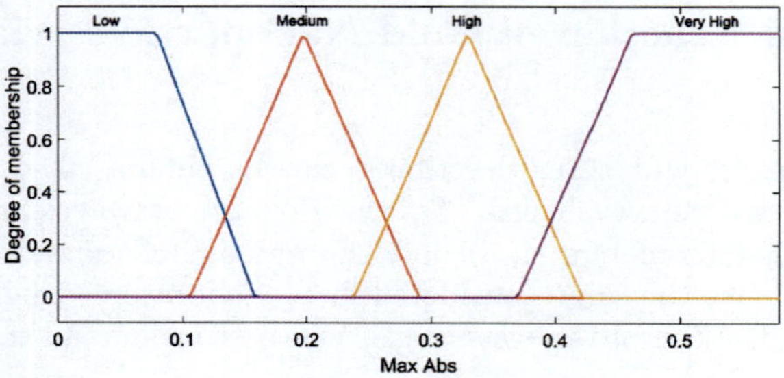

Fig. 7. Fuzzy sets *Low* (L), *Medium* (M), *High* (H) and *Very High* (VH) for the classification of the maximum absorbance of gold nanoparticles.

Table 1. Fuzzy rules for the estimative of wavelength × maximum absorbance. The fuzzy sets of the antecedents and the consequents can be seen in Figs. 6 and 7, respectively.

If wavelength is ____	then maximum absorbance is ____
L	H
ML	M
M	H
MH	VH
H	M
VH	L

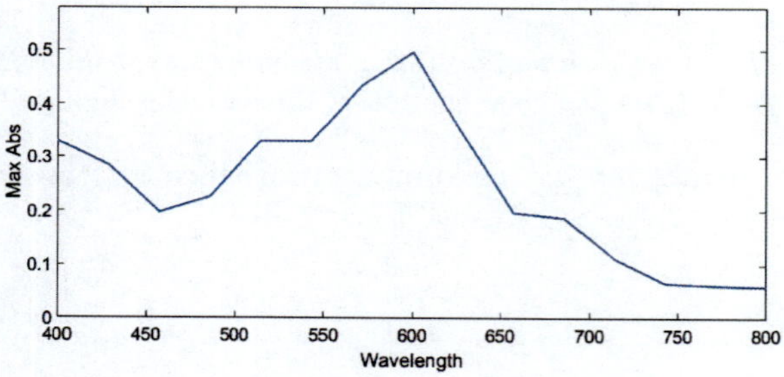

Fig. 8. Curve of wavelength × maximum absorbance generated by the proposed FRBS (Table 1) with Mamdani method.

Now the next step is to estimate the stability point (0.508) reached in the experiments, which can be seen in Fig. 5. This estimation will be given by the p-fuzzy system, where the FRBS that feeds the p-fuzzy system consists of: the input is the maximum absorbance value and the output is the slope of the absorbing curve.

The classification of the maximum absorbance by fuzzy sets will be the same as before (see Fig. 7). The classification of the slope is given by:

- negative *N2*;
- null *N1*;
- positive *P1*;
- high positive *P2*,

whose fuzzy sets are provided in Fig. 9. The classification of the slope is motivated by the work proposed by Wasques et al. [5].

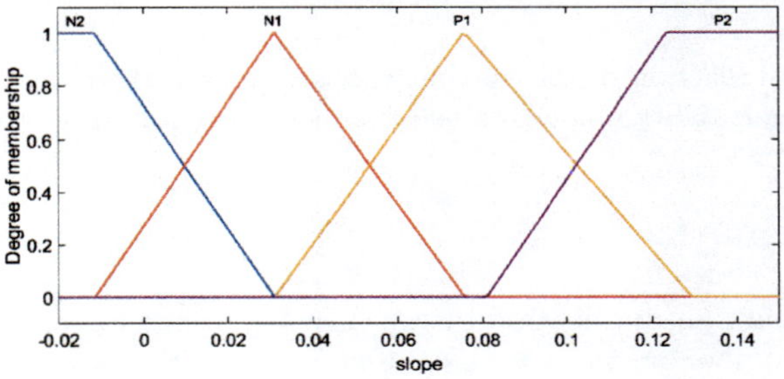

Fig. 9. Fuzzy sets negative *N2*, null *N1*, positive *P1* and high positive *P2* for the classification of the slope of the absorbing curve.

The fuzzy rules that feeds the *p*-fuzzy system are given in Table 2.

Table 2. Fuzzy rules that feeds the *p*-fuzzy system that simulates the behaviour of the chemical reaction presented in Fig. 5

If maximum absorbance is ___	then the slope of the absorbing curve is ___
L	*P*2
M	*P*1
H	*N*1
VH	*N*2

Figure 10 depicts the simulation of the gold nanoparticles formation via *p*-fuzzy system. Note that the qualitative behaviour is very similar to the experiments, as one can observe in Fig. 5. Moreover, the stability point obtained via *p*-fuzzy system was 0.5087, which is very close to the value obtained experimentally (0.508).

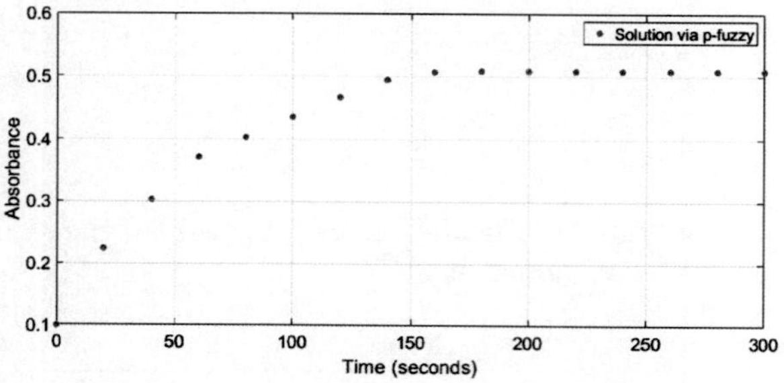

Fig. 10. Chemical reaction of the maximum absorbance with respect to time, via p-fuzzy system. The stability point obtained from the simulation is 0.5087.

5 Final Remarks

This work presents an application of a p-fuzzy system to predict the behavior of some chemical phenomena, such as the growth kinetics of gold nanoparticles. For this purpose, two fuzzy rule-based systems were constructed, the first was used to calibrate the fuzzy sets that model the wavelength by absorbance, and the second to estimate the growth slope of gold nanoparticles, in which the latter is set to feed the p-fuzzy system.

The first fuzzy rule-based system was built and calibrated from a dataset obtained in the laboratory of Ilum School of Science. After that, the fuzzy sets proposed in the fuzzyfication step were reused to build a second fuzzy rule-based system, in order to feed the p-fuzzy system. For the construction of the fuzzy sets of the slope and the fuzzy rules the authors adapted the modeling proposed in [5].

Through the simulations presented here, it was possible to reproduce the behavior of the curves of wavelength by absorbance. Moreover, it was possible to estimate satisfactorily the equilibrium point of the growth of gold nanoparticles.

This study reveals the potential of p-fuzzy systems to predict the kinetics of gold nanostructures and it can also be extended to other plasmonic structures, such as core-shell and anisotropic nanoparticles. Understanding and being able to predict the reaction in the synthesis of plasmonic nanostructures may open new avenues to design more efficient synthesis approaches.

Acknowledgment. The authors thank the financial support and the dataset given by Ilum School of Science - Brazilian Center for Research in Energy and Materials (CNPEM). The first author thanks the support of FAPESP n° 2023/03927-0 and the second and third authors thank the support of FAPESP n° 2023/02558-0.

References

1. Sardar, R., Funston, A.M., Mulvaney, P., Murray, R.W.: Gold nanoparticles: past, present, and future. Langmuir **25**(24), 13840–13851 (2009)
2. Bansal, S.A., Kumar, V., Karimi, J., Singh, A.P., Kumar, S.: Role of gold nanoparticles in advanced biomedical applications. Nanoscale Adv. **2**, 3764–3787 (2020)
3. Polte, J., et al.: Mechanism of gold nanoparticle formation in the classical citrate synthesis method derived from coupled in situ XANES and SAXS evaluation. J. Am. Chem. Soc. **132**, 1296–1301 (2010)
4. Sanchez, D.E., Wasques, V.F., Esmi, E., Barros, L.C.: Consecutive chemical reactions models via P-fuzzy systems. Proc. Ser. Braz. Soc. Comput. Appl. Math. **7**(1), 1–7 (2020)
5. Wasques, V.F., Santo Pedro, F., Esmi, E., Barros, L.C.: Synthesis chemical reaction model via P-fuzzy systems. In: Dick, S., Kreinovich, V., Lingras, P. (eds.) NAFIPS 2022. LNNS, vol. 500, pp. 336–347. Springer, Cham (2022). https://doi.org/10.1007/978-3-031-16038-7_32
6. Zadeh, L.A.: Fuzzy sets. Inf. Control **8**, 338–353 (1965)
7. Gomide, F., Pedrycz, W.: Introduction to Fuzzy Sets: Analysis and Design (Complex Adaptive Systems). MIT Press, Cambridge (1998)
8. Barros, L.C., Bassanezi, R.C., Lodwick, W.A.: A First Course in Fuzzy Logic, Fuzzy Dynamical Systems, and Biomathematics. Springer, Heidelberg (2017). https://doi.org/10.1007/978-3-662-53324-6

Optimization of Artificial Potential Field Using Genetic Algorithm for Human-Aware Navigation of Autonomous Mobile Robots

Shurendher Kumar Sampathkumar(✉), Anirudh Chhabra, Daegyun Choi, and Donghoon Kim

University of Cincinnati, Cincinnati, OH, USA
{sampatsr,chhabrad,choidg}@mail.uc.edu, Donghoon.Kim@uc.edu

Abstract. Path planning is an indispensable capability for an autonomous mobile robot to formulate a path from the initial position to the desired position while maneuvering safely from obstacles. In traditional path planning methods, the task constraint might be to generate a minimum energy or minimum distance path. However, in human-rich environments, apart from avoiding obstacles, the mobile robot must also navigate by recognizing and acting according to the social norms of humans. Hence, this study proposes an artificial potential field (APF)-based human-aware navigation framework for an autonomous mobile robot. Also, the genetic algorithm is utilized to optimize the scaling factors and the orders of potential functions of the APF. As a result, the shortest path from the start to the desired position is found for autonomous mobile robots without violating human factor constraints.

1 Introduction

Path planning and collision avoidance form the base for navigation of autonomous mobile robots (AMRs) and are a popular area of mobile robot research [1]. Generally, path planning methods aim to create a minimum time or minimum energy path for the AMRs to travel from the start position to the desired goal while avoiding collisions with obstacles in the environment [2]. Traditionally, AMRs are deployed in factory settings where human presence is limited. However, in recent times, AMRs are being utilized in more dynamically human-rich settings such as restaurants, airports, and shopping malls [3–5]. In these cases, the capability of the AMRs to navigate safely in such crowded environments becomes crucial. Unlike a static or dynamic obstacle, defined in conventional collision avoidance problems, the presence of humans requires an alteration to the path planning approaches by considering human factors. This is known as human-aware navigation (HAN).

HAN can be defined as "to operate and be accepted in the same environment as people, a robot must not only be able to avoid collisions with them but also

K. Cohen et al. (Eds.): NAFIPS 2023, LNNS 751, pp. 160–171, 2023.
https://doi.org/10.1007/978-3-031-46778-3_15

recognize and act accordingly to the social behavior of humans" [6]. Unlike a traditional path planning method, the goal of HAN is to ease discomfort and not cause harm to humans. Perception, prediction, and path planning play key roles in HAN. Mateus et al. proposed a vision-based perception for pedestrian detection using deep convoluted neural networks [7]. Bruckschen et al. [8] predicted the human's navigation goal based on the AMR's observation and prior knowledge about typical human transitions between objects. Then, a time-dependent cost map is generated with the human's future time steps, which is then solved to generate the shortest path for the AMRs. In this work, as a first step, we would like to investigate the artificial potential field (APF) as a HAN path planner without considerations for perception and prediction. This is because the principle of the APF is simple and straightforward for hardware implementation and offers outstanding real-time performance [9–11].

However, the APF suffers from certain limitations. At certain locations, the repulsive potential field generated by the obstacle might be equal to the attractive potential field generated by the desired position. In such locations, the agent tends to get stuck at that position due to local minima and will not be able to reach the desired position [11]. Another limitation is the goal non-reachable with obstacles nearby (GNRON) [9]. The agent might not be able to reach the desired position when it is very close to the obstacle. These problems would also be observed when implementing HAN-APF. For instance, in a shopping mall scenario, the shelves are arranged such that they offer a narrow passage between them. When a human is present in this passage, depending upon the dimensions of the passage, the HAN factors might cause local minimal issues. This would lead to the AMR getting stuck at that particular location. Proper selection of scaling parameters is significant to overcome such scenario. Hence, to resolve this, we propose using a genetic algorithm (GA) to optimize the scaling parameters of the HAN-APF subject to minimizing the path length and avoiding collisions.

Overall, the primary contribution of this work is to introduce a new path planning method for HAN based on the APF called HAN-APF. Moreover, the human factors included in HAN are prioritized based on their importance and optimized with the GA to find a shortest path to the desired position for the AMR. The paper is structured as follows: Sect. 2 introduces the conventional APF, gives an overview of the AMR and the human model used in this study, and introduces the concepts of HAN and GA; Sect. 3 presents the proposed method of HAN-APF and the optimization using the GA; Sect. 4 focuses on the description of the simulation study and the corresponding results; and Sect. 5 concludes the work and discusses future work.

2 Preliminaries

This work aims to navigate the AMR from the start to the desired location using the GA-optimized HAN-APF approach while considering human factors. For simplification, it is assumed that the position of the static obstacles and humans are known to the AMRs. Also, it is considered that the AMR is familiar with the pose of the human. The simulation is set up in a two-dimensional environment and the HAN planner doesn't consider interactions of humans with the objects in the environment. Furthermore, the human is assumed to be static in this work and is assumed to be a rigid body while the AMR is considered as a point mass.

2.1 Path Planning

The AMR aims to generate a path from its initial position to the desired location such that the AMR avoids potential collisions with obstacles in the environment as it traverses along the path. Usually, a search algorithm based on a specific strategy is defined to incrementally explore the workspace for generating a feasible path for the AMR [12]. One of the most popular strategies for approaching such problems is the APF [13].

The APF operates by defining attractive and repulsive potential fields in the robot's environment. These fields originate from objects in the environment based on the goal of the robot. Generally, the target position generates an attractive field while the obstacles generate a repulsive field, allowing the robot to traverse toward the goal while avoiding the obstacles. The attractive, V_a, and repulsive, V_r, potential functions are then combined to obtain the following total potential function, V, thereby governing the motion of the AMR [9,13]:

$$V(\mathbf{p}) = V_a(\mathbf{p}) + V_r(\mathbf{p}), \tag{1}$$

where

$$V_a(\mathbf{p}) = \frac{1}{2} k_a d(\mathbf{p}, \mathbf{p}_f)^{n_a}, \tag{2}$$

$$V_r(\mathbf{p}) = \begin{cases} \frac{1}{2} k_r \left(\frac{1}{d(\mathbf{p}, \mathbf{p}_h)} - \frac{1}{d_0} \right)^{n_r} & d(\mathbf{p}, \mathbf{p}_h) \leq d_0, \\ 0 & d(\mathbf{p}, \mathbf{p}_h) > d_0. \end{cases} \tag{3}$$

Here, $\mathbf{p} = [x_r \ y_r]^T \in \mathbb{R}^2$ represents the position of the AMR, $\mathbf{p}_f \in \mathbb{R}^2$ and $\mathbf{p}_i \in \mathbb{R}^2$ are the final position and the initial positions of the AMR, $\mathbf{p}_h = [x_h \ y_h]^T \in \mathbb{R}^2$ is the position of the human (obstacle), $d(\mathbf{a}, \mathbf{b})$ is the relative distance between $\mathbf{a} \in \mathbb{R}^2$ and $\mathbf{b} \in \mathbb{R}^2$, and d_o denote the distance of influence (DOI) of the repulsive potential. Additionally, k_a and k_r represent the scaling factors for attractive and repulsive potential functions while n_a and n_r denote the order of the attractive and repulsive functions.

The attractive ($\mathbf{f}_a(\mathbf{p}) \in \mathbb{R}^2$) and repulsive ($\mathbf{f}_r(\mathbf{p}) \in \mathbb{R}^2$) potential fields at a given position ($\mathbf{p}$) are expressed using the negative gradient of the individual potential functions as follows [9]:

$$\mathbf{f}_a(\mathbf{p}) = -\nabla V_a(\mathbf{p}) = k_a d(\mathbf{p}_f - \mathbf{p}), \tag{4}$$

$$\mathbf{f}_r(\mathbf{p}) = -\nabla V_r(\mathbf{p}) = \begin{cases} k_r \left(\frac{1}{d(\mathbf{p}, \mathbf{p}_h)} - \frac{1}{d_o} \right) \frac{1}{d(\mathbf{p}, \mathbf{p}_h)^3} (\mathbf{p} - \mathbf{p}_h) & d(\mathbf{p}, \mathbf{p}_h) \leq d_0, \\ 0 & d(\mathbf{p}, \mathbf{p}_h) > d_0. \end{cases} \tag{5}$$

The total potential field generated is then obtained by adding the attractive and the repulsive potential fields.

2.2 Kinematics

2.2.1 Human Model

The human motion is modeled based on a simplified unicycle model as follows:

$$\begin{bmatrix} x_{\mathrm{h}}(t+1) \\ y_{\mathrm{h}}(t+1) \\ \psi_{\mathrm{h}}(t+1) \end{bmatrix} = \begin{bmatrix} \begin{bmatrix} x_{\mathrm{h}}(t) + v_{\mathrm{h}_x}(t)\Delta t \cos\phi(t) \end{bmatrix} \\ \begin{bmatrix} y_{\mathrm{h}}(t) + v_{\mathrm{h}_y}(t)\Delta t \sin\phi(t) \end{bmatrix} \\ \psi_{\mathrm{h}}(t) + \omega_{\mathrm{h}}(t)\Delta t \end{bmatrix}, \tag{6}$$

where x_{h} and y_{h} indicate the Cartesian position, ψ_{h} is the heading angle, v_{h_x} and v_{h_y} denote the linear velocity in x and y directions, ω_{h} is the angular velocity of the human, and Δt denotes the time step.

2.2.2 Autonomous Mobile Robot (AMR) Model

In this work, the AMR is assumed to be a point mass moving under the influence of potential fields generated by the APF. The corresponding position and velocity of the AMR are expressed as

$$\begin{bmatrix} x_{\mathrm{r}}(t+1) \\ y_{\mathrm{r}}(t+1) \\ v_x(t+1) \\ v_y(t+1) \end{bmatrix} = \begin{bmatrix} x_{\mathrm{r}}(t) + f_x\Delta t \\ y_{\mathrm{r}}(t) + f_y\Delta t \\ f_x \\ f_y \end{bmatrix}, \tag{7}$$

where x_{r} and y_{r} denote the Cartesian position of the AMR, v_x and v_y are the respectively linear velocities, and Δt denotes the time step. Note that f_x and f_y are the total potential field vector $\mathbf{f}$ in the x and y directions, respectively. Since the AMR is assumed to be a point mass, the angular velocity and the heading angle are not considered in this case (Fig. 1).

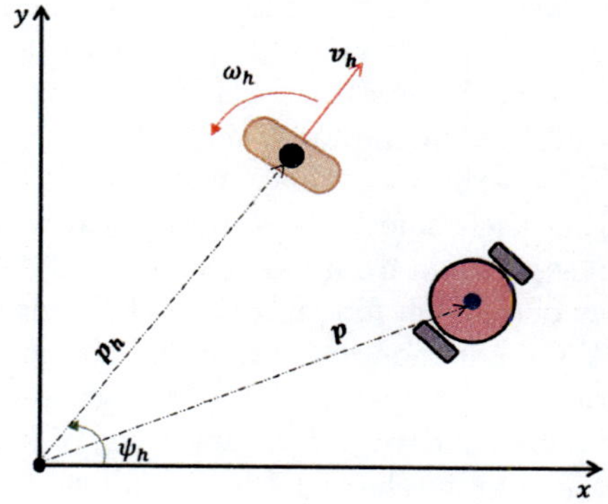

Fig. 1. Representation of the human and the AMR in the Cartesian frame

2.3 Human Aware Navigation (HAN)

The HAN is the overlap of robot motion planning and human-robot interaction [14]. As mentioned in Sect. 1, the primary goal of HAN is the reduction of stress

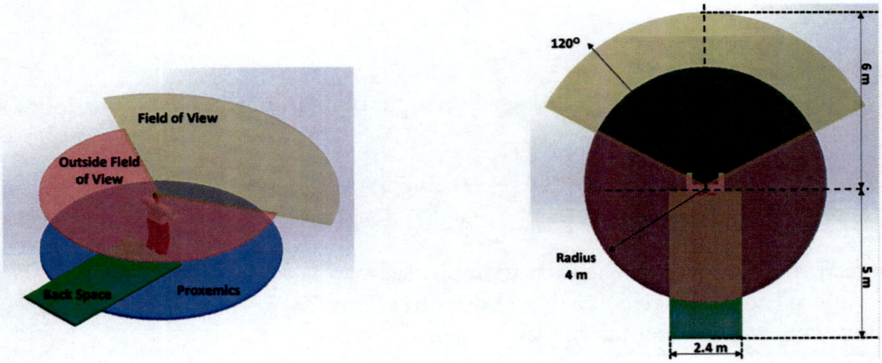

Fig. 2. Pictorial representation of HAN factors: the yellow region represents the field of view (FOV), the blue region highlights the proxemics, and the green region depicts the backspace

and cause no harm to humans. Thus, we define HAN as "navigation of the AMRs based on human-centered potential fields causing least discomfort with no harm". There are many different goals that have been considered for HAN with the common criteria of improving acceptance of AMRs. Some of these goals include minimizing annoyance and stress for humans, ensuring that the AMRs behave more naturally (similar to humans), and AMRs follow cultural norms. Kruse et al. classify the research on HAN into three major categories: comfort, naturalness, and sociability [14]. The 'safety factor' is considered as part of 'comfort' in Ref. [14] as being safe is comfortable for humans. However, this work considers human safety as a separate category from comfort as safety is a basic necessity and takes higher precedence than comfort.

Apart from these, there are other HAN factors like the acceptable noise made by AMRs, the appearance of the AMRs, etc. [14] This work only considers two factors: human safety and human comfort.

The primary goal of the human safety factor is to avoid collisions with humans, thereby ensuring their safety. This is achieved when the AMRs maintain a certain minimum separation from humans. Many HAN strategies consider the proxemics theory as one of the fundamental HAN rules [7,8,15]. The proxemics theory classifies the separation maintained by humans from other individuals based on certain rules and navigates the AMR accordingly. However, these specifications are lenient [16] and vary with age, type of social interaction, and culture. This work considers a circle region of a radius of 4 m with the human at the center as shown in Fig. 2 [16].

The human comfort factor aims to subside the annoyance and stress for humans during interaction with AMRs [14]. One way to give humans comfort is to avoid the surprise of navigating in the region behind the human [8,14]. This is called the region of backspace (BS). Generally, BS is considered to be a rectangular region behind the human with a length of 5 m and a width of 2.4 m as shown in Fig. 2. This dimension for BS is chosen based on survey results [8]. Another way is for the AMRs to avoid maneuvering in FOV of the human. This ensures that AMR does not navigate in the path of humans, thereby not being a potential obstacle for humans and causing discomfort. It is known that beyond 120° from the human center visual line, monocular vision area is reached which limits three-dimensional perception, and a fifth of the field of vision is lost causing difficulty in viewing [17]. Therefore, based on this, the FOV of humans is assumed to be 120 deg in this work [18].

2.4 Genetic Algorithm (GA)

The GA is an evolutionary algorithm inspired by the idea of natural selection [19]. Natural selection implies that given a population of individuals and a fitness function, the fittest individual is selected. The goal of the GA is to optimize individuals based on a specified fitness function. The idea is that in a population group, individuals (called parents) are selected for reproduction resulting in children. These children represent the next generation in which children with higher fitness scores are selected for further reproduction. This process is continued based on criteria such as obtaining an optimal (or near optimal) solution.

The first step in GA is creating a random set of individuals. These individuals are called *chromosomes* and the set of chromosomes is called a *population*. Each *chromosome* represents a potential solution for a set of parameters that need to be optimized. These parameters are set up in each chromosome as *genes*. Each *chromosome* is then evaluated through a fitness function to obtain a fitness score. The fitness score shows the ability of the particular individual to compete. The *chromosomes* with the best fitness scores are selected for reproduction and thereby sought as solutions to the optimization problem. Using techniques such as crossover and mutation, a diverse new *population* is created in every generation. However, the elitism technique is also employed to preserve the best solutions from accidental modification of the fittest *chromosomes*.

3 Proposed Approach

In this section, the APF-based HAN path planner is designed, followed by the definition of optimization criteria for the selected parameters.

3.1 HAN-APF

In traditional APF, the potential functions are limited to the attraction toward the goal and repulsion from obstacles. However, for considering the HAN factors, individual potentials for each factor are created. Thus, the total repulsive potential functions change prompting the AMR to adapt a path according to the HAN factors. The potential function for HAN-APF is redefined as follows:

$$V_{\mathrm{h}}(\mathbf{p}) = V_{\mathrm{a}}(\mathbf{p}) + \Sigma V_{\mathrm{i}}(\mathbf{p}), \tag{8}$$

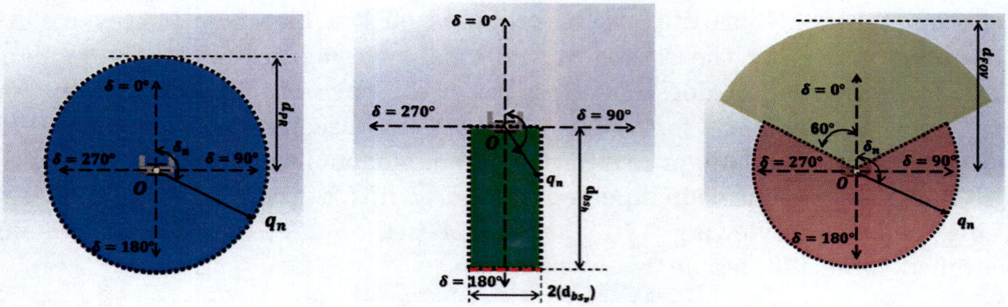

Fig. 3. Calculation of DOI for the HAN factors: proxemics (left), BS (center), and FOV (right)

where

$$V_{\mathrm{i}}(\mathbf{p}) = \begin{cases} \frac{1}{2}k_{\mathrm{i}}\left(\frac{1}{d(\mathbf{p},\mathbf{p}_{\mathrm{h}})} - \frac{1}{d_{\mathrm{i}}}\right)^{n_{\mathrm{i}}} & d(\mathbf{p},\mathbf{p}_{\mathrm{h}}) \leq d_{\mathrm{i}}, \\ 0 & d(\mathbf{p},\mathbf{p}_{\mathrm{h}}) > d_{\mathrm{i}}, \end{cases} \tag{9}$$

where i = $\left[\mathrm{PR, BS, FOV}\right]$ denote the additional potential functions based on the considered HAN factors. Here, k_i is the respective scaling factor, n_i is the corresponding order, and d_i denotes the DOI.

Although the potential functions for HAN-APF are similar to traditional APF, the geometry of DOI varies as shown in Fig. 2. Hence, unlike a constant value representing the radius of the circle, we define the DOI by calculating the distance from the human position and a series of points, q_n generated along the boundary of corresponding human factor. For instance, BS is defined with 34 points along the boundary of the BS factor, while FOV and PR are defined with 56 and 120 points respectively. The location of these points are significant because the DOI is obtained by considering the closest point along the boundary of human factor from the AMR. The DOI for each HAN factor is obtained using δ as shown in Fig. 3. In the case of BS, for each point (q_n) along the immediate edges of the back-space factor region. However, for the distant edges, the d_{BS} is constant and set as 5 m. For the FOV case, the d_{FOV} remains constant for each point (q_n) along the circumference of the FOV region and is equal to 6 m and for points outside this region, $d_{\mathrm{FOV}} = 0$. Note that the DOI for PR remains constant across all angles and is similar to the repulsive potential function from conventional APF.

3.2 GA Optimization

This work considers the use of GA to optimize the scaling factors and the order of the PR, BS and FOV potential functions. These parameters are defined as genes of a chromosome (Ω) as follows:

$$\Omega = \left[k_{\mathrm{a}}\ k_{\mathrm{i}}\ n_{\mathrm{a}}\ n_{\mathrm{i}}\ \right], \tag{10}$$

where k_a and n_a represent the scaling factor and the order for the attractive potential function and k_i and n_i represent the scaling factors and orders for the HAN-based repulsive potential functions. Once the parameterized chromosome is defined, then a population of randomized solutions is created to initialize the algorithm. These potential solutions are then evaluated using the fitness function to determine fitness and proceed with the optimization process. The choice of the fitness function is crucial to the optimization and is defined as follows:

$$\mathcal{J} = -(l_{\mathrm{AMR}} + \Sigma \rho_{\mathrm{i}}) \tag{11}$$

where l_{AMR} is the length of the path traversed by the AMR, and ρ_i represent the penalties applied for violating the corresponding human factor constraints. Note that the minus sign indicates the minimization of the specified parameters as the GA naturally maximizes the fitness function. Here, the penalties are chosen as $\rho_{\mathrm{PR}} \gg \rho_{\mathrm{FOV}} > \rho_{\mathrm{BS}}$ to prioritize violations of the space near humans. This is due to the existing priority of the spaces around humans. Since human safety takes precedence over comfort, the penalty for proxemics is the largest so that it is never violated. The penalty for the FOV is kept higher than the penalty for back-space due to the fact that the AMR has more chance of being noticed by the human in the FOV than in the back-space. Also, the AMR has a higher chance of being a potential obstacle and disturbing the path of humans when navigating in the FOV when compared with the back-space. This structure of the fitness function provides a good way to allow prioritization of space around human obstacles while being able to generate a path with the least distance by minimizing the path length and the penalties applied to the solution.

4 Simulation Study

The simulation environment is defined as a two-dimensional space of 130 m × 130 m area. The performance of the proposed APF-based HAN path planner is tested in this environment for PR, FOV, and BS HAN factors. The dimensions for d_{PR}, $d_{\mathrm{BS_v}}$, $d_{\mathrm{BS_h}}$, and d_{FOV} are 4 m, 1.2 m, 5 m, and 6 m, respectively. The corresponding simulation parameters are given in Table 1. Here, training refers to the optimization of the scaling factors and the orders of potential functions for the respective HAN factors. Furthermore, the minimization of path length without violation of human factors is considered as the optimization criteria.

Table 1. Simulation parameters

Parameter	Symbol	Training	Testing
Start position of AMR	$(x_{\mathrm{r}}, y_{\mathrm{r}})$	(50 m, 120 m)	(50 m, 120 m)
Desired position of AMR	$(x_{\mathrm{d}}, y_{\mathrm{d}})$	(50 m, 20 m)	(50 m, 20 m)
Human pose	$(x_{\mathrm{h}}, y_{\mathrm{h}}, \psi_{\mathrm{h}})$	(50 m, 80 m, 350°)	(50 m, 80 m, 10°)
Human velocities	$(v_{\mathrm{h_x}}, v_{\mathrm{h_y}}, \omega_{\mathrm{h}})$	(0 m/s, 0 m/s, 0°/s)	(0 m/s, 0 m/s, 0°/s)

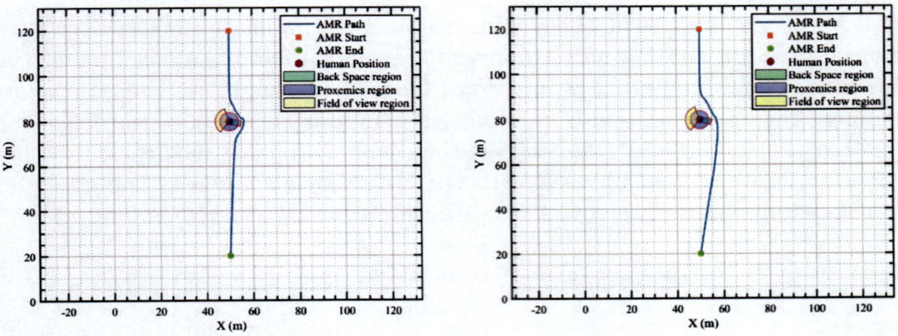

Fig. 4. AMR path generation with non-optimal (*left*) vs optimal (*right*) factors: Training

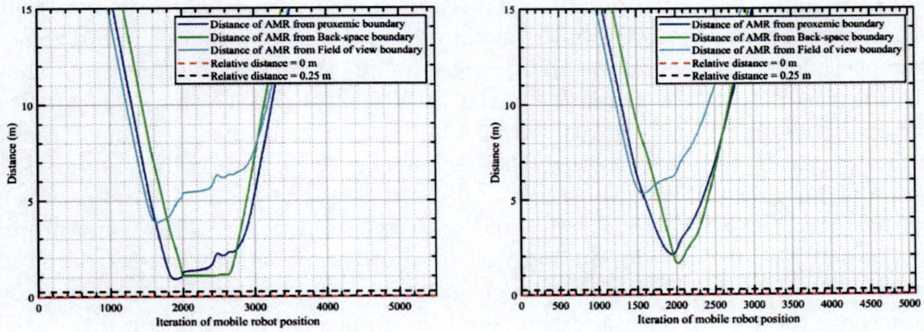

Fig. 5. Relative distance plots considering non-optimal (*left*) vs optimal (*right*) factors: Training

For conventional HAN-APF, the scaling factors ($\mathbf{k}_{\text{ut}}$) and the order of potentials ($\mathbf{n}_{\text{ut}}$) are chosen manually such that AMR reaches the desired position for a particular heading angle without violating the HAN factors. The scaling factors are, $\mathbf{k}_{\text{ut}}$= [0.01, 200, 0.005, 10] and orders are, $\mathbf{n}_{\text{ut}}$= [2, 2, 1, 1]. The resulting path length for the AMR is 109.98 m. Also, the maximum velocity for the AMR is constrained to be 2 m/s which is the common maximum velocity for AMR.

The HAN-APF parameters are then optimized using GA for minimizing path length while not violating HAN factors. The GA is implemented over 1000 generations with a population size of 10, a crossover rate of 85%, mutation rate of 20%, and 10% elitism rate. For selecting the best chromosomes in each generation, the tournament selection technique is implemented with a tournament size of 4. Moreover, the search space for the GA is set within the lower and upper bounds defined in Table 2.

The optimized orders and scaling factors for the potential functions are obtained as $\mathbf{n}_{\text{t}}$= [1, 2, 1, 1] and $\mathbf{k}_{\text{t}} = \left[.523\ 9.921\ 0.088\ 7.075\right] \times 10^3$, respectively, generating a path length of 101.48 m for the AMR in the training scenario. It can

Table 2. GA Search Space

Parameter	k_a	k_{PR}	k_{FOV}	k_{BS}	n_a	n_{PR}	n_{FOV}	n_{BS}
Lower bounds	0	0	0	0	1	1	1	1
Upper bounds	1000	1000	1000	1000	4	3	3	3

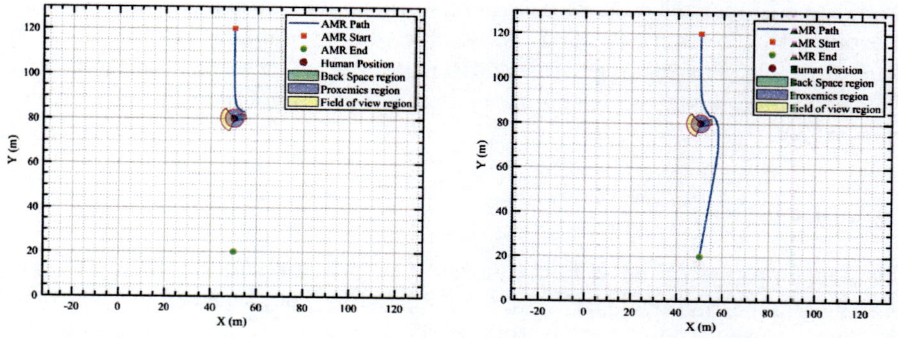

Fig. 6. AMR path generation with non-optimal (*left*) vs optimal (*right*) factors: Testing.

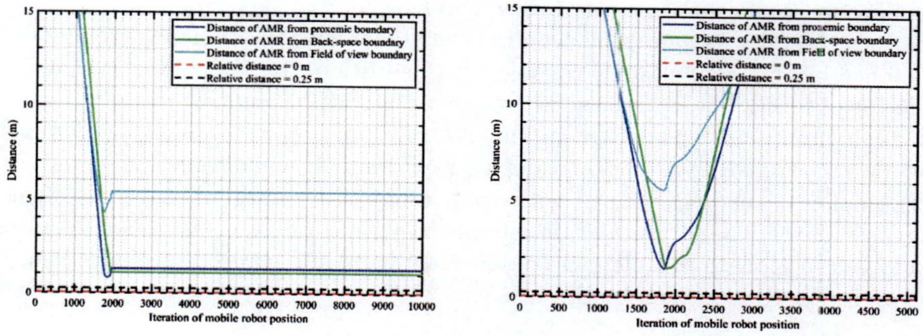

Fig. 7. Relative distance plots considering non-optimal (*left*) vs optimal (*right*) factors: Testing.

be observed in Fig. 4 that although for both non-optimal parameters and optimal parameters, the AMR reaches the desired position, the path length is considerably shorter in the optimized case. This is because while using non-optimal parameters, the HAN-APF generates a turbulent path, whereas using the optimal factors, the HAN-APF generates a smoother curve around the human obstacle as shown in Fig. 4. Therefore, by properly tuning the parameters, the AMR does not need to travel additional distance around the human obstacle. Moreover, as shown in Fig. 5, the AMR never violates any HAN factors.

The optimized parameters are then tested in different environmental conditions where the human obstacle has a different heading angle. While the

GA-optimized HAN-APF path planner was able to find a path to the desired position with a path length of 102.28 m, the AMR isn't able to find a path beyond the human obstacle when non-optimal parameters are used to model the HAN-APF. As shown in Fig. 6 and Fig. 7, the AMR gets stuck in local minima when non-optimal HAN-APF is used whereas the AMR successfully reaches the desired position using the GA-optimized HAN-APF. This further signifies that through the proper selection of parameters for the APF, the local minima could be avoided to a certain extent. This, however, highly depends on the geometry of the obstacle. For instance, the AMR has more chances of getting trapped in local minima when the shape of the obstacle is symmetrical or "U" shaped obstacle [20].

5 Conclusion

This paper proposes a novel human-aware navigation (HAN) framework based on artificial potential field (APF) called HAN-APF. The proposed method takes into consideration the human factors, such as proxemics, backspace, and field of view for the path planning of an autonomous mobile robot (AMR). The human factors are modeled as repulsive potential fields with a varying distance of influence. The proposed method is validated while taking into consideration the combination of proxemics, field of view, and backspace. The results show that the AMR successfully navigates by avoiding the human and the avoidance maneuver is defined by the human factor considered. In addition, the scaling factor and the order of potential are optimized using genetic algorithm and successfully tested for robustness and scalability. In the future, the dynamic nature of humans will be considered with noise to depict randomness, while also analyzing the performance of the path planner with other kinds of obstacles in the environment. Furthermore, HAN-APF could be integrated with a range measurement sensor model and a global path planner to detect humans in advance thereby reducing the overall path traversed by the AMR.

References

1. Chen, W., Zhang, T., Zou, Y.: Int. J. Adv. Rob. Syst. **15**(3), 1729881418776183 (2018)
2. Szczepanski, R., Tarczewski, T., Erwinski, K.: IEEE Access **10**, 39729 (2022)
3. Bear Robotics. Resturant robots (2023). https://www.bearrobotics.ai/servi. Accessed 14 Mar 2023
4. Robotics Tomorrow. Airport robots (2022). https://www.roboticstomorrow.com/story/2022/11/ottonomyio-partners-with-pittsburgh-international-airports-xbridge-to-innovate-customer-experiences-at-the-airport/19763/l. Accessed 14 Mar 2023
5. Korea times. Shopping mall robots (2022). https://www.koreatimes.co.kr/www/tech/2023/03/129_337566.html. Accessed 14 Mar 2023
6. Story, M., Jaksic, C., Fletcher, S.R., Webb, P., Tang, G., Carberry, J.: Paladyn J. Behav. Robot. **12**(1), 379 (2021)

7. Mateus, A., Ribeiro, D., Miraldo, P., Nascimento, J.C.: Robot. Auton. Syst. **113**, 23 (2019)
8. Bruckschen, L., Bungert, K., Dengler, N., Bennewitz, M.: In: 2020 IEEE/RSJ International Conference on Intelligent Robots and Systems (IROS), pp. 11032–11037. IEEE (2020)
9. Choi, D., Chhabra, A., Kim, D.: In: AIAA SCITECH 2022 Forum, p. 0272 (2022)
10. Lin, X., Wang, Z.Q., Chen, X.Y.: In: 2020 27th Saint Petersburg International Conference on Integrated Navigation Systems (ICINS), pp. 1–5 (2020). https://doi.org/10.23919/ICINS43215.2020.9134006
11. Omar, R., Sabudin, E., Che Ku Melor, C.K., et al.: ARPN J. Eng. Appl. Sci. **11**(18), 10801 (2016)
12. Spong, M.W., Hutchinson, S., Vidyasagar, M., et al.: Robot Modeling and Control, vol. 3. Wiley, New York (2006)
13. Khatib, O.: In: Proceedings of the 1985 IEEE International Conference on Robotics and Automation, vol. 2, pp. 500–505. IEEE (1985)
14. Kruse, T., Pandey, A.K., Alami, R., Kirsch, A.: Robot. Auton. Syst. **61**(12), 1726 (2013)
15. Smith, T., Chen, Y., Hewitt, N., Hu, B., Gu, Y.: Int. J. Soc. Robot. 1–18 (2021)
16. Rios-Martinez, J., Spalanzani, A., Laugier, C.: Int. J. Soc. Robot. **7**, 137 (2015)
17. Robotics Tomorrow. Human vision (2022). https://www.bristol.gov.uk/sensory-support-service/for-families/vision-support-for-families/eye-conditions/monocular-vision. Accessed 14 Mar 2023
18. Wang, Z., Nagai, Y., Zhu, D., Liu, J., Zou, N.: In: IOP Conference Series: Materials Science and Engineering, vol. 573, p. 012093. IOP Publishing (2019)
19. Mirjalili, S.: Evolutionary Algorithms and Neural Networks, pp. 43–55. Springer, Heidelberg (2019). https://doi.org/10.1007/978-3-319-93025-1
20. Jiang, Q., Cai, K., Xu, F.: Mech. Sci. **14**(1), 87 (2023)

Numerical Solutions of Fuzzy Population Models: A Case Study for Chagas' Disease Dynamics

Beatriz Laiate[1(✉)], Felipe Longo[2], José Ronaldo Alves[2,3], and João Frederico C. A. Meyer[2]

[1] Federal Institute of Education, Science and Technology of Espírito Santo, Serra, ES, Brazil
beatrizlaiate@gmail.com, beatriz.laiate@ifes.edu.br

[2] Institute of Mathematics, Statistics and Scientific Computing, University of Campinas, Campinas, Brazil
longo@ime.unicamp.br, j261237@dac.unicamp.br, jmeyer@unicamp.br

[3] CCINAT, Federal University of Vale do São Francisco, Senhor do Bonfim, Brazil

Abstract. This manuscript aims to investigate numerical solutions of biological models given by first-order systems of nonlinear fuzzy differential equations when the populations are given by specific fuzzy functions called $\mathcal{S}$–linearly correlated fuzzy processes. To this end, the notion of ψ–numerical methods is introduced and a discussion on the *fuzzy basal number* of the model is made. Lastly, numerical simulations of a within-host model of Chagas' disease are presented.

Keywords: Fuzzy differential equations · Fuzzy modeling · Strong linear independence · Fuzzy basal number · Chagas' Disease

1 Introduction

Mathematical modeling of infectious diseases is a recognized tool of comprehension of biological features of several types of infection in human body [16]. In particular, tropical weather diseases, whose prevalence indexes are higher in developing countries, continue without a proper understanding of their biological functioning and evolution.

Chagas' disease, or *american-trypanosomiasis* is the disease caused by the infection of *Trypanosoma Cruzi*, a protozoan transmitted by the barber. Specially common in regions with predominance of mud houses, it is still considered a neglected disease in the world. Studies reveal that the natural history of the infection is characterized by an incubation period with approximated duration of 7 days. After this phase, several cell types are invaded by the *T. Cruzi*. After the incubation period, the pathogenesis of Chagas' disease can be classified into acute and chronic phases, according to time of the infection and serological characteristics [17]. The immune response to the *T. Cruzi* infection have already been

K. Cohen et al. (Eds.): NAFIPS 2023, LNNS 751, pp. 172–183, 2023.
https://doi.org/10.1007/978-3-031-46778-3_16

studied by several authors [3,4,9,19,20]. The adaptive immune system embraces both humoral and cellular immunity, and it consists on one of the biological response to the T. Cruzi infection.

In the last decade, countless papers have appeared concerning biological applications of fuzzy sets theory spanning a vast quantity of situations, be they from a macroscopic point of view as [2] and [15], for example, to a microscopic approach [13]. Physiological measurements, such as the immune response to an infection, are particularly endowed with enormous uncertainties [1]. When dealing with physiological models described by differential equations it is expected that embracing uncertainty from the epistemic point of view can bring additional understanding information when compared to the classical approach.

This manuscript aims to study a within-host Chagas' disease predicting adaptive immune response to the T. Cruzi infection as a fuzzy population model. To this end, we recall some concepts on fuzzy calculus in Banach spaces of $\mathbb{R}_{\mathcal{F}}$ [5–7] so that we describe the populations involved as fuzzy-number-valued functions called $\mathcal{S}$–linearly correlated fuzzy processes. We introduce the concept of *fuzzy basal number* as a notion of characterization to the uncertainty intrinsic to a given population dynamics. The remainder of this paper provides the following content: Sect. 2 recalls some basic notions on fuzzy sets theory and fuzzy calculus. Section 3 introduces the notion of ψ–numerical methods of a fuzzy differential equation under the ψ–derivative. Section 4 presents a within-host model of Chagas' disease in a fuzzy environment. It also provides some numerical simulations to the solution. Section 5 brings some final remarks on the present paper.

2 Basic Notions

A fuzzy number A is a convex, bounded and compactly supported fuzzy subset of $\mathbb{R}$. The class of all fuzzy numbers is denoted by $\mathbb{R}_{\mathcal{F}}$. For all $A \in \mathbb{R}_{\mathcal{F}}$ given and $\alpha \in [0,1]$, there exist $a_\alpha^-, a_\alpha^+ \in \mathbb{R}$ with $a_\alpha^- \le a_\alpha^+$ such that $[A]_\alpha = \left[a_\alpha^-, a_\alpha^+\right]$. The Pompeiu-Hausdorff distance between two fuzzy numbers is the operator $\mathcal{D}_\infty : \mathbb{R}_{\mathcal{F}} \times \mathbb{R}_{\mathcal{F}} \to \mathbb{R}^+$ given by

$$\mathcal{D}_\infty (A, B) = \sup_{\alpha \in [0,1]} \inf \left\{\left|a_\alpha^- - b_\alpha^-\right|, \left|a_\alpha^+ - b_\alpha^+\right|\right\}, \tag{1}$$

for all $A, B \in \mathbb{R}_{\mathcal{F}}$ given levelwise by $[A]_\alpha = \left[a_\alpha^-, a_\alpha^+\right]$ and $[B]_\alpha = \left[b_\alpha^-, b_\alpha^+\right]$, $\alpha \in [0,1]$, respectively. The space $(\mathbb{R}_{\mathcal{F}}, \mathcal{D}_\infty)$ is a complete, but not separable, metric space. The core of a fuzzy number A is the set of all elements x such that $A(x) = 1$. The set of all fuzzy numbers whose core is a unitary set is denoted by $\mathbb{R}_{\mathcal{F}}^{\wedge} = \{A \in \mathbb{R}_{\mathcal{F}} \mid [A]_1 = \{a\} \text{ for some } a \in \mathbb{R}\}$.

A fuzzy number $A \in \mathbb{R}_{\mathcal{F}}$ is said to be symmetric with respect to $x \in \mathbb{R}$ if $A(x - y) = A(x + y)$ holds for all $y \in \mathbb{R}$, and in this case it is denoted $(A|x)$. If there is no $x \in \mathbb{R}$ such that A is symmetric, they it is said to be non-symmetric [7].

In this paper, we deal with fuzzy population models given by fuzzy numbers in vector spaces of $\mathbb{R}_{\mathcal{F}}$. To this end, we recall the concepts presented in [5,6,12,14].

Let $\mathcal{A} = \{A_1, \ldots, A_n\} \subset \mathbb{R}_{\mathcal{F}}$ be a finite set of fuzzy numbers. We denote the subset $\mathcal{S}(\mathcal{A})$ of $\mathbb{R}_{\mathcal{F}}$ as the set given by

$$\mathcal{S}(\mathcal{A}) = \{q_1 A_1 + \ldots + q_n A_n \mid q_1, \ldots, q_n \in \mathbb{R}\}, \tag{2}$$

where "+" and $q_i A_i$ represent the standard operations of sum and scalar multiplication in $\mathbb{R}_{\mathcal{F}}$, for all $i = 1, \ldots, n$.

The set $\mathcal{S}(\mathcal{A})$ can be seen as the set of all Minkowski combinations of the fuzzy numbers $A_1, \ldots, A_n$.

Definition 1 [5]. A set of fuzzy numbers $\mathcal{A} = \{A_1, \ldots, A_n\}$ is said to be strongly linearly independent (SLI, for short) if for all $B \in \mathcal{S}(\mathcal{A})$ given by $B = q_1 A_1 + \ldots + q_n A_n$, the implication $(B|0) \Rightarrow q_1 = \ldots, q_n = 0$ holds.

The next theorem characterizes the set $\mathcal{A} \subseteq \mathbb{R}_{\mathcal{F}}$ when it has the property of *strong linear independence*.

Theorem 1 [5]. *Let $\mathcal{A} = \{A_1, \ldots, A_n\} \subset \mathbb{R}_{\mathcal{F}}$ be given. The set $\mathcal{A}$ is SLI if, and only if, the function $\psi : \mathbb{R}^n \to \mathcal{S}(\mathcal{A})$ given by*

$$\psi(x_1, \ldots, x_n) = x_1 A_1 + \ldots + x_n A_n$$

for all $(x_1, \ldots, x_n) \in \mathbb{R}^n$, is a bijection.

If $B \in \mathcal{S}(\mathcal{A})$, then there exist unique $q_1, \ldots, q_n \in \mathbb{R}$ such that $B = \psi(q_1, \ldots, q_n) = q_1 A_1 + \ldots + q_n A_n$ whenever $\mathcal{A} \subseteq \mathbb{R}_{\mathcal{F}}$ is SLI. In addition, we can write levelwise

$$[B]_\alpha = q_1 [A_1]_\alpha + \ldots + q_n [A_n]_\alpha, \ \forall \alpha \in [0, 1],$$

and we call B an *$\mathcal{S}$–linearly correlated fuzzy number*. The next theorem provides two practical methods to generate specific SLI sets of fuzzy numbers, which will be the object of study of this paper.

Theorem 2 *(Adapted from [5]). Let $A \in \mathbb{R}_{\mathcal{F}}$ be a non-symmetric triangular fuzzy number. The sets*

$$\mathcal{A} = \{A^i\}_{i=0,1,\ldots,n} \quad \text{and} \quad \mathcal{A} = \left\{\hat{f}_i(A)\right\}_{i=0,1,\ldots,n}$$

are SLI for all $n \geq 1$.

Theorem 2 reveals that the SLI sets generated via fuzzy modifiers and the Zadeh's extension of polynomial functions are based on a unique non-symmetric fuzzy number $A \in \mathbb{R}_{\mathcal{F}}$.

Let $\mathcal{A} = \{A_1, \ldots, A_n\} \subset \mathbb{R}_{\mathcal{F}}$ be an SLI set satisfying $\mathbb{R} \subset \mathcal{S}(\mathcal{A}) \subset \mathbb{R}_{\mathcal{F}}^{\wedge}$. Note that the SLI sets given by Theorem 2 satisfy this condition. It is a well-known fact that $A_i \in \mathbb{R}$ for some $i = 1, \ldots, n$, in this case [12]. Suppose that $A_1 = \chi_{\{x\}}$ for some $x \in \mathbb{R}$ and let $B, C \in \mathcal{S}(\mathcal{A})$ be given by $B = q_1 A_1 + \ldots + q_n A_n$ and $C = p_1 A_1 + \ldots + p_n A_n$. The ψ–arithmetic operations in $\mathcal{S}(\mathcal{A})$ are defined as:

- $\lambda \cdot_\psi B = (\lambda q_1)A_1 + \ldots + (\lambda q_n)A_n$, $\forall \lambda \in \mathbb{R}$;
- $B +_\psi C = (q_1 + p_1)A_1 + \ldots + (q_n + p_n)A_n$;
- $B -_\psi C = (q_1 - p_1)A_1 + \ldots + (q_n - p_n)A_n$;
- $B \odot_\psi C = cB +_\psi bC -_\psi bc$;
- $B \div_\psi C = B \odot_\psi C_\psi^{-1}$,

where $[B]_1 = \{b\}$, $[C]_1 = \{c\}$ and $C_\psi^{-1} = \left(\frac{2}{c} - \frac{p_1 x}{c^2}\right) - \frac{p_2}{c^2}A_2 - \ldots - \frac{p_n}{c^2}A_n \in \mathcal{S}(\mathcal{A})$.

It is noteworthy that the operations of sum and scalar multiplication in $\mathcal{S}(\mathcal{A})$ (and, consequently, the difference) are operations induced from the isomorphism $\psi : \mathbb{R}^n \to \mathcal{S}(\mathcal{A})$, so that $(\mathcal{S}(\mathcal{A}), +_\psi, \cdot_\psi)$ turns to be a vector space of dimension n [5]. Moreover, the space $\mathcal{S}(\mathcal{A})$ is closed under all the ψ–operations, that is, $B \otimes_\psi C \in \mathcal{S}(\mathcal{A})$ for all $\otimes_\psi \in \{+_\psi, -_\psi, \odot_\psi, \div_\psi\}$ and $B, C \in \mathcal{S}(\mathcal{A})$. Additionally, $\left[B \otimes_\psi C\right]_1 = b \otimes c$, that is, the ψ-arithmetic operations are direct extensions of the arithmetic operations on real numbers.

The operator $\|\cdot\|_\psi : \mathcal{S}(\mathcal{A}) \to \mathbb{R}^+$ given by $\|B\|_\psi = \max_{i=1,\ldots,n} |q_i|$, for all $B \in \mathcal{S}(\mathcal{A})$, defines a norm in $\mathcal{S}(\mathcal{A})$, so that $\left(\mathcal{S}(\mathcal{A}), +_\psi, \cdot_\psi, \|\cdot\|_\psi\right)$ is a Banach space isometric to the Euclidean space $(\mathbb{R}, \|\cdot\|)$, where $\|\cdot\| : \mathbb{R}^n \to \mathbb{R}$ stands for the max norm. Moreover, the operator $\mathcal{D}_\psi : \mathcal{S}(\mathcal{A}) \times \mathcal{S}(\mathcal{A}) \to \mathbb{R}^+$ given by $\mathcal{D}_\psi(B, C) = \left\|B -_\psi C\right\|_\psi$ for all $B, C \in \mathcal{S}(\mathcal{A})$ defines a metric in $\mathcal{S}(\mathcal{A})$. The next theorem ensures that there is a close relation between the operation $\odot_\psi$ and the operator $\|\cdot\|_\psi$ (resp. $\mathcal{D}_\psi$):

Theorem 3 [10]. *Let $\mathcal{A} = \{A_1, \ldots, A_n\}$ be an SLI set satisfying $\mathbb{R} \subseteq \mathcal{S}(\mathcal{A}) \subseteq \mathbb{R}_\mathcal{F}^\wedge$. For all $A, B \in \mathcal{S}(\mathcal{A})$, the following inequality holds:*

$$\left\|A \odot_\psi B\right\|_\psi \leq K \|A\|_\psi \|B\|_\psi ,$$

where $K = K(\mathcal{A})$ is a constant. In addition, for all $C \in \mathcal{S}(\mathcal{A})$, the inequality

$$\mathcal{D}_\psi (A \odot_\psi C, B \odot_\psi C) \leq K \|C\|_\psi \mathcal{D}_\psi (A, B)$$

holds.

For a given SLI set $\mathcal{A} \subset \mathbb{R}_\mathcal{F}$, a fuzzy-number-valued function of the form $f : [a, b] \to \mathcal{S}(\mathcal{A})$ is called an *$\mathcal{S}$–linearly correlated fuzzy process*. In this case, there exists $q : [a, b] \to \mathbb{R}^n$ given by $q(t) = (q_1(t), \ldots, q_n(t))$ such that

$$f(t) = q_1(t)A_1 + \ldots + q_n(t)A_n, \ \forall t \in [a, b].$$

Thus, in this case, f is completely determined by the real-valued functions $q_i(\cdot)$, coordinates of f in the space $\mathcal{S}(\mathcal{A})$.

Since an $\mathcal{S}$–linearly correlated fuzzy process is defined in a Banach space, the Fréchet derivative of f is well-defined [18]. In fact, according to [6], f is Fréchet differentiable at $t \in [a, b]$ if, and only if, q is differentiable at $t \in [a, b]$. Moreover, the equality

$$f'[t](h) = \left(q_1'(t)h\right) A_1 + \ldots + \left(q_n'(t)h\right) A_n \tag{3}$$

holds for all $h \in \mathbb{R}$. The induced operations of sum, subtraction and scalar multiplication in $\mathcal{S}(\mathcal{A})$ whenever $\mathcal{A} \subset \mathbb{R}_\mathcal{F}$ is SLI allow the following definition:

Definition 2 [6]. Let $\mathcal{A} \subset \mathbb{R}_{\mathcal{F}}$ be SLI. The fuzzy function $f : [a, b] \to \mathcal{S}(\mathcal{A})$ is said to be ψ–differentiable at $t \in [a, b]$ if there exists a fuzzy number $f'(t) \in \mathcal{S}(\mathcal{A})$ such that

$$\lim_{h\to 0} \frac{1}{h} \cdot_{\psi} \left(f(t+h) -_{\psi} f(t)\right) = f'(t) \tag{4}$$

where the limit is given w.r.t. $\mathcal{D}_{\psi}$. In this case, we say that $f'(t)$ is the ψ–derivative of f at t. Moreover, f is said to be ψ–differentiable in $[a, b]$ if $f'(t)$ exists for all $t \in [a, b]$.

Definition 2 provides a notion of derivative defined in terms of the induced metric $\mathcal{D}_{\psi}$ in $\mathcal{S}(\mathcal{A})$, instead of the Hausdorff metric $\mathcal{D}_{\infty}$. In addition, the ψ–derivative is associated with the Fréchet derivative of f when $h = 1$, so that the equality $f'(t) = f'[t](1)$ holds (see [6] for details). Thus, for a given SLI set $\mathcal{A} \subset \mathbb{R}_{\mathcal{F}}$, a function $f : [a, b] \to \mathcal{S}(\mathcal{A})$ is ψ–differentiable at $t \in [a, b]$ if, and only if, it is Fréchet differentiable at t. Additionally,

$$f'(t) = q_1'(t)A_1 + \ldots + q_n'(t)A_n. \tag{5}$$

holds. Thus, the ψ–derivative of f is completely determined by the derivative of its coordinates $q_i(\cdot)$, for $i = 1, \ldots, n$. As an immediate consequence, FIVPs and systems of FIVPs for $\mathcal{S}$–linearly correlated fuzzy processes boil down to systems of IVPs, whose initial conditions are associated to the fuzzy initial conditions given by $\mathcal{S}$–linearly correlated fuzzy numbers [6,11].

In this paper we focus on SLI sets of the form $\{A_i\}_{i=0,1,\ldots,n} = \{A^i\}_{i=0,1,\ldots,n}$, as described by Theorem 2. Thus, the populations involved in this study are described as fuzzy-number-valued functions of the form

$$f(t) = q_1(t) + q_2(t)A + \ldots + q_n(t)A^n, \tag{6}$$

where $A \in \mathbb{R}_{\mathcal{F}}$ is a non-symmetric fuzzy number. Therefore, we consider that in specific population dynamics, we can describe all populations in terms of a unique fuzzy number.

3 Fuzzy Population Models in $\mathcal{S}(\mathcal{A})$

If $\mathcal{A} \subset \mathbb{R}_{\mathcal{F}}$ is an SLI set generated as in Theorem 2, then it is completely determined by a given $A \in \mathbb{R}_{\mathcal{F}}$. Then, the vector space $\left(\mathcal{S}(\mathcal{A}), +_{\psi}, \cdot_{\psi}\right)$ can model population dynamics whose uncertainty is directly related to the fuzzy number A. Moreover, if $A \in \mathbb{R}_{\mathcal{F}}^{\wedge}$, then the ψ–cross operations are well defined in $\mathcal{S}(\mathcal{A})$.

Such considerations motivate the following:

Definition 3. Let $\mathcal{A} \subset \mathbb{R}_{\mathcal{F}}$ be an SLI set given as in Theorem 2, for some $A \in \mathbb{R}_{\mathcal{F}}$, and consider the first-order system of FIVPs

$$\begin{cases} X_i'(t) = f_i\left(t, X_1, \ldots, X_p\right) \\ X_i(t_0) = X_0 \in \mathcal{S}(\mathcal{A}) \end{cases}, \tag{7}$$

where $X_i : [a, b] \to \mathcal{S}\left(\mathcal{A}\right)$ for $i = 1, \ldots, p$. Then the fuzzy number $A \in \mathbb{R}_{\mathcal{F}}$ is called a *fuzzy basal number* of (7).

Note that for a FIVP whose underlying SLI set is given by $\mathcal{A} = \{A^i\}_{i=0,\ldots,n}$ for some non-symmetric $A \in \mathbb{R}_{\mathcal{F}}$, the fuzziness of the solutions is directly determined by the fuzzy basal number, since

$$[X_i(t)]_0 = (q_1(t) + q_2(t) + \ldots + q_n(t))\, [A]_0 \tag{8}$$

holds for all $t \geq 0$.

3.1 The ψ–Runge-Kutta Method

In the classical case, the fourth order R-K method has its convergence ensured under some appropriate conditions. In the case of FIVP under the ψ–derivative, the iterative steps for the adapted methods are given in terms of the ψ–arithmetic operations. Moreover, the initial conditions are given by $\mathcal{S}$–linearly correlated fuzzy numbers. We will call this numerical method a *ψ–type numerical method.*

Consider the FIVP (7), where $f_i : [a, b] \times \mathcal{S}(\mathcal{A})^p \to \mathcal{S}(\mathcal{A})$ is continuous w.r.t. $\mathcal{D}_\psi$, $i = 1, \ldots, p$. Here, $\mathcal{S}(\mathcal{A})^p$ denotes the cartesian product $\mathcal{S}(\mathcal{A}) \times \ldots \times \mathcal{S}(\mathcal{A})$, which is a fuzzy subset of the p-dimensional Euclidean space $\mathbb{R}^p$.

Consider that the interval $[a, b] \subseteq \mathbb{R}$ is divided into the aforementioned subintervals. The *ψ–Runge Kutta* method proposes that for each iteration k,

$$X_i^{k+1} = X_i^k +_\psi \frac{h}{6} \cdot \left(K_1^k +_\psi 2 \cdot K_2^k +_\psi 2 \cdot K_3^k +_\psi K_4^k\right),$$

where

$$\begin{aligned}
K_1^k &= f_i\left(t^k, X_1^k, \ldots, X_n^k\right) \\
K_2^k &= f_i\left(t_k +_\psi \frac{h}{2}, X_1^k +_\psi \frac{h}{2} \cdot K_1^k, \ldots, X_n^k +_\psi \frac{h}{2} \cdot K_1^k\right) \\
K_3^k &= f_i\left(t_k +_\psi \frac{h}{2}, X_1^k +_\psi \frac{h}{2} \cdot K_2^k, \ldots, X_n^k +_\psi \frac{h}{2} \cdot K_2^k\right) \\
K_4^k &= f_i\left(t_k +_\psi h, X_1^k +_\psi h \cdot K_3^k, \ldots, X_n^k +_\psi h \cdot K_3^k\right)
\end{aligned} \tag{9}$$

It is noteworthy that since the ψ–arithmetic operations define the ψ–numerical methods, the ψ–Runge-Kutta method produces iterative steps which boil down to the classical corresponding methods into the coordinates of the populations involved.

The next section presents an application of the fourth order ψ–Runge Kutta method to a within-host Chagas' disease model.

3.2 Chagas' Disease Dynamics

T. cruzi (T) penetrates the body of the human host in the trypomastigote form, in which a process of protein release occurs. The host cell invasion process (C) is a result of alterations in the cell membrane caused by the substances release by the protozoan [17]. Already inside the target cell, *T. cruzi* is now in the amastigote form, which takes advantage of the cellular structure, rich in nutrients, to replicate itself. With the rupture of the cell and consequent release of its contents into the extracellular environment, this entire population of new parasites is relaunched in the blood flow and goes in search of infecting new target cells.

Throughout this process, the host's organism begins to show the first activities of an immune response (both innate and adaptive), with regard specifically to humoral immunity (represented in the model). This acts mainly through B lymphocytes (B), which after being activated (B_p), begin to secrete antibodies that act by attacking the circulating forms of *T. cruzi*. Thus, the humoral immunity mechanism is activated when free *T. Cruzi* antigens in the bloodstream activate B lymphocytes, which are responsible for producing antibodies after a differentiation process.

The following model considers the *T. Cruzi*, target cells, B lymphocytes, and activated B lymphocytes, represented by T, C, B and B_p, respectively [17].

$$\begin{cases} \dfrac{dT}{dt}(t) = \tau\alpha T(t)C(t) - \mu_t T(t) - \varepsilon T(t)B_p(t) \\ \dfrac{dC}{dt}(t) = k_c - \mu_c C(t) - \alpha T(t)C(t) \\ \dfrac{dB}{dt}(t) = k_b - \mu_t B(t) - \beta T(t)B(t) \\ \dfrac{dB_p}{dt}(t) = \beta T(t)B(t) - \mu_p B_p(t) + \gamma B_p(t)T(t) \end{cases}, \tag{10}$$

whose initial conditions are given by $T(0) = T_0$, $C(0) = C_0$, $B(0) = B_0$, $B_p(0) = B_{p_0} \in \mathbb{R}$. The within-host model (10) described as a FIVP under the ψ–derivative is

$$\begin{cases} T'(t) = \tau\alpha \cdot_\psi T(t) \odot_\psi C(t) -_\psi \mu_t \cdot_\psi T(t) -_\psi \varepsilon \cdot_\psi T(t) \odot_\psi B_p(t) \\ C'(t) = k_c -_\psi \mu_c \cdot_\psi C(t) -_\psi \alpha \cdot_\psi T(t) \odot_\psi C(t) \\ B'(t) = k_b -_\psi \mu_t \cdot_\psi B(t) -_\psi \beta \cdot_\psi T(t) \odot_\psi B(t) \\ B'_p(t) = \beta \cdot_\psi T(t) \odot_\psi B(t) -_\psi \mu_p \cdot_\psi B_p(t) +_\psi \gamma \cdot_\psi B_p(t) \odot_\psi T(t) \end{cases}, \tag{11}$$

whose fuzzy initial conditions are given by $T(0) = T_0$, $C(0) = C_0$, $B(0) = B_0$, $B_p(0) = B_{p_0} \in \mathcal{S}(\mathcal{A})$. We are assuming that the populations involved are described by $\mathcal{S}$-linearly correlated fuzzy processes of the form $P(t) = p_1(t) + p_2(t)A + p_3(t)A^2$, where $A \in \mathbb{R}_\mathcal{F}$ is a non-symmetric fuzzy number given, that is, we are considering that the underlying SLI set is given by $\mathcal{A} = \{1, A, A^2\}$. Hence, we are assuming that there exist real-vector functions $q, c, b, b_p : [0, T] \to \mathbb{R}^3$ such that

$$\begin{cases} T(t) &= q_1(t) + q_2(t)A + q_3(t)A^2 \\ C(t) &= c_1(t) + c_2(t)A + c_3(t)A^2 \\ B(t) &= b_1(t) + b_2(t)A + b_3(t)A^2 \\ B_p(t) &= b_{p_1}(t) + b_{p_2}(t)A + b_{p_3}(t)A^2 \end{cases} \tag{12}$$

for all $t \geq 0$. Since the FIVP (11) is a 4–dimensional system of fuzzy differential equations, it can be rewritten as

$$\begin{cases} T'(t) = F_1(T, C, B, B_p) \\ C'(t) = F_2(T, C, B, B_p) \\ B'(t) = F_3(T, C, B, B_p) \\ B'_p(t) = F_4(T, C, B, B_p) \end{cases}, \tag{13}$$

where $T_0, C_0, B_0, B_{p_0} \in \mathcal{S}(\mathcal{A})$. Let us denote

$$\mathcal{D}_{\psi_p} : \mathcal{S}(\mathcal{A})^p \times \mathcal{S}(\mathcal{A})^p \to \mathbb{R}^+$$

given by

$$\mathcal{D}_{\psi_p}(X, Y) = \max_{i=1,\dots,p} \mathcal{D}_{\psi}(X_i, Y_i), \quad \forall X, Y \in \mathcal{S}(\mathcal{A})^p.$$

The operator $\mathcal{D}_{\psi_p}$ defines a metric in $\mathcal{S}(\mathcal{A})^p$ for every $p = 1, 2, \dots$. Moreover, $\left(\mathcal{S}(\mathcal{A})^p, \mathcal{D}_{\psi_p}\right)$ is a complete metric space [8]. In this case, we consider the vector-valued fuzzy function $F : [a, b] \to \mathcal{S}(\mathcal{A})^4$ given by $F = (F_1, F_2, F_3, F_4)$.

Proposition 1. *The function $F : \mathcal{S}(\mathcal{A})^4 \to \mathcal{S}(\mathcal{A})^4$ given by the FIVP (7) is locally Lipschitz continuous.*

Proof. It is worth noting that for each $i = 1, 2, 3, 4$, the function $F_i : \mathcal{S}(\mathcal{A})^4 \to \mathcal{S}(\mathcal{A})$ is continuous w.r.t. $\mathcal{D}_\psi$ [10]. In addition, from Theorem 3, it follows that

$$\begin{aligned} \mathcal{D}_\psi(A \odot_\psi C, B \odot_\psi D) &\leq \mathcal{D}_\psi(A \odot_\psi C, B \odot_\psi C) + \mathcal{D}_\psi(B \odot_\psi C, B \odot_\psi D) \\ &\leq K\left(\|C\|_\psi \mathcal{D}_\psi(A, B) + \|B\|_\psi \mathcal{D}_\psi(C, D)\right). \end{aligned} \tag{14}$$

for any $A, B, C, D \in \mathcal{S}(\mathcal{A})$.

Note that $F : [a, b] \to \mathcal{S}(\mathcal{A})^4$, given by $F = (F_1, F_2, F_3, F_4)$, is locally Lipschitz continuous in $\left(\mathcal{S}(\mathcal{A})^4, \mathcal{D}_{\psi_4}\right)$ if, and only if, F_1, F_2, F_3 and F_4 are locally Lipschitz continuous in $\left(\mathcal{S}(\mathcal{A}), \mathcal{D}_\psi\right)$. Below, we prove that F_1 is Lipschitz (the proof for F_2, F_3 and F_4 is analogous). In fact, since $\mathcal{D}_{\psi_4} : \mathcal{S}(\mathcal{A})^p \times \mathcal{S}(\mathcal{A})^p \to \mathbb{R}^+$ is a metric, and so, by Eq. 14, we have that

$$\begin{aligned} &\mathcal{D}_{\psi_4}(F(X), F(Y)) \leq \mathcal{D}_\psi(F_1(T, C, B, B_p), F_1(X, Y, Z, W)) \\ &\leq |\tau\alpha| \mathcal{D}_\psi(T \otimes_\psi C, X \otimes_\psi Y) + |\mu_T| \mathcal{D}_\psi(T, X) + |\varepsilon| \mathcal{D}_\psi(T \otimes_\psi B_p, X \otimes_\psi W) \\ &\leq |\tau\alpha| K\Big(\|C\|_\psi \mathcal{D}_\psi(T, X) + \|X\|_\psi \mathcal{D}_\psi(C, Y)\Big) + |\mu_T| \mathcal{D}_\psi(T, X) \\ &+ |\varepsilon| K\Big(\|B_p\|_\psi \mathcal{D}_\psi(T, X) + \|X\|_\psi \mathcal{D}_\psi(B_p, W)\Big) \\ &\leq |\tau\alpha| K \max\{\|T\|, \|C\|, \|B\|, \|B_p\|\}\, \mathcal{D}_\psi(T, X) \\ &+ |\tau\alpha| K \max\{\|X\|, \|Y\|, \|Z\|, \|W\|\}\, \mathcal{D}_\psi(C, Y) \\ &+ |\mu_T| \mathcal{D}_\psi(T, X) + |\varepsilon| K \max\{\|T\|, \|C\|, \|B\|, \|B_p\|\}\, \mathcal{D}_\psi(T, X) \\ &+ |\varepsilon| K \max\{\|X\|, \|Y\|, \|Z\|, \|W\|\}\, \mathcal{D}_\psi(B_p, W) \\ &\leq \Big((|\tau\alpha| + |\varepsilon|) K (\max\{\|T\|, \|C\|, \|B\|, \|B_p\|\} + \max\{\|X\|, \|Y\|, \|Z\|, \|W\|\}) + |\mu_T|\Big) \\ &\mathcal{D}_{\psi_4}((T, C, B, B_p), (X, Y, Z, W)) \end{aligned}$$

Then, for some $R > 0$, it follows that $F : [a,b] \times \mathcal{B}(0;R) \subset (\mathcal{S}(\mathcal{A}))^p \to (\mathcal{S}(\mathcal{A}))^p$ satisfies

$$\mathcal{D}^{\infty}_{\psi_4}(F_1(t,T,C,B,B_p), F_1(t,X,Y,Z)) \leq M\mathcal{D}_{\psi_4}\left((T,C,B,B_p),(X,Y,Z,W)\right),$$

where $M = (|\tau\alpha| + |\varepsilon|)2KR + |\mu_T|$, from which we conclude that F_1 is locally Lipschitz continuous. □

Theorem 4. *The FIVP* (13) *has a solution* $X(\cdot) = (T(\cdot), C(\cdot), B(\cdot), B_p(\cdot))$ *defined in an interval* $[t_0, t_0 + k] \subseteq [a,b]$, *where* $k > 0$ *is sufficiently small. Moreover,* X *is unique.*

Proof. The proof follows straightforward from Banach's fixed point theorem and from Proposition 1. □

We shall consider the FIVP (11) with fuzzy initial conditions given by $T_0 = 8{-}A{+}A^2$, $C_0 = 0.3{+}0.2A{-}0.2A^2$, $B_0 = 9.4{+}0.1A{-}0.1A^2$, $B_p = 0.8{+}0.75A{-}0.75A^2 \in \mathcal{S}(\mathcal{A})$, where $A = (0.5, 1, 2) \in \mathbb{R}_{\mathcal{F}}$ is the fuzzy basal number of the model. The injectiveness of $\psi : \mathbb{R}^3 \to \mathcal{S}(\mathcal{A})$ assures that solving (11) is equivalent to solve a family of differential equations, given in terms of q, c, b, b_p, whose initial conditions are given completely determined by the fuzzy initial conditions of (11).

Figure 1 depicts the numerical solution of the IVPs when $q_0 = (0.8, -1, 1)$, $c_0 = (0.3, 0.2, -0.2)$, $b_0 = (9.4, 0.1, -0.1)$, $b_{p0} = (0.8, 0.75, -0.75)$, for which the ψ–RK method was applied. The corresponding fuzzy solutions to (11) are represented by Fig. 2.

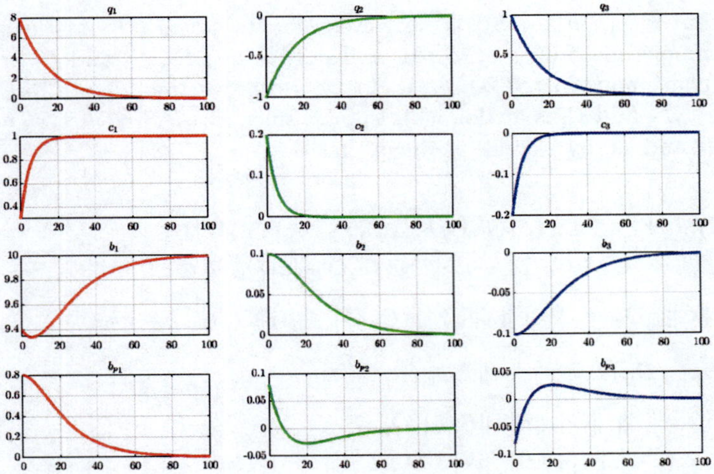

Fig. 1. From top to bottom and from left to right: numerical solutions to the classical vector-real-valued functions $q, c, b, b_p : [0, 100] \to \mathbb{R}^3$. The red, green and blue lines correspond to the i-th coordinates of each vector-valued function, where $i = 1, 2, 3$, respectively. Note the *T. Cruzi* population coordinates decay at the top of graphical depiction.

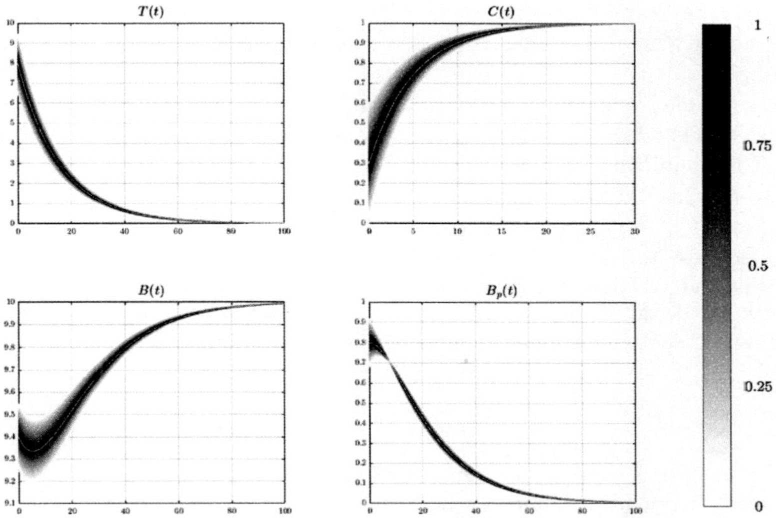

Fig. 2. From top to bottom and from left to right: top view of the numerical solutions to the FIVP (11) at the interval $[0, 100]$, given by the populations T, C, B and B_p, respectively. The gray-scale lines represent the α–levels of the solutions, where α varies from 0 to 1. Note that the classical solution of (10) coincides with the core of the fuzzy solutions.

4 Final Remarks

This paper presented a case study on a within-host Chagas' disease model considering humoral immunity response in a fuzzy environment. To this end, we considered a four-population model in which each population is described as a special type of fuzzy-number-valued function given by Minkowski combinations of SLI sets of fuzzy numbers. Since each population is given by a fuzzy process embedded in a Banach space of $\mathbb{R}_{\mathcal{F}}$, we described the Chagas' model as a FIVP under the ψ–derivative, which is based on the notion of Fréchet derivative of those functions.

When considering the populations as $\mathcal{S}$–linearly correlated fuzzy processes, we can apply the notion of *fuzzy basal number*, introduced in Sect. 3, which correspond to the main source of the uncertainty of the model. The bigger the diameter of the fuzzy basal number, the bigger the diameter of each population involved.

We introduced the notion of ψ–type numerical methods, which consist on extensions of the classical numerical methods embracing the ψ-arithmetic operations. The fuzzy numerical solutions to the FIVP were provided using the fourth order ψ–Runge Kutta method, which consist on the classical RK method applied to the corresponding IVPs of the coefficients of each population. As it was expected, there has been observed a T. Cruzi decay, as well as an activated B lymphocyte count decay with time. On the other hand, the B lymphocytes and target cells count have increased with time. All populations had its fuzziness vanishing with time, since the associated coordinates into the space $\mathcal{S}(\mathcal{A})$ approximates 0.

Lastly, we can observe the classical solution of (10) coincides with the fuzzy solutions of (11) when the initial conditions of (10) coincide with the core of the fuzzy initial conditions of (11). In future works, we aim to study other immune responses to the *T. Cruzi* infection.

Acknowledgments. The authors would like to thank the Brazilian National Council for Scientific and Technological Development under grant n. 140692/2020-7, the Federal University of Vale do São Francisco, Bahia, Brazil, and the Institute of Mathematics, Statistics and Scientific Computing of the University of Campinas, Brazil.

References

1. Ayres, P., Lee, J.Y., Paas, F., van Merriënboer, J.J.: The validity of physiological measures to identify differences in intrinsic cognitive load. Front. Psychol. **12**, 702538 (2021)
2. Barros, L.C., Bassanezi, R.C., Tonelli, P.A.: Fuzzy modelling in population dynamics. Ecol. Model. **128**(1), 27–33 (2000)
3. Cardoso, M.S., Reis-Cunha, J.L., Bartholomeu, D.C.: Evasion of the immune response by *Trypanosoma cruzi* during acute infection. Front. Immunol. **6**, 659 (2016)
4. de Freitas, L.M., Maioli, T.U., de Ribeiro, H.A.L., Tieri, P., Castiglione, F.: A mathematical model of Chagas disease infection predicts inhibition of the immune system. In: 2018 IEEE International Conference on Bioinformatics and Biomedicine (BIBM), pp. 1374–1379. IEEE (2018)
5. Esmi, E., de Barros, L.C., Santo Pedro, F., Laiate, B.: Banach spaces generated by strongly linearly independent fuzzy numbers. Fuzzy Sets Syst. **417**, 110–129 (2021)
6. Esmi, E., Laiate, B., Santo Pedro, F., Barros, L.C.: Calculus for fuzzy functions with strongly linearly independent fuzzy coefficients. Fuzzy Sets Syst. **436**, 1–31 (2021)
7. Esmi, E., Santo Pedro, F., de Barros, L.C., Lodwick, W.: Fréchet derivative for linearly correlated fuzzy function. Inf. Sci. **435**, 150–160 (2018)
8. Kreyszig, E.: Introductory Functional Analysis with Applications, vol. 17. Wiley, Hoboken (1991)
9. Kumar, S., Tarleton, R.L.: The relative contribution of antibody production and $CD8^+$ T cell function to immune control of *Trypanosoma cruzi*. Parasite Immunol. **20**(5), 207–216 (1998)
10. Laiate, B.: On the properties of fuzzy differential equations under cross operations. Submitted for publication
11. Laiate, B., Esmi, E., Pedro, F.S., Barros, L.C.: Solutions of systems of linear fuzzy differential equations for a special class of fuzzy processes. In: Rayz, J., Raskin, V., Dick, S., Kreinovich, V. (eds.) NAFIPS 2021. LNNS, vol. 258, pp. 217–228. Springer, Cham (2022). https://doi.org/10.1007/978-3-030-82099-2_20
12. Laiate, B., Watanabe, R.A., Esmi, E., Santo Pedro, F., Barros, L.C.: A cross product of $\mathcal{J}$-linearly correlated fuzzy numbers. In: 2021 IEEE International Conference on Fuzzy Systems (FUZZ-IEEE), pp. 1–6. IEEE (2021)

13. Liu, H., Zhang, F., Mishra, S.K., Zhou, S., Zheng, J.: Knowledge-guided fuzzy logic modeling to infer cellular signaling networks from proteomic data. Sci. Rep. **6**(1), 1–12 (2016)
14. Longo, F., Laiate, B., Pedro, F.S., Esmi, E., Barros, L.C., Meyer, J.F.C.A.: A-cross product for autocorrelated fuzzy processes: the hutchinson equation. In: Rayz, J., Raskin, V., Dick, S., Kreinovich, V. (eds.) NAFIPS 2021. LNNS, vol. 258, pp. 241–252. Springer, Cham (2022). https://doi.org/10.1007/978-3-030-82099-2_22
15. Massad, E., Ortega, N.R.S., de Barros, L.C., Struchiner, C.J.: Fuzzy Logic in Action: Applications in Epidemiology and Beyond, vol. 232. Springer, Heidelberg (2009). https://doi.org/10.1007/978-3-540-69094-8
16. Murray, J.D.: Mathematical Biology: I. An Introduction. Springer, New York (2002). https://doi.org/10.1007/b98868
17. Oliveira, L.S.: Modelando a interaçao entre o Sistema Imunológico Humano e Trypanosoma cruzi. Ph.D. thesis, Dissertaçao de Mestrado em Matemática Aplicada, IMECC-Unicamp (2010)
18. Penot, J.-P.: Calculus Without Derivatives, vol. 266. Springer, New York (2012). https://doi.org/10.1007/978-1-4614-4538-8
19. Vega-Royero, S., Sibona, G.: Mathematical modeling of the inmune response to the chagasic trypanosomiasis. Ciencia en Desarrollo **10**(2), 177–184 (2019)
20. Yang, H.M.: A mathematical model to assess the immune response against *trypanosoma cruzi* infection. J. Biol. Syst. **23**(01), 131–163 (2015)

Fuzzy Logic++: Towards Developing Fuzzy Education Curricula Using ACM/IEEE/AAAI CS2023

Christian Servin[1(✉)], Brett A. Becker[2], Eric Eaton[3], and Amruth Kumar[4]

[1] Information Technology Systems Department, El Paso Community College, 919 Hunter Dr., El Paso, TX 79915-1908, USA
cservin1@epcc.edu

[2] School of Computer Science and Informatics, University College Dublin, Belfield, Dublin 4, Ireland
brett.becker@ucd.ie

[3] Computer and Information Science Department, University of Pennsylvania, 3330 Walnut Street, Philadelphia, PA 19104-6309, USA
eeaton@seas.upenn.edu

[4] Computer Science, Ramapo College of New Jersey, 505 Ramapo Valley Road, Mahwah, NJ 07430-1680, USA
amruth@ramapo.edu

Abstract. Fuzzy Logic education has been considered part of artificial intelligence, which in many cases, can be found in the *electives* curricula or, inclusively, at the graduate studies for the past decades. Because of the persistent emergence of computing areas, such as Artificial Intelligence and Machine Learning, Robotics, and others, there is an appetite for recognizing concepts on intelligent systems in undergraduate education. The ACM/IEEE/AAAI Computer Science Curricula, also known as CS2023, has proposed significant changes from the last version CS2013, particularly in artificial intelligence. These proposed recommendations will have an impact in the next ten years in computer science undergraduate education, including fundamentals of computer programming. This work presents the changes that may impact computer science education curricula, particularly the knowledge areas and competencies model that will influence fuzzy logic education and computing. We also present the intersection of knowledge areas in fuzzy logic, especially for explainable AI.

1 Introduction

Fuzzy logic is a valuable tool for addressing the challenges of uncertainty and imprecision in AI applications. Its ability to model complex systems and make decisions based on incomplete or uncertain information makes it a valuable component of many AI systems, e.g., [2, 16, 17]. Therefore, there is a need to recognize concepts from fuzzy logic that can be intersected by curricula recommendations and then offer them in computer science programs.

Another important direction of artificial intelligence that has high-impact in society nowadays, specially with ideas on ethics, diversity, equity, inclusion, and justice is

K. Cohen et al. (Eds.): NAFIPS 2023, LNNS 751, pp. 184–193, 2023.
https://doi.org/10.1007/978-3-031-46778-3_17

Explainable AI (xAI) [6,19]. AI techniques such as deep learning, face challenges in proving explanations on their final results, specially when these generated conclusions and recommendations are not necessary the optimal ones or they rely in a set of rules from a "black box" algorithm. Therefore, it is desirable to provide explanation to these solutions through xAI.

Fuzzy logic concepts and topics have been offered at higher-level computer science courses, such as special topics, electives, or even graduate studies. Recognizing these topics in other courses has been challenging unless it is a specialized study area, such as Artificial Intelligence (AI) or intelligent system study track. Due to the significant changes in technology and computing, such as the introduction of Large Language Models (LLM), intelligent systems, recognition, society, ethics, professionalism processing information, uncertainty, and ambiguity, these concepts on AI are now part of the CS curricula.

In this work, authors recognize several *knowledge areas* that are part of the CS2023 curricular guideline that highlights areas of interest in the fuzzy logic community. These areas promise topics, learning outcomes and professional competencies that potentially can increase the concepts in fuzzy logic through knowledge areas in artificial intelligence, specialized platforms development, and society, ethics, and professionalism.

2 About ACM/IEEE/AAAI CS2023: Updates and Concentrations

In the Spring of 2021, the joint work of the Association of Computing Machinery (ACM), the Institute of Electrical and Electronics Engineering – Computer Society (IEEE-CS), and the American Association of Artificial Intelligence (AAAI) started the effort to revise undergraduate Computer Science curricular guidelines issued once every ten years, currently referred to as CS2023. The work comprises 18 Knowledge Areas (KA), covering significant concepts in the most influential computer science areas: Algorithms and Complexity (AL), Architecture and Organization (AR), Artificial Intelligence (AI), Data Management (DM), Graphics and Interactive Techniques (GIT), Human-Computer Interaction (HCI), Mathematical and Statistical Foundations (MSF), Modeling (MOD), Networking and Communication (NC), Operating Systems (OS), Parallel and Distributed Computing (PDC), Programming Languages (PL), Security (SEC), Society, Ethics, and Professionalism (SEP), Software Development Fundamentals (SDF), Software Engineering (SE), Specialized Platform Development (SPD), and Systems Fundamentals (SF).

CS2023: Selected Updates

Several updates are considered in the CS2023 version; many of these details can be found at [15]. This section discusses selected changes/updates that impact the fuzzy logic education aspects.

Knowledge Area Renamed. A total of eight knowledge areas have been renamed to emphasize their focus that contemporary contextualizes in computer science. Three of these areas are:

- Intelligent Systems (IS), now is called *Artificial Intelligence (AI)*
- Platform Based Development (PBD), now is called *Specialized Platform Development (SPD)*
- Social Issues and Professional Practice (SP), now is called *Society, Ethics, and Professionalism (SEP)*

Core and Knowledge Hours. In the previous CS2013 work, core hours were defined in *Tier I* and *Tier II*, implying that computer science programs were expected to cover 100% of Tier I (165 h covered) core topics and at least 80% of Tier II (143 h) topics. In the CS2023 adopted the definition of the Computer Science (CS) Core and the Knowledge Area (KA) core, where

- CS core - topics that every Computer Science graduate must know
- KA core - topics that any coverage of the Knowledge Area must include.

Several KAs have grown significantly in CS Core, including Artificial Intelligence, Specialized Platform Development, Society, Ethics, and Professionalism.

Incorporating Society, Ethics, and Professionalism as a Knowledge Unit. Due to the emerging changes in society and technology, the computer science society recognizes the need to incorporate the SEP considerations and responsibilities in each knowledge area from this work. These aspects influence each knowledge area depending on the competency area to which they belong, i.e., *software*, *systems*, *applications*, and the *theoretical foundations of computer science*.

Characteristics of Computer Science Graduates. Based on the computer science education community and industry's input, the CS2023 lists several characteristics that CS Graduates should have. Some of the ones include:

- Algorithmic problem-solver – Good solutions to common problems at an appropriate level of abstraction
- Cross-disciplinary – understanding of non-computing disciplines
- Handle ambiguity and uncertainty
- Strong mathematical and logical skills

The fuzzy logic community shares similar characteristics to computer science graduates, especially since: many of the applications in fuzzy require an understanding of cross-disciplinary work; how to incorporate effective solutions to cover different levels of abstraction; the fact that fuzzy logic requires an understanding of mathematical concepts such as fuzzy sets and, perhaps the most significant one that is, handle uncertainty, since fuzzy systems are designed to process ambiguity and uncertainty.

3 Knowledge Areas Intersecting Fuzzy

Although there are 18 KA in CS2023 that have impact in fuzzy logic education, there are three main ones that have a thoroughgoing impact in fuzzy: Artificial Intelligence, Specialized Platform Development, and SEP.

Fuzzy Logic Intersecting Artificial Intelligence Education. Fuzzy logic is related to artificial intelligence (AI) in several ways. Fuzzy logic is a mathematical framework for dealing with uncertainty and imprecision, an important component of many AI applications. Fuzzy logic can be used to:

- model complex systems and make decisions based on incomplete or uncertain information. This is crucial in many AI applications, such as expert systems, decision support systems, and autonomous agents, e.g., see [29];
- basis for machine learning algorithms. For example, fuzzy clustering algorithms can be used to group data points based on their similarity. In contrast, fuzzy decision trees can classify data based on a set of rules, e.g., see [13,22,28];
- model human reasoning and decision-making. This can be useful in developing intelligent systems that are more natural and intuitive for humans to interact with, e.g., see [1,4,32];
- control complex systems like robots or industrial processes. These systems can be made more flexible and adaptive to changing conditions by using fuzzy logic controllers, e.g., see [3,5,9,31];

Specialized Platform Development is Important to Support Intelligent Computing Curricula: Because of the new emerging technology and rapid growth of technology and ubiquitous computing, there is a need for:

- platforms that support mobile and web development to provide accessibility to many applications and software available;
- platforms that support iterative development, such as Large Language Models (LLM) and tools that employ deep generative models (e.g., *ChatGPT, DALL-E, Midjourney*);
- platforms that support Robotics/Drone development focus on hardware, constraints/considerations, and software architectures.

From the education viewpoint, there are several best practices that use specialized platforms to disseminate and demonstrate computing curricula. For example, in [21], authors have successfully demonstrate software development principles using various low-cost and accessible environments, such as *Raspberry PIs* and *Arduinos* for Robotics and Drone programming. In [7], authors demonstrate computational thinking principles using Robotics in a Middle School environment. In [11], authors used a mobile platform called *App Inventor* to teach CS principles.

Concepts in SEP are Important to Explainable Fuzzy (xAI). The KA Society, Ethics, and Professionalism is now part of every KA for CS2023 as a KU. Since AI is ubiquitous, some important ethical considerations and responsibilities are essential for computer science education. The Intelligent computing considerations include *transparency*, *privacy*, *accountability*, *safety*, *human oversight*, and *fairness*. The last two, in particular, partly motivate this section. Humans should be able to intervene and correct errors or biases in AI systems when necessary. This will lead to recognizing aspects on **j**ustice, **e**quity, **d**iversity, and **i**nclusion, often referred to as JEDI. This AI ubiquity has reactivated ethical dilemmas that have been in humanity for decades or centuries, such as the *trolley dilemma* [10,25,27], in addition, to introduce challenges in the engineering, biometric technologies, and algorithmic decision-making systems, in civic life through recognizing human characteristics including facial recognition, skin colors, bringing concerns on diversity, equity, and inclusion, e.g., see: [12,20,30]. The xAI introduces flexibility in determining solutions in a more *humanistic* perspective. The concept of *justice* is quite often hard to define in intelligent systems, especially since the main concept is completely *ambiguous* and *"blind"*, it has a degree of matter and is defined based on expert consideration. In other words, it is *fuzzy*.

Therefore, this work is an opportunity to incorporate educational recommendations into multidisciplinary areas. Mainly, this work focuses on the following knowledge areas of interest: Artificial Intelligence, Specialized Platform Development and Society, Ethics, and Professionalism.

4 Fuzzy Education: Beyond Knowledge Areas

Educators have proposed strategies and techniques for teaching fuzzy concepts/topics in higher education, such as [14,24,26]. Authors have even proposed a "fuzzy" flavor on concepts covered in computer science fundamentals, such as Data Structures and Algorithms courses, e.g., [8] or [18]. These works introduce fuzzy in the first two years of computer science. Additionally, there is existing *mapping* to learning outcomes proposed to the previous CS2013 version [23] that can help educators introduce fuzzy-based concepts in computer science fundamentals.

These are several of the topics proposed in the current version of the CS2023 project that impacts fuzzy. In this section, we concentrate on the following three areas.

Artificial Intelligence (AI)

Artificial intelligence has become commonplace in many areas, such as businesses, news articles, and everyday conversation, for years, primarily driven by high-impact machine learning applications. These advances were made possible by the widespread availability of large datasets, increased computational power, and algorithmic improvements. In particular, a shift from engineered representations to representations learned automatically through optimization over large datasets. The resulting advances have put such terms as "neural networks" and "deep learning" into everyday vernacular. These are some of the knowledge units with their corresponding learning outcomes under Artificial Intelligence KA that are now part of the computer science core and that can be applied to topics of fuzzy logic.

- AI: Fundamental Issues
  - **AI-FUN-01** Additional depth on problem characteristics with examples
  - **AI-FUN-02** Additional depth on nature of agents with examples
  - **AI-FUN-03** Additional depth on AI Applications, growth, and Impact (economic, societal, ethics)
- AI: Algorithms Search/Fundamental Data Structures and Algorithms
  - **AI-SEA-01** Design the state space representation for a puzzle (e.g., N-queens or 3-jug problem)

  - **AI-SEA-02** Select and implement an appropriate uninformed search algorithm for a problem (e.g., tic-tac-toe), and characterize its time and space complexities.
  - **AI-SEA-03** Select and implement an appropriate informed search algorithm for a problem after designing a helpful heuristic function (e.g., a robot navigating a 2D gridworld).
  - **AI-SEA-04** Evaluate whether a heuristic for a given problem is admissible/can guarantee an optimal solution
  - **AI-SEA-05** Design and implement a genetic algorithm solution to a problem.
  - **AI-SEA-06** Design and implement a simulated annealing schedule to avoid local minima in a problem.
  - **AI-SEA-07** Apply minimax search with alpha-beta pruning to prune search space in a two-player adversarial game (e.g., connect four).
- AI: Fundamental Knowledge Representation and Reasoning
  - **AI-REP-01** Given a natural language problem statement, encode it as a symbolic or logical representation.
  - **AI-REP-02** Explain how we can make decisions under uncertainty, using concepts such as Bayes theorem and utility.
  - **AI-REP-03** Make a probabilistic inference in a real-world problem using Bayes' theorem to determine the probability of a hypothesis given evidence.
  - **AI-REP-04** Apply Bayes' rule to determine the probability of a hypothesis given evidence.
  - **AI-REP-05** Compute the probability of outcomes and test whether outcomes are independent.
- AI: Machine Learning
  - **AI-ML-01** Describe the differences among the three main styles of learning: supervised, reinforcement, and unsupervised.
  - **AI-ML-02** Differentiate the terms of AI, machine learning, and deep learning.
  - **AI-ML-03** Frame an application as a classification problem, including the available input features and output to be predicted (e.g., identifying alphabetic characters from pixel grid input).
  - **AI-ML-04** Apply two or more simple statistical learning algorithms (such as k-nearest-neighbors and logistic regression) to a classification task and measure the classifiers' accuracy.
  - **AI-ML-05** Identify overfitting in the context of a problem and learning curves and describe solutions to overfitting.
  - **AI-ML-06** Explain how machine learning works as an optimization/search process.

Specialized Platforms Development (SPD)

Emerging Computing Areas such as data science/analytics - use multi-platforms to retrieve sensing data. Cybersecurity - involves protecting specific data extraction, recognizing protocols to protect network transfer ability, and manipulating it. Artificial intelligence and machine learning - use artifacts that retrieve information for robotics, drones to perform specific tasks, and other platforms that perform data analysis and visualizations. These are some of the learning outcomes from selected knowledge units under SPD KA that are now part of the computer science core and that can be applied to topics of fuzzy logic.

- Describe how the state is maintained in web programming
- Implement a location-aware mobile application that uses data APIs.
- Implement a sensor-driven mobile application that logs data on a server
- Compare robot-specific languages and techniques with those used for general-purpose software development
- Discuss the constraints a given robotic platform imposes on developers
- Interactively analyze large datasets
- Create a program that performs a task using LLM systems
- Contrast a program developed by an AI platform and by a human

Society Ethics and Professionalism (SEP)

The SEP KA is extensible. Each KA in CS2023 possesses a KU that focuses on SEP. These knowledge units under SEP KA are now part of the computer science core and can potentially be topics of fuzzy logic, particularly Explainable Fuzzy AI.

- Social implications of computing in a hyper-networked world where the capabilities of artificial intelligence are rapidly evolving
- Impact of involving computing technologies, particularly artificial intelligence, biometric technologies and algorithmic decision-making systems, in civic life (e.g. facial recognition technology, biometric tags, resource distribution algorithms, policing software)
- Interpret the social context of a given design and its implementation.
- Articulate the implications of social media use for different identities, cultures, and communities.
- Ethical theories and decision-making (philosophical and social frameworks)
- Define and distinguish equity, equality, diversity, and inclusion
- From AI: Applications and Societal Impact
  – **AI-SEP-01** Given a real-world application domain and problem, formulate an AI solution to it, identifying proper data/input, preprocessing, representations, AI techniques, and evaluation metrics/methodology.
  – **AI-SEP-02** Analyze the societal impact of one or more specific real-world AI applications, identifying issues regarding ethics, fairness, bias, trust, and explainability
  – **AI-SEP-03** Describe some of the failure modes of current deep generative models for language or images, and how this could affect their use in an application.

5 Summary and Discussion

This paper describes selected knowledge areas of fuzzy logic from the new ACM/IEEE/AAAI CS2023 curricula project. The CS2023 project proposes significant changes from its previous version CS2013. Significant changes include increasing computer science core hours in knowledge areas related to fuzzy logic, such as artificial intelligence, specialized platform development, society, ethics, and professionalism, impacting xAI areas. Another feature of the project is the recognition of the ability to process uncertainty/ambiguity as one of the characteristics of graduate students. This characteristic governs the philosophy behind fuzzy logic. The Artificial Intelligence KA recognizes areas and topics that impact curricular considerations in computer science. Many of these topics cover fundamentals knowledge representation and reasoning and machine learning – topics that are often used in the fuzzy community. The specialized platform development KA is used to provide vehicles for development. Software and applications are now developed in specialized platforms such as mobile, robotics, and interactive ones. Therefore, this area connects learning outcomes and platforms for the future development of fuzzy systems. Finally, society, ethics, and professionalism cover topics that impact society and humankind. Explainable AI frameworks help systems to understand and interpret predictions made by machine learning models, especially the ones that are a matter of degree, i.e., concerns/inquiries provided under the SEP. Future work includes: packaging a course in Fuzzy Logic or Explainable Fuzzy Logic that can cover several knowledge units from various KAs, emphasizing fuzzy challenges. Fuzzy logic aligned to Artificial Intelligence recognizes real-world applications and challenges for the following years; it is of interest to this curricular practice to capture many aspects that infuse into the computing curriculum. Finally, one of the ultimate objectives of this work is to create best practices for these KAs and discover competencies that can impact areas in fuzzy logic.

Acknowledgments. Partial support for this work was provided by the National Science Foundation under grant DUE-2231333.

References

1. Abbasi, S.H.R., Shabaninia, F.: A research on employing fuzzy composite concepts based on human reasoning through singleton and non-singleton fuzzification. In: 2011 IEEE International Conference on Information Reuse & Integration, pp. 500–501 (2011)
2. Bělohlávek, R., Dauben, J.W., Klir, G.J.: Fuzzy Logic and Mathematics: A Historical Perspective. Oxford University Press, Oxford (2017)
3. Biglarbegian, M., Melek, W.W., Mendel, J.M.: Design of novel interval type-2 fuzzy controllers for modular and reconfigurable robots: theory and experiments. IEEE Trans. Ind. Electron. **58**(4), 1371–1384 (2011)
4. Bouchon-Meunier, B.: Fuzzy models in analogy and case-based reasoning. In: NAFIPS 2009 - 2009 Annual Meeting of the North American Fuzzy Information Processing Society, pp. 1–2 (2009)
5. Chiu, S., Cheng, J.J.: Automatic generation of fuzzy rulebase for robot arm posture selection. In: NAFIPS/IFIS/NASA 1994. Proceedings of the First International Joint Conference of The North American Fuzzy Information Processing Society Biannual Conference. The Industrial Fuzzy Control and Intellige, pp. 436–440 (1994)
6. Cohen, K., Bokati, L., Ceberio, M., Kosheleva, O., Kreinovich, V.: Why fuzzy techniques in explainable AI? Which fuzzy techniques in explainable AI? In: Rayz, J., Raskin, V., Dick, S., Kreinovich, V. (eds.) NAFIPS 2021. LNNS, vol. 258, pp. 74–78. Springer, Cham (2022). https://doi.org/10.1007/978-3-030-82099-2_7

7. Junior, A.O.C., e Silva, J.P.F.L., Rivera, J.A., Guedes, E.B.: ThinkCarpet: potentializing computational thinking with educational robotics in middle school. In: 2022 Latin American Robotics Symposium (LARS), 2022 Brazilian Symposium on Robotics (SBR), and 2022 Workshop on Robotics in Education (WRE), pp. 372–377 (2022)
8. Deng, Y., Chen, Y., Zhang, Y., Mahadevan, S.: Fuzzy Dijkstra algorithm for shortest path problem under uncertain environment. Appl. Soft Comput. **12**, 1231–1237 (2012)
9. Feng, G.: A survey on analysis and design of model-based fuzzy control systems. IEEE Trans. Fuzzy Syst. **14**(5), 676–697 (2006)
10. Gogoll, J., Müller, J.F.: Autonomous cars: in favor of a mandatory ethics setting. Sci. Eng. Ethics **23**, 681–700 (2016). https://doi.org/10.1007/s11948-016-9806-x
11. Gray, J., Abelson, H., Wolber, D., Friend, M.: Teaching CS principles with app inventor. In: Proceedings of the 50th Annual Southeast Regional Conference, ACM-SE 2012, pp. 405–406. Association for Computing Machinery, New York (2012)
12. Jora, R.B., Sodhi, K.K., Mittal, P., Saxena, P.: Role of artificial intelligence (AI) in meeting diversity, equality and inclusion (DEI) goals. In: 2022 8th International Conference on Advanced Computing and Communication Systems (ICACCS), vol. 1, pp. 1687–1690 (2022)
13. Kasabov, N.K., Song, Q.: DENFIS: dynamic evolving neural-fuzzy inference system and its application for time-series prediction. IEEE Trans. Fuzzy Syst. **10**(2), 144–154 (2002)
14. Kosheleva, O., Villaverde, K.: How Interval and Fuzzy Techniques Can Improve Teaching. Springer, Heidelberg (2018). https://doi.org/10.1007/978-3-662-55993-2
15. Kumar, A.N., Raj, R.K.: Computer science curricula 2023 (CS2023): community engagement by the ACM/IEEE-CS/AAAI joint task force. In: Proceedings of the 54th ACM Technical Symposium on Computer Science Education V. 2, SIGCSE 2023, pp. 1212–1213. Association for Computing Machinery, New York (2023)
16. Mendel, J.M.: Uncertain Rule-Based Fuzzy Systems, vol. 684. Springer, Cham (2017). https://doi.org/10.1007/978-3-319-51370-6
17. Nguyen, H.T., Walker, C.L., Walker, E.A.: A First Course in Fuzzy Logic. CRC Press, Boca Raton (2019)
18. Olaru, C., Wehenkel, L.: A complete fuzzy decision tree technique. Fuzzy Sets Syst. **138**(2), 221–254 (2003)
19. Phillips, P.J., Hahn, C.A., Fontana, P.C., Broniatowski, D.A., Przybocki, M.A.: Four principles of explainable artificial intelligence, Gaithersburg, Maryland, p. 18 (2020)
20. Popescu, A., Stefan, L.-D., Deshayes-Chossart, J., Ionescu, B.: Face verification with challenging imposters and diversified demographics. In: 2022 IEEE/CVF Winter Conference on Applications of Computer Vision (WACV), pp. 1151–1160 (2022)
21. Rani, A., Chaudhary, A., Sinha, N., Mohanty, M., Chaudhary, R.: Drone: the green technology for future agriculture. Harit Dhara **2**(1), 3–6 (2019)
22. Sanchez, R., Servin, C., Argaez, M.: Sparse fuzzy techniques improve machine learning. In: 2013 Joint IFSA World Congress and NAFIPS Annual Meeting (IFSA/NAFIPS), pp. 531–535 (2013)
23. Servín, C.: Fuzzy information processing computing curricula: a perspective from the first two-years in computing education. In: Rayz, J., Raskin, V., Dick, S., Kreinovich, V. (eds.) NAFIPS 2021. LNNS, vol. 258, pp. 391–399. Springer, Cham (2022). https://doi.org/10.1007/978-3-030-82099-2_35
24. Servin, C., Kreinovich, V.: Towards efficient algorithms for approximating a fuzzy relation by fuzzy rules: case when "and"-and "or"-operation are distributive. In: 2014 IEEE Conference on Norbert Wiener in the 21st Century (21CW), pp. 1–7 (2014)
25. Servin, C., Kreinovich, V., Shahbazova, S.: Ethical dilemma of self-driving cars: conservative solution. In: Shahbazova, S.N., Abbasov, A.M., Kreinovich, V., Kacprzyk, J., Batyrshin, I.Z. (eds.) Recent Developments and the New Directions of Research, Foundations,

and Applications. STUDFUZZ, vol. 423, pp. 93–98. Springer, Cham (2023). https://doi.org/10.1007/978-3-031-23476-7_9

26. Servin, C., Muela, G.: A gentle introduction to fuzzy logic in CS I course: using PLTL as a vehicle to obliquely introduce the concept of fuzzy logic (2016)
27. Thomson, J.J.: The trolley problem. Yale Law J. **94**(6), 1395–1415 (1985)
28. Tizhoosh, H.R.: Opposition-based learning: a new scheme for machine intelligence. In: International Conference on Computational Intelligence for Modelling, Control and Automation and International Conference on Intelligent Agents, Web Technologies and Internet Commerce (CIMCA-IAWTIC 2006), vol. 1, pp. 695–701 (2005)
29. Wang, L.-X., Mendel, J.M.: Generating fuzzy rules by learning from examples. IEEE Trans. Syst. Man Cybern. **22**(6), 1414–1427 (1992)
30. Weber, C.: Engineering bias in AI. IEEE Pulse **10**(1), 15–17 (2019)
31. Yung-Jen Hsu, J., Lo, D.-C., Hsu, S.-C.: Fuzzy control for behavior-based mobile robots. In: NAFIPS/IFIS/NASA 1994. Proceedings of the First International Joint Conference of The North American Fuzzy Information Processing Society Biannual Conference. The Industrial Fuzzy Control and Intellige, pp. 209–213 (1994)
32. Zadeh, L.A.: Fuzzy logic = computing with words. IEEE Trans. Fuzzy Syst. **4**(2), 103–111 (1996)

Associative Property of Interactive Addition for Intervals: Application in the Malthusian Model

Vinícius F. Wasques[1,2](✉), Allan Edley Ramos de Andrade[3], and Pedro H. M. Zanineli[4]

[1] Ilum School of Science - CNPEM, Campinas, Brazil
[2] São Paulo State University - UNESP, Rio Claro, Brazil
vwasques@outlook.com
[3] Federal University of Mato Grosso do Sul - UFMS, Três Lagoas, Brazil
allan.andrade@ufms.br
[4] Ilum School of Science - CNPEM, Campinas, Brazil
pedro220068@ilum.cnpem.br

Abstract. This paper presents a study about the properties of the interactive arithmetic sum $+_{0.5}$ for intervals. This particular arithmetic operator is defined by a family of joint possibility distributions, which gives raise to the notion of interactivity. This work shows that the class of intervals $\mathcal{I}$ under the sum $+_{0.5}$ is a commutative monoid, that is, the operation $+_{0.5}$ satisfies the commutative and associative properties and $\mathcal{I}$ has an identity element. In order to illustrate the properties of this arithmetic sum, the Malthusian problem is investigated from the perspective of numerical methods, such as the Euler's and Runge-Kutta methods. Finally, a comparison between the interactive arithmetic $+_{0.5}$ and the standard arithmetic is presented in order to provide the advantages of the sum $+_{0.5}$.

Keywords: Fuzzy Arithmetic · Interval Differential Equation · Fuzzy Numerical Method · Fuzzy Interactivity

1 Introduction

Several phenomena of nature can be described by mathematical tools such as differential equations. For example, Malthus stated that a population grows proportional to itself. The differential equation that models this problem is given by $x' = \lambda x$, where λ is the growth rate. So, differential equations can be used to analyze how certain populations evolve over time. Approaches like this require information such as growth rates and initial population value. However, such parameters are not always precisely described, since there are several uncertainties in the phenomena. These uncertainties can be modeled by fuzzy numbers

V. F. Wasques—FAPESP ° 2023/03927-0.

K. Cohen et al. (Eds.): NAFIPS 2023, LNNS 751, pp. 194–206, 2023.
https://doi.org/10.1007/978-3-031-46778-3_18

or intervals. Hence, the fuzzy set theory or interval theory may provide a more realistic overview of the studied phenomenon.

On the other hand, solving differential equations is not a simple task. This task becomes even more complex when we incorporate fuzzy or interval variables. Thus, numerical methods are useful to study these type of problems. In this sense, it is desirable that the numerical methods satisfy some requirements, such as the order in which the method is computed. That is, the calculation from left to right should be the same as from right to left.

The usual (Minkowski) interval sum holds the property of commutativity and associativity, making it consistent in that sense. On the other hand, the sum between two intervals produces a new interval that is larger in size than the previous ones. Here we are referring to the size by the width of the interval. In the context of fuzzy set theory, the larger the width, the greater the uncertainty. With that in mind, we are interested in other arithmetics that have good properties like the ones mentioned above, so that we can have a better control of their widths.

Different types of sum between fuzzy numbers (or intervals) have been proposed in the literature, but not all of them satisfy the associative property for example. Here the focus will be the interactive sum [1]. The idea of interactivity is connected with a fuzzy relation called joint possibility distribution, which has a similar interpretation to distributions in probability theory. Among the different types of interactive sum, we will discuss the sum $+_{0.5}$ that is associated with the distribution $J_{0.5}$ proposed in [2]. We will analyze its properties from an algebraic point of view, and from there, provide interval numerical solutions to the population problem proposed by Malthus [3] considering an interval initial condition.

Next we present the preliminaries concepts in the classical and fuzzy sets theory, for a better understanding of the work.

2 Preliminaries

Let us present some definitions of abstract algebra that are necessary for understanding the paper. A *groupoid* is defined by a non-empty set S and a binary operation $\sigma : S \times S \to S$ and it is denoted by (S, σ). If σ is associative, that is, for all $x, y, z \in S$ it follows that $\sigma(\sigma(x, y), z) = \sigma(x, \sigma(y, z))$, then (S, σ) is called by a *semigroup*. Moreover, if σ is commutative, that is, $\sigma(x, y) = \sigma(y, x)$, for all $x, y \in S$, and $\sigma(x, 1_S) = \sigma(1_S, x) = x$, for all $x \in S$, where 1_S is the identity element of S, then S is called a commutative monoid.

The Euler's and Runge-Kutta methods are mathematical tools to provide numerical solutions for ordinary differential equations given in form of an initial value problem

$$\begin{cases} x' = f(t, x(t)) \\ x(t_0) = x_0 \in \mathbb{R} \end{cases}.$$

The numerical method proposed by Euler is defined by

$$x_{n+1} = x_n + hf(t_n, x_n),$$

where h is the size of each interval $[t_n, t_{n+1}]$ and x_{n+1} is the approximation of $x(t)$ at $t = t_{n+1}$. As an extension of the Euler's method, the Runge-Kutta method is given by

$$x_{n+1} = x_n + \frac{h}{6}(k_1 + 2k_2 + 2k_3 + k_4),$$

where $k_1 = f(t_n, x_n)$, $k_2 = f(t_n + \frac{h}{2}, x_n + \frac{h}{2}k_1)$, $k_3 = f(t_n + \frac{h}{2}, x_n + \frac{h}{2}k_2)$ and $k_4 = f(t_n + h, x_n + hk_3)$.

2.1 Fuzzy Sets Theory

Here the Fuzzy Sets Theory will be presented as an extension of the Interval Theory. Recall that the Interval Theory studies the structure, properties and applications of elements in the form of $[a, b] = \{x \in \mathbb{R} : a \leq x \leq b\}$. A fuzzy subset A of a universe X is defined by a generalization of the characteristic function $\chi_A : X \to \{0, 1\}$, that is, a function $\varphi_A : X \to [0, 1]$ called by a membership function.

For example, the interval $[1, 2]$ can be defined by the function $\chi_{[1,2]}$ which satisfies $\chi_{[1,2]}(x) = 1$, if $x \in [1, 2]$ or $\chi_{[1,2]}(x) = 0$, if $x \notin [1, 2]$. On the other hand, $[1, 2]$ can be generalized by a trapezoidal fuzzy set A, such as the one defined by $\varphi_A(x) = 1$, if $x \in [1, 2]$, $\varphi_A(x) = x$, if $x \in [0, 1]$, $\varphi_A(x) = 3 - x$, if $x \in [2, 3]$ and $\varphi_A(x) = 0$ otherwise.

For each $0 < \alpha \leq 1$, the α-cut of A is defined by $[A]^\alpha = \{u \in X : \varphi_A(u) \geq \alpha\}$, and supposing that X is a topological space, then the 0-cut of A is defined by $[A]^0 = cl\{u \in X : \varphi_A(u) > 0\}$, where cl denotes the closure of a subset of X [4]. The width of a fuzzy set is computed by the size of its 0-cut and as larger is the width, the greater is the uncertainty that it models.

In modelling and applications the concept of a fuzzy number is considered, since it extends the definition of a real number. A fuzzy subset A of X is a fuzzy number if the topological space is given by $X = \mathbb{R}$ and every α-cut of A is a nonempty and close interval with bounded support ($suppA = \{u \in \mathbb{R} : \varphi_A(u) > 0\}$) [2]. The class of fuzzy numbers is denoted by $\mathbb{R}_\mathcal{F}$.

In the space of the fuzzy numbers the arithmetic operations is defined in terms of the extension principles, such as Zadeh's and sup-J extension principles,

$$(A \otimes B)(z) = \sup_{x \otimes y = z} \min\{A(x), B(y)\} \quad \text{(Zadeh's extension principle)}$$

$$(A \otimes_J B)(z) = \sup_{x \otimes y = z} J(A(x), B(y)) \quad \text{(Sup-}J\text{ extension principle)},$$

where $\otimes$ is an arithmetic operation and J is a relationship called joint possibility distribution (JPD). Recall that J is a JPD between the fuzzy numbers A and B if $A(x) = \sup_{u \in \mathbb{R}} J(x, u)$ and if $B(y) = \sup_{v \in \mathbb{R}} J(v, y)$, simultaneously [1].

The arithmetic via Zadeh's and sup-J extensions are called *standard* (or non-interactive) and *interactive* arithmetics, respectively. The main difference between these two types of arithmetic is that sup-J extension produces outputs more specific than the Zadeh's extension, meaning that $A \otimes_J B$ has width less or equal than the width of $A \otimes B$ [5]. For intervals the addition via Zadeh's extension is given by $[\underline{a}, \overline{a}] + [\underline{b}, \overline{b}] = [\underline{a} + \underline{b}, \overline{a} + \overline{b}]$, which is equivalent to the Minkowski addition [6]. On the other hand, the addition via sup-J extension produces other results, for example, the distribution $J = J_0$ proposed by [7,8], produces $[\underline{a}, \overline{a}] +_{J_0} [\underline{b}, \overline{b}] = [\min(\underline{a} + \overline{b}, \underline{b} + \overline{a}), \max(\underline{a} + \overline{b}, \underline{b} + \overline{a})]$ [9]. For example, let be $A = [1, 2]$ and $B = [3, 7]$, thus $A + B = [1 + 3, 2 + 7] = [4, 9]$ and $A +_{J_0} B = [\min(2+3, 1+7), \max(2+3, 1+7)] = [5, 8]$. Note that $width(A +_{J_0} B) = 3 \leq 5 = width(A + B)$.

Arithmetic operations such as $\otimes_{J_0}$ are important because with them is possible to control the growing of the width, in contrast to the standard arithmetic. The growing of the width means a propagation of uncertainty, which leads to inconsistencies in modelling [10]. Other interactive arithmetics can be considered in order to avoid this problem, such as the family of JPDs proposed by Esmi et al. [2], denoted by J_γ. The construction of this family is very elaborated, so here we focus only in the expression of the sum, obtained from the sup-J extension principle. The sum $A +_\gamma B$ is provided in the next theorem.

Theorem 1 [2]. *Let $A, B \in \mathbb{R}_{\mathcal{F}_C}$, whose α-cuts are $[A]^\alpha = [a_\alpha^-, a_\alpha^+]$ and $[B]^\alpha = [b_\alpha^-, b_\alpha^+]$. For each $\gamma \in [0, 1]$, it follows that the α-cuts of $A +_\gamma B$ are given by*

$$[A +_\gamma B]^\alpha = [c_\alpha^-, c_\alpha^+] + \{\overline{a} + \overline{b}\} \tag{1}$$

where,

$$c_\alpha^- = \inf_{\beta \geq \alpha} h_{(A+B)}^-(\beta, \gamma) \quad e \quad c_\alpha^+ = \sup_{\beta \geq \alpha} h_{(A+B)}^+(\beta, \gamma). \tag{2}$$

with

$$\begin{aligned} h_{(A+B)}^-(\beta,\gamma) = \min\{ & (a^{(\overline{a})})_\beta^- + (b^{(\overline{b})})_\beta^+ + \gamma((b^{(\overline{b})})_\beta^- - (b^{(\overline{b})})_\beta^+), \\ & (a^{(\overline{a})})_\beta^+ + (b^{(\overline{b})})_\beta^- + \gamma((a^{(\overline{a})})_\beta^- - (a^{(\overline{a})})_\beta^+), \\ & \gamma((a^{(\overline{a})})_\beta^- + (b^{(\overline{b})})_\beta^-)\} \end{aligned}$$

and

$$\begin{aligned} h_{(A+B)}^+(\beta,\gamma) = \max\{ & (a^{(\overline{a})})_\beta^- + (b^{(\overline{b})})_\beta^+ + \gamma((a^{(\overline{a})})_\beta^+ - (a^{(\overline{a})})_\beta^-), \\ & (a^{(\overline{a})})_\beta^+ + (b^{(\overline{b})})_\beta^- + \gamma((b^{(\overline{b})})_\beta^+ - (b^{(\overline{b})})_\beta^-), , \\ & \gamma((a^{(\overline{a})})_\beta^+ + (b^{(\overline{b})})_\beta^+)\} \end{aligned}$$

where $(a^{(\overline{a})})_\beta^-, (a^{(\overline{a})})_\beta^+, (b^{(\overline{b})})_\beta^-$ and $(b^{(\overline{b})})_\beta^+$ represent the translated endpoints of the fuzzy numbers A and B, respectively, by their middle points.

In the context of Interval Theory a similar result can be obtained. This result is enunciated in Theorem 2.

Theorem 2 [11]. *Let $A = [\underline{a}, \overline{a}]$ and $B = [\underline{b}, \overline{b}]$ be two intervals and $\gamma \in [0, 1]$. The interactive sum $C = A +_\gamma B$ for any $\gamma \in [0, 1]$ is given by $C = [\underline{c}, \overline{c}]$, where*

$$\overline{c} = \max\{\overline{a} + \underline{b} + \gamma(\overline{b} - \underline{b}), \underline{a} + \overline{b} + \gamma(\overline{a} - \underline{a})\}$$

and

$$\underline{c} = \min\{\underline{a} + \overline{b} - \gamma(\overline{b} - \underline{b}), \overline{a} + \underline{b} - \gamma(\overline{a} - \underline{a})\}.$$

For example, let be $A = [1, 2]$ and $B = [2, 5]$. The interactive sums $+_\gamma$ are given by $A +_0 B = [4, 6]$, $A +_{0.25} B = [3.75, 6.25]$, $A +_{0.5} B = [3.5, 6.5]$, $A +_{0.75} B = [3.25, 6.75]$ and $A +_1 B = [3, 7]$.

This paper focuses on a specific joint possibility distribution of this family, the $J_{0.5}$ which will be discussed in the next section. From now on, the discussion will be provided only for the class of intervals, which is denoted by $\mathcal{I}$.

3 $J_{0.5}$-Joint Possibility Distribution

The joint possibility distribution $J_{0.5}$ has a very technical construction, which can be consulted in [2] for more details. This relation restricts the domain of the Cartesian products in which the elements of the intervals are associated. The graphical representation of this joint possibility distribution for intervals is given in Fig. 1.

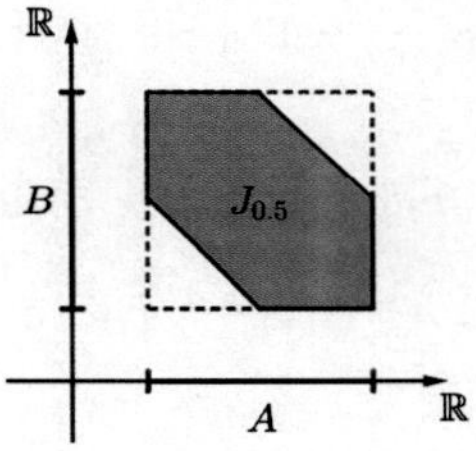

Fig. 1. Representation of the joint possibility distribution $J_{0.5}$ for intervals

The sum between interactive fuzzy numbers based on $J_{0.5}$ can be obtained from Theorem 2, by applying $\gamma = 0.5$. The expression of $+_{0.5}$ is given by the following corollary.

Corollary 1. *Let $A = [\underline{a}, \overline{a}]$ and $B = [\underline{b}, \overline{b}]$ be two intervals. The interactive sum $C = A +_{0.5} B$ is given by $C = [\underline{c}, \overline{c}]$, where*

$$\underline{c} = \underline{a} + 0.5(\overline{b} + \underline{b}) \text{ and } \overline{c} = \overline{a} + 0.5(\overline{b} + \underline{b})$$

or

$$\underline{c} = \underline{b} + 0.5(\overline{a} + \underline{a}) \text{ and } \overline{c} = \overline{b} + 0.5(\overline{a} + \underline{a}).$$

From the above corollary, a connection between the width of $A +_{0.5} B$ and its operands can be established, which is provided in the next lemma.

Lemma 1. *Let $A = [\underline{a}, \overline{a}]$ and $B = [\underline{b}, \overline{b}]$ be two intervals. Thus,*

1. $A +_{0.5} B = A + \boldsymbol{b}$, *if* $width(A) \geq width(B)$ *or* $A +_{0.5} B = B + \boldsymbol{a}$, *if* $width(B) \geq width(A)$, *where* $\boldsymbol{a}$ *and* $\boldsymbol{b}$ *are the middle points of* A *and* B, *respectively.*

2. $width(A +_{0.5} B) = \max\{width(A), width(B)\}$.

Proof: 1. From Corollary 1, the sum $A +_{0.5} B$ is given by

$$\text{Case 1: } A +_{0.5} B = [\underline{a} + 0.5(\overline{b} + \underline{b}), \overline{a} + 0.5(\overline{b} + \underline{b})] \text{ or}$$
$$\text{Case 2: } A +_{0.5} B = [\underline{b} + 0.5(\overline{a} + \underline{a}), \overline{b} + 0.5(\overline{a} + \underline{a})].$$

If $width(A) \leq width(B)$, then

$$\begin{aligned}\overline{a} - \underline{a} \leq \overline{b} - \underline{b} &\Rightarrow 0.5(\overline{a} - \underline{a}) \leq 0.5(\overline{b} - \underline{b}) \\ &\Rightarrow 0.5\overline{a} + 0.5\underline{b} \leq 0.5\overline{b} + 0.5\underline{a}.\end{aligned}$$

Adding $0.5\overline{a} + 0.5\overline{b}$ in both sides, it follows that

$$\begin{aligned}&0.5\overline{a} + 0.5\overline{b} + 0.5\overline{a} + 0.5\underline{b} \leq 0.5\overline{a} + 0.5\overline{b} + 0.5\overline{b} + 0.5\underline{a} \\ &\Rightarrow \overline{a} + 0.5(\overline{b} + \underline{b}) \leq \overline{b} + 0.5(\overline{a} + \underline{a}).\end{aligned}$$

Hence, the right endpoint of $A +_{0.5} B$ is given by $\overline{b} + 0.5(\overline{a} + \underline{a})$. Similarly, it follows that the left endpoint is given by $\underline{b} + 0.5(\overline{a} + \underline{a})$. Consequently,

$$\begin{aligned}A +_{0.5} B &= [\underline{b} + 0.5(\overline{a} + \underline{a}), \overline{b} + 0.5(\overline{a} + \underline{a})] \\ &= B + 0.5(\overline{a} + \underline{a}) \\ &= B + \mathbf{a}.\end{aligned}$$

If $width(B) \leq width(A)$, then

$$\begin{aligned}A +_{0.5} B &= [\underline{a} + 0.5(\overline{b} + \underline{b}), \overline{a} + 0.5(\overline{b} + \underline{b})] \\ &= A + 0.5(\overline{b} + \underline{b}) \\ &= A + \mathbf{b}.\end{aligned}$$

2. Suppose that $\max\{width(A), width(B)\} = width(A)$, that is, $width(B) \leq width(A)$. From item 1., $A +_{0.5} B = [\underline{a} + 0.5(\overline{b} + \underline{b}), \overline{a} + 0.5(\overline{b} + \underline{b})]$. Thus, $width(A +_{0.5} B) = \overline{a} - \underline{a} = width(A)$. The other case is analogous and the proof is complete.

Next, the above lemma will be considered to proof the associative property of intervals for the sum $+_{0.5}$. In fact, this result reveals that the class $\mathcal{I}$ under the binary operation $+_{0.5}$ is a semigroup.

Theorem 3. *The set* $(\mathcal{I}, +_{0.5})$ *is a semigroup.*

Proof: First note that $(\mathcal{I}, +_{0.5})$ is a groupoid, since $\mathcal{I}$ is non-empty and $+_{0.5} : \mathcal{I} \times \mathcal{I} \to \mathcal{I}$ is a binary operation. Now let us show that $+_{0.5} : \mathcal{I} \times \mathcal{I} \to \mathcal{I}$ is an associative operation. To this end, consider $A = [\underline{a}, \overline{a}]$, $B = [\underline{b}, \overline{b}]$ and $C = [\underline{c}, \overline{c}]$ elements of $\mathcal{I}$. Hence, we must analyze all the possible cases,

$$\begin{aligned}
i)&\ width(A) \geq width(B) \geq width(C)\\
ii)&\ width(A) \geq width(C) \geq width(B)\\
iii)&\ width(B) \geq width(A) \geq width(C)\\
iv)&\ width(B) \geq width(C) \geq width(A)\\
v)&\ width(C) \geq width(A) \geq width(B)\\
vi)&\ width(C) \geq width(B) \geq width(A)
\end{aligned}$$

Here we will proof only the first case and the others can be checked in a similar way.

From Lemma 1, if $width(A) \geq width(B)$, then $A +_{0.5} B = A + \mathbf{b}$. Moreover, if $width(B) \geq width(C)$, then $B +_{0.5} C = B + \mathbf{c}$. Since, $width(A) \geq width(C)$, it follows that

$$\begin{aligned}
(A +_{0.5} B) +_{0.5} C &= (A + \mathbf{b}) +_{0.5} C\\
&= A +_{0.5} C + \mathbf{b}\\
&= A + \mathbf{c} + \mathbf{b}.
\end{aligned}$$

On the other hand,

$$\begin{aligned}
A +_{0.5} (B +_{0.5} C) &= A +_{0.5} (B + \mathbf{c})\\
&= (A +_{0.5} B) + \mathbf{c}\\
&= A + \mathbf{b} + \mathbf{c}.
\end{aligned}$$

Therefore, $(A +_{0.5} B) +_{0.5} C = A +_{0.5} (B +_{0.5} C)$ and by checking the other five cases we conclude that $+_{0.5}$ is an associative operation. Hence, $(\mathcal{I}, +_{0.5})$ is a semigroup.

Note that $(\mathcal{I}, +_{0.5})$ is also a commutative monoid, since $A +_{0.5} B = B +_{0.5} A$, for all $A, B \in \mathcal{I}$ and $0_{\mathcal{I}} = [0, 0] \in \mathcal{I}$ is the identity element under $+_{0.5}$. This result is enunciated in the next proposition.

Proposition 1. *The semigroup* $(\mathcal{I}, +_{0.5})$ *is a commutative monoid.*

Here the main property of this semigroup will be explored in numerical methods. Note that in methods such as Euler and Runge-Kutta, the associativity is a fundamental property to ensure the consistency of the final numerical solution, since the calculation can not depend on the order in which it is computed. In the next section we provide an application of this interactive sum in a growth population model called Malthusian problem.

4 Application in the Malthusian Model

Malthus proposed the first model to study the dynamics of population growth. The assumptions of his model was the population growths proportionally as itself, which gives raise to the following ordinary differential equation $x' = \lambda x$, where $0 < \lambda < 1$ is the growth rate. So, from an information about the initial value of the population $x(0) = x_0$ the solution for this problem is given by $x(t) = x_0 e^{\lambda t}$, which is coherent with the hypothesis stated by Malthus.

However, it is not simple to accurately determine the initial value $x(0)$. So, it is more reasonable to model $x(0)$ by intervals or fuzzy numbers, in order to consider an inaccuracy around the initial value of the population. Hence, the Malthusian problem studied here is defined by

$$\begin{cases} x' = \lambda x \\ x(0) = X_0 \in \mathcal{I} \end{cases},$$

where $X_0 = [\underline{x_0}, \overline{x_0}]$ and $0 < \lambda < 1$.

In terms of Zadeh's (or sup-J) extension principle for one input, the interval solution to this problem is given by $\hat{X}(t) = [\underline{x_0}e^{\lambda t}, \overline{x_0}e^{\lambda t}]$. Now, from the perspective of numerical methods, Wasques et al. [10] proposed numerical solutions for fuzzy values by considering classical numerical methods and extending the classical arithmetic operations involved in the method. For example, the Euler's method for the fuzzy case is given by $X_{n+1} = X_n \oplus hf(t_n, X_n)$, where h is the size of the intervals $[t_n, t_{n+1}]$, X_n is a fuzzy number for all n and $\oplus$ is a fuzzy sum. Analogously for the Runge-Kutta method [5].

Here, it will be considered the methodology proposed by Wasques et al. [5,10], where X_n will be given by intervals and $\oplus$ will be given by the interactive sum $+_{0.5}$. In order to provide the advantages of $+_{0.5}$, the numerical solution via Minkowski sum will be also provided.

4.1 Numerical Solution via Euler's Method

The numerical solutions via Euler's method based on the sums $+$ and $+_{0.5}$ are respectively given by

$$X_{n+1} = X_n + h(\lambda X_n) \tag{3}$$

and

$$X_{n+1} = X_n +_{0.5} h(\lambda X_n). \tag{4}$$

The numerical solutions (3) and (4) are depicted in Figs. 2 and 3, respectively.

Note that both solutions have the qualitatively behaviour of the classical solution proposed by Malthus, that is, the solution has a exponential growth. Figure 2 reveals that the width of the numerical solution via standard sum is increasing, which implies that the uncertainty is propagated along time, as expected [5]. However, the numerical solution via interactive sum $+_{0.5}$ does not have the same behaviour. Figure 3 suggests that the width remains constant along time, which implies that the uncertainty could be controlled. Proposition 2 proves this fact.

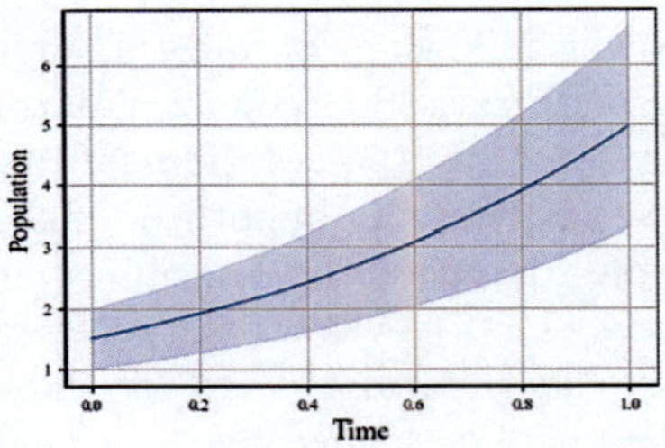

Fig. 2. Numerical solution via Euler's method considering the Minkowski sum. The blue shadow represents the interval numerical solution and the blue line represent the deterministic solution for initial value 1.5. The parameters are $h = 0.01, \lambda = 0.2$ and $x_0 = [1, 2]$.

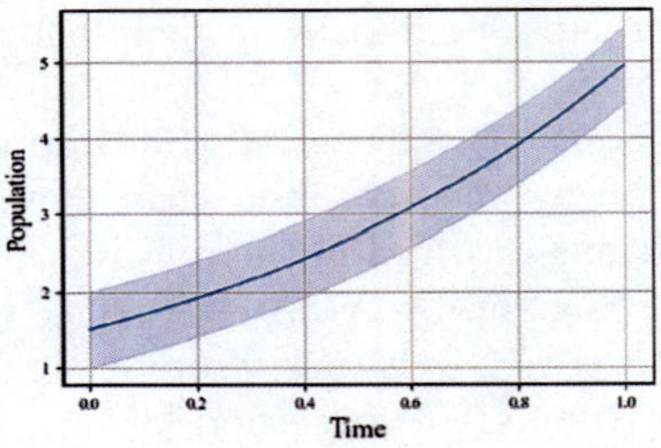

Fig. 3. Numerical solution via Euler's method considering the interactive sum $+_{0.5}$. The blue shadow represents the interval numerical solution and the blue line represent the deterministic solution for initial value 1.5. The parameters are $h = 0.01, \lambda = 0.2$ and $x_0 = [1, 2]$.

Proposition 2. *The numerical solution* (4) *to the Malthusian problem has constant width.*

Proof: First note that for sufficient small value of h, it follows that $0 < h\lambda < 1$, since $0 < \lambda < 1$. Consequently, if $X_n = [\underline{x_n}, \overline{x_n}]$ and $h\lambda X_n = [h\lambda\underline{x_n}, h\lambda\overline{x_n}]$, then $width(X_n) = \overline{x_n} - \underline{x_n} \geq h\lambda(\overline{x_n} - \underline{x_n}) = width(h\lambda X_n)$, for all $n \in \mathbb{N}$. Thus, from item 2 of Lemma 1, it follows that $width(X_{n+1}) = \max\{width(X_n), width(h\lambda X_n)\} = width(X_n)$, for all $n \in \mathbb{N}$. Therefore, we conclude that the numerical solution via Euler's method with interactive arithmetic $+_{0.5}$ has constant width.

It is important to observe here that both numerical solutions are consistent in the following sense, each element x_{n+1} was obtained in a unique way, since the commutative property holds for both sums. In the next section it will be explored the associative property as well.

4.2 Numerical Solution via Runge-Kutta Method

The numerical solutions via Runge-Kutta method based on the sums $+$ and $+_{0.5}$ are respectively given by

$$X_{n+1} = X_n + \frac{h}{6}(K_1 + 2K_2 + 2K_3 + K_4), \tag{5}$$

where

$$\begin{aligned} K_1 &= \lambda X_n, \\ K_2 &= \lambda\left(X_n + \frac{h}{2}K_1\right), \\ K_3 &= \lambda\left(X_n + \frac{h}{2}K_2\right), \\ K_4 &= \lambda\left(X_n + hK_3\right) \end{aligned}$$

and

$$X_{n+1} = X_n +_{0.5} \frac{h}{6}(K_1 +_{0.5} 2K_2 +_{0.5} 2K_3 +_{0.5} K_4), \tag{6}$$

where

$$\begin{aligned} K_1 &= \lambda X_n, \\ K_2 &= \lambda\left(X_n +_{0.5} \frac{h}{2}K_1\right), \\ K_3 &= \lambda\left(X_n +_{0.5} \frac{h}{2}K_2\right), \\ K_4 &= \lambda\left(X_n +_{0.5} hK_3\right). \end{aligned}$$

The numerical solutions (5) and (6) are depicted in Figs. 4 and 5, respectively.

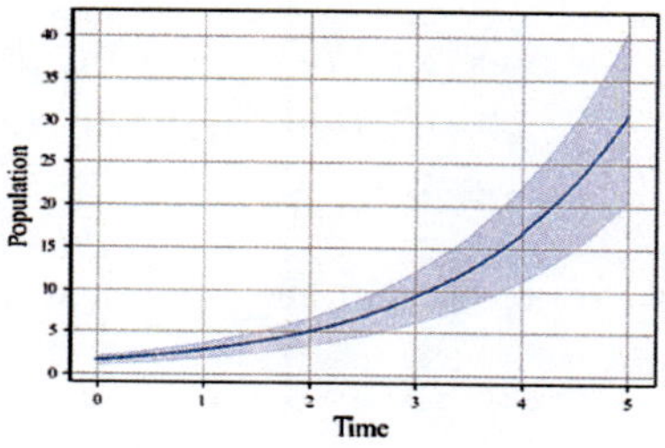

Fig. 4. Numerical solution via Runge-Kutta method considering the Minkowski sum. The blue shadow represents the interval numerical solution and the blue line represent the deterministic solution for initial value 1.5. The parameters are $h = 0.01, \lambda = 0.2$ and $x_0 = [1, 2]$.

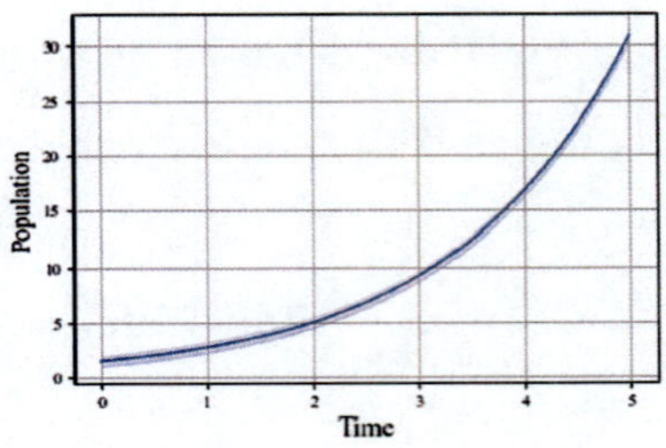

Fig. 5. Numerical solution via Runge-Kutta method considering the interactive sum $+_{0.5}$. The blue shadow represents the interval numerical solution and the blue line represent the deterministic solution for initial value 1.5. The parameters are $h = 0.01$, $\lambda = 0.2$ and $x_0 = [1, 2]$.

It is important to observe that both numerical solutions are consistent in terms of the associative property, since each element X_{n+1} was obtained uniquely, since the order of arithmetic operations is independent.

As well as the Euler's method, the Runge-Kutta method provided numerical solutions qualitatively similar to the deterministic solution. For the Runge-Kutta method, it is clear how much it is possible to have greater control of the uncertainty throughout the phenomenon by comparing the widths of the interval solutions seen in Figs. 4 and 5. In fact, the width of the numerical solution (6) is also constant, as Proposition 3 reveals.

Proposition 3. *The numerical solution* (6) *to the Malthusian problem has constant width.*

Proof: First, let us analyze K_1, K_2, K_3 and K_4. Note that $width(K_1) = width(\lambda X_n)$. Also, from item 2 of Lemma 1, $width(K_2) = \lambda width(X_n)$, since $width(X_n) \geq width(\frac{h}{2}K_1)$. Similarly, $width(K_4) = width(K_3) = width(K_2) = width(K_1)$. Now, let be $A = K_1 +_{0.5} 2K_2 +_{0.5} 2K_3 +_{0.5} K_4$. So, $width(A) = \max\{width(K_1), width(2K_2),\ width(2K_3), width(K_4)\} = width(2K_2)$. Consequently,

$$\begin{aligned} width(X_{n+1}) &= \max\{width(X_n), (h/6)width(2K_2)\} \\ &= \max\{width(X_n), (h\lambda/3)width(X_n)\} \\ &= width(X_n), \end{aligned}$$

for all $n \in \mathbb{N}$. Therefore, we conclude that the numerical solution (6) has constant width.

5 Final Remarks

This papers studied the algebraic properties of the set $\mathcal{I}$ under the sum $+_{0.5}$. We showed that $(\mathcal{I}, +_{0.5})$ is a commutative monoid. From this result, the Euler and

Runge-Kutta methods were considered to provide a numerical solution to the Malthusian problem, where the initial condition is given by an interval value. These numerical methods are consistent in the following sense, the arithmetic operations are independent of computation order.

Section 4 provided a comparison between the numerical solutions via Euler's and Runge-Kutta methods considering two different arithmetics, the standard (Minkowski) and interactive ($+_{0.5}$). The simulations corroborate the theoretical results that the standard arithmetic propagates uncertainty, while in the interactive arithmetic there is a better control of this propagation. In fact, this paper proved that the numerical solution via interactive sum $+_{0.5}$ has constant width in both numerical methods, which means that the uncertainty remains constant along time. For future works, we intend to investigate the algebraic properties of the arithmetic operation $+_{0.5}$ in the space of fuzzy numbers.

Acknowledgment. The first author thanks the support of V. F. Wasques - FAPESP n° 2023/03927-0, the second author thanks the support of Federal University of Mato Grosso do Sul (UFMS/MEC) and first and third authors thank the support of Ilum School of Science (CNPEM).

References

1. Fuller, R., Majlender, P.: On interactive fuzzy numbers. Fuzzy Sets Syst. **143**(3), 355–369 (2004)
2. Esmi, E., Wasques, V.F., Barros, L.C.: Addition and subtraction of interactive fuzzy numbers via family of joint possibility distributions. Fuzzy Sets Syst. **424**, 105–131 (2021)
3. Edelstein-Keshet, L.: Mathematical Models in Biology. Random House (1988)
4. Barros, L.C., Bassanezi, R.C., Lodwick, W.A.: A First Course in Fuzzy Logic, Fuzzy Dynamical Systems, and Biomathematics. Springer, Heidelberg (2017). https://doi.org/10.1007/978-3-662-53324-6
5. Wasques, V.F., Esmi, E., Barros, L.C., Bede, B.: Comparison between numerical solutions of fuzzy initial-value problems via interactive and standard arithmetics. In: Kearfott, R.B., Batyrshin, I., Reformat, M., Ceberio, M., Kreinovich, V. (eds.) IFSA/NAFIPS 2019 2019. AISC, vol. 1000, pp. 704–715. Springer, Cham (2019). https://doi.org/10.1007/978-3-030-21920-8_62
6. Moore, R.E., Kearfott, R.B., Cloud, M.J.: Introduction to Interval Analysis. SIAM (2009)
7. Esmi, E., Barroso, G., Barros, L.C., Sussner, P.: A family of joint possibility distributions for adding interactive fuzzy numbers inspired by biomathematical models. In: Proceedings of the 2015 Conference of the International Fuzzy Systems Association and the European Society for Fuzzy Logic and Technology (EUSFLAT 2015) (2015)
8. Wasques, V.F., Esmi, E., Barros, L.C., Sussner, P.: The generalized fuzzy derivative is interactive. Inf. Sci. **519**, 93–109 (2020)
9. Wasques, V.F., Pinto, N.J.B., Esmi, E., de Barros, L.C.: Consistence of interactive fuzzy initial conditions. In: Bede, B., Ceberio, M., De Cock, M., Kreinovich, V. (eds.) NAFIPS 2020. AISC, vol. 1337, pp. 143–155. Springer, Cham (2022). https://doi.org/10.1007/978-3-030-81561-5_13

10. Wasques, V.F., Esmi, E., Barros, L.C., Sussner, P.: Numerical solution for fuzzy initial value problems via interactive arithmetic: application to chemical reactions. Int. J. Comput. Intell. Syst. **13**(1), 1517–1529 (2020)
11. Esmi, E., Sacilotto, C., Wasques, V.F., Barros, L.C.: Numerical solution for interval initial value problems based on interactive arithmetic. Iran. J. Fuzzy Syst. **19**(6), 1–12 (2022)

Genetic Fuzzy Passivity-Based Control Applied to a Robust Control Benchmark Problem

Jared Burton(✉) and Kelly Cohen

Department of Aerospace Engineering, University of Cincinnati, Cincinnati, OH, USA
burtonjd@mail.uc.edu, cohenky@ucmail.uc.edu

Abstract. Second-order systems are incredibly common in many aspects of engineering, especially where structures and mechanics play a major role. Multi-degree-of-freedom, 2nd-order systems are those with dynamics described by a system of coupled, ordinary differential equations of degree 2. Passivity-based control is a common scheme used in controlling these systems due to its physical inspiration and robustness. In this work, a Fuzzy Inference System (FIS) is developed to augment a passivity-based controller for improved robustness. The FIS is trained by Genetic Algorithm and applied to a robust control benchmark problem.

1 Introduction

1.1 Second-Order Systems

Second-order systems are incredibly common in many aspects of engineering, especially where structures and mechanics play a major role. Multi-degree-of-freedom, 2nd-order systems are those with dynamics described by a system of coupled, ordinary differential equations of degree 2. As a basic consequence of conservation of momentum, many aircraft, spacecraft, land vehicles, underwater vehicles, robots, and structures are modeled in this way, to name a few. Recently, an interest in highly flexible variants of such systems has further motivated the study of 2nd-order systems [1]. Flexible dynamic models may more accurately represent a system's true behavior and can be used to develop control schemes which benefit from the 2nd-order dynamics. For aircraft and spacecraft design, significant reductions in weight and power usage can be achieved by relaxing the rigid-body assumption and adopting flexible alternatives [2]. In fact, the treatment of flexible bodies as rigid is responsible for many past mistakes, some of which led to catastrophic failure [3]. In grounded robotics too, weight and power savings motivate the study of second-order systems and control laws which take advantage of their structure. For example, Disney Research has investigated the use of flexible robotics in animatronics for film production and theme-park attractions [4]. Mixing position and force-based manipulation for tactile robotics is another example where 2nd-order systems are simulated, allowing for improved control [5].

K. Cohen et al. (Eds.): NAFIPS 2023, LNNS 751, pp. 207–218, 2023.
https://doi.org/10.1007/978-3-031-46778-3_19

1.2 Passivity-Based Control

Den Hartog, in his seminal work on vibration absorption, introduced a means by which a simple spring-mass-damper system could be attached to a structure to modify its natural frequency and absorb and dissipate vibrational energy [6]. This passive, mechanical appendage effectively reduced the amplitude of forced vibrations and subsequently was used in many applications such as aircraft, for example [7]. Since then, many control approaches based on the concept of passivity have been developed for application to 2nd-order systems [8, 9]. 'Passivity,' as it is used here, refers to control schemes which operate without an external source of energy. Passivity-based control may be implemented mechanically, via electrical networks, or both, and has been used to control satellites, structures, etc. [10, 11]. These controllers come with natural stability guarantees because they are always energy dissipating or at least do not increase the total mechanical energy [8]. Consequently, they have been implemented as stand-alone solutions as well as alongside existing control laws to improve performance and robustness.

Even still, basic passivity-based control approaches are limited as they may require bulky mechanical components or extensive electrical networks to properly implement. Truly passive systems are also often collocated, meaning that sensors and actuators measure and affect the same location on a structure. These are limitations which may be overcome by introducing virtual-passive controllers. These are controllers which emulate passivity-based control via a microcontroller or integrated circuit, support non-collocated sensors and actuators, and may rely on an external source of energy for control input. This may provide further-improved performance by producing greater forcing. Virtual controllers do not guarantee all the benefits of true passivity-based control but are able to overcome some of the major limitations while retaining many of the advantages.

In practice, a microcontroller may be used to implement the control loop of a virtual passive system. This consists of measurement, state estimation, integration of the dynamics of the virtual passive system, control calculation, and actuator output. The dynamics of the virtual passive system are typically chosen to reflect a simple physical system such as a basic mass-spring-damper or other, low degree-of-freedom structure. These are also typically 2nd-order systems with constant coefficients. In this work, variable (nonlinear) damping is leveraged to improve performance robustness and stability robustness for a benchmark control problem.

1.3 Robust Control Benchmark

A benchmark problem for robust control was introduced at the American Control Conference in the year 1990 and updated in the following years [12]. The problem consists of controlling two carts connected by a spring (Fig. 1). Each cart has a nominal mass of $m_1 = m_2 = 1$. The spring stiffness has nominal value $\hat{k} = 1$ and the true value is allowed to vary between 0.5 and 2. Appropriate units are assumed where necessary. For a given value of stiffness k, the open-loop system is linear and exhibits both a single rigid-body mode and a single vibrational mode. Disturbances are applied to the second mass only. Measurements y of the second mass are subject to noise and used for control feedback. The control input to the system is provided as a force on the first mass.

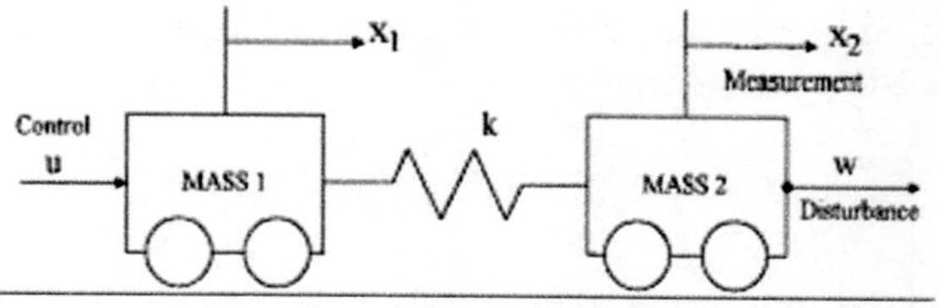

Fig. 1. Two-cart Robust Control Benchmark Problem (image from [13])

The goal of the robust control problem is to find a control law such that stability and several performance requirements are met despite variation in the spring stiffness. The performance requirements are specified for a scenario in which the second mass is subject to a unit impulse disturbance at time $t = 0$. The closed-loop system should have a settling time of 15 s or less, be 'insensitive to high-frequency sensor noise,' and have peak control effort magnitude less than 1 [12].

Reference [14] catalogues 10 controllers designed by various researchers using a variety of methods including variations of loop-transfer recovery, H-infinity control, and LQG regulation. In this work, a fuzzy virtual passivity-based controller is applied to the robust control benchmark problem. This serves as a continuation of work done by Cohen et al. Which used a Fuzzy Inference System (FIS) to vary the virtual damping of such a controller [11, 13]. In reference [13], Cohen et al. manually tuned a FIS to mimic the optimal control solution in which the damping coefficient is the controlled variable. Varying the damping of a truly passive system does not void stability and can improve performance and robustness.

1.4 Genetic Fuzzy Approach

In this work, Fuzzy Inference System tuning is performed by a Genetic Algorithm (GA) to improve the solution obtained in [13]. The genetic tuning strategy was chosen for its ease of implementation, incredible generality, and its ability to search a large space for high-quality solutions. These have been illustrated by the application of Genetic Fuzzy solutions to many problems in recent years [15, 16]. The Genetic Fuzzy approach is expected to improve upon the solution obtained via manual tuning by Cohen et. al [13]. Also, it provides a partial answer to the question of whether the optimal-control inspired solution remains optimal in this robust-control context.

2 Methodology

2.1 System Model

The open-loop model of the two-cart system is given in Eq. 1 in state-space form where x_1 and x_2 are the positions of the 1st and 2nd masses and x_3 and x_4 are their velocities. Measurements of the position of the 2nd mass are corrupted by noise: $y = x_2 + \upsilon$. For sake of simplicity, the measurement noise υ is modeled as a 2 Hz sinusoid: $\upsilon = 0.01\sin(4\pi t)$. This approximately accounts for the performance requirement that the system operate in the presence of high-frequency noise. The control input is modeled by u and the disturbance to cart 2 is modeled by w.

$$\begin{bmatrix} \dot{x}_1 \\ x_2 \\ x_3 \\ x_4 \end{bmatrix} = \begin{bmatrix} 0 & 0 & 1\,0 \\ 0 & 0 & 0\,1 \\ -k/m_1 & k/m_1 & 0\,0 \\ k/m_2 & -k/m_2 & 0\,0 \end{bmatrix} \begin{bmatrix} x_1 \\ x_2 \\ x_3 \\ x_4 \end{bmatrix} + \begin{bmatrix} 0 \\ 0 \\ 1/m_1 \\ 0 \end{bmatrix} u + \begin{bmatrix} 0 \\ 0 \\ 0 \\ 1/m_2 \end{bmatrix} w\# \tag{1}$$

To control this system, a virtual passivity-based solution is developed. The structure of the solution is adopted from [13] which mimics a mass attached to cart 1 by a spring and damper (Fig. 2). This additional mass is also attached to a fictitious wall by another spring such that it will drive the displacements of the carts to 0. This fictitious system is shown in Fig. 2. Cohen et al. picked values for the virtual system $m_v = 600$, $k_{v1} = 0.35$, and $k_{v2} = 2.9$ [13]. These values produced acceptable results and are reused in this work for sake of comparison. The virtual damping coefficient d_v can assume values within range [0, 2]. In reference [13], d_v at time t is determined by a FIS which takes the relative position and velocity between the virtual cart and cart 1 as input. These inputs were chosen such that the FIS would mimic the bang-bang optimal control solution. In this work, d_v is determined by a FIS which takes the estimated positions of carts 1 and 2 as inputs.

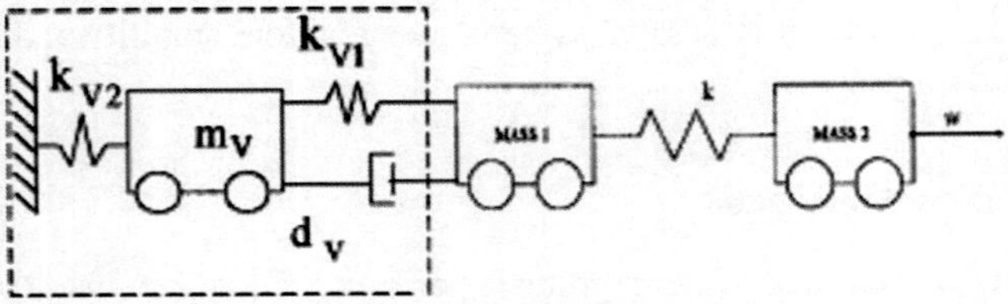

Fig. 2. Two-cart System with Virtual Passive Controller in dashed box (image from [13])

Only the position of cart 2 is measured directly; the position of cart 1 must be estimated. Cohen et al. used a Luenberger Observer to obtain the state estimate [13, 17]. The observer is modeled in continuous time in Eq. 2 where $\hat{x}$ is the state vector estimate. The gain matrix $M = \begin{bmatrix} 18\ 6\ 0\ 16 \end{bmatrix}^T$ was obtained by pole placement [13]. Once again, the same values are used here for comparison.

$$\dot{\hat{x}} = A\hat{x} + Bu + M\left(y - C\hat{x}\right)\# \tag{2}$$

The full system block diagram is shown in Fig. 3. It accounts for the carts, the state observer, the FIS and virtual passive components, disturbances, and noise.

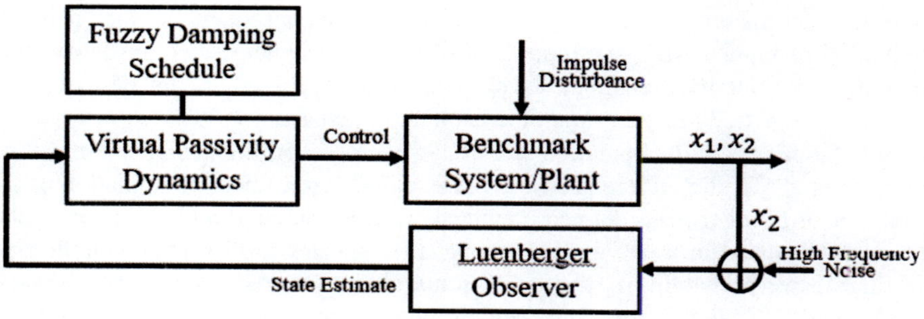

Fig. 3. Controller and Plant Diagram

2.2 Performance Metrics

Linear, constant-coefficient systems admit analytical tests for stability and sometimes allow closed-form representations of performance metrics such as settling time and maximum control effort. In general, for nonlinear and time-varying dynamics, this is not always the case. This is true for the nonlinear virtual damping used here. Instead, Monte-Carlo analysis may be used to estimate the probability of violating a system requirement or the probability distribution of a quantity of interest. A variation of this technique called Stochastic Robustness Analysis (SRA) takes advantage of the properties of binomial random variables to provide confidence bounds for these probabilities [18, 19]. Most importantly, confidence bounds may be calculated without integrating over probability distributions [19]. This motivates the use of binomial performance metrics. For the robust control benchmark problem, three performance metrics are considered: probability of instability (P_I), probability of exceeding a 15 s settling-time (P_{TS}), and probability of exceeding a maximum control magnitude of 1 (P_u) [14].

Mathematically verifying nonlinear system stability is difficult, if not impossible for the general case. In reference [13], Cohen et al. developed an approximate map which gave bounds for 'stable' damping values in terms of the unknown spring stiffness. If the observer or virtual parameters (other than damping) were changed, this map would cease to be applicable, and an alternative check would be required. In this work, a map of stable regions is not used because it is not easily generated or understood for the general case. Instead, the more general and readily applicable SRA is used. Settling time was defined here to mean the time after which both the 1st and 2nd cart's positions remained less than 0.1 for a 20 s time horizon. Both settling time and control exceedance are easily checked.

2.3 Genetic Training

Both the membership functions (MFs) and rule base of the damping coefficient FIS are trained via GA. MFs defined over the 2 inputs ($\hat{x}_1$ and $\hat{x}_2$) are triangular or shoulder type with domain [-1, 1]. The edges of the leftmost and rightmost MF for each input are fixed at the boundaries of the domain. The left and right edges of a MF coincide with the centers of neighboring MFs to simplify training and avoid degenerate cases. This leaves 1 tunable parameter per MF. The FIS is Takagi-Sugeno type so no MFs are specified for the output [20]. The rule base provides a consequent for each pairwise combination of input 1 MFs and input 2 MFs. Rule consequents are constant values. With 5 MFs specified for each input, the chromosome is specified as a list of 35 floating point numbers. With 7 MFs per input, the list length increases to 63.

The GA operates with population size 40, 95% crossover probability, 1% mutation probability, and 5% elitism rate [16]. Fitness is defined as a linear combination of the squared performance metrics defined previously (Eq. 3). Squared probability estimates were originally used in reference [21] to penalize greater probabilities significantly more than lesser probabilities. The coefficients of the squared probability estimates were obtained by trial-and-error and found to perform well. The probability estimates used during training are obtained by Monte Carlo simulations with 100 runs per solution. The unknown stiffness parameter is sampled randomly from a uniform distribution in each simulation. Training is carried out as in Fig. 4.

$$fitness = -10^5 P_I^2 - 10^3 P_{TS}^2 - 10^4 P_u^2 \# \quad (3)$$

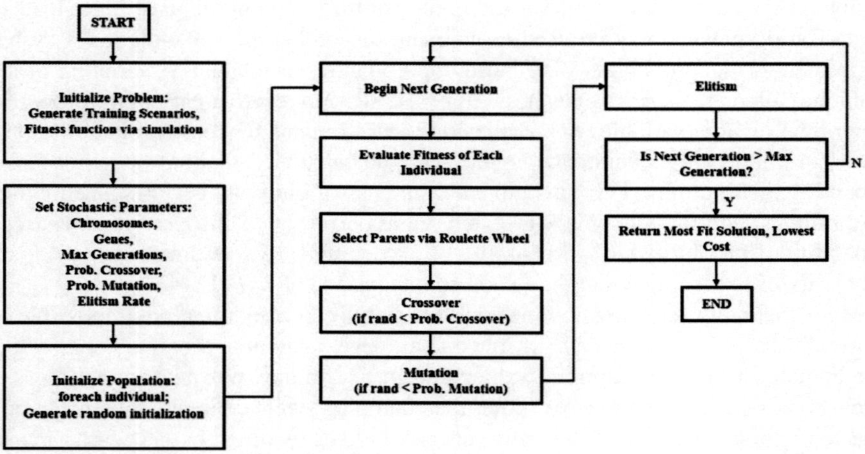

Fig. 4. Genetic Training Operation

3 Results and Discussion

Initially, a FIS is trained with 5 MFs defined for each input variable. The GA converged to a good enough solution after less than 100 generations. The trained membership functions and control surface are shown in Figs. 5 and 6. The rulebase is shown in Table 1. The control surface is jagged and could be further refined but was found to be acceptable for the purpose of illustration. Simulated displacement responses and control histories for an impulse disturbance are shown in Fig. 7 for several values of the spring stiffness. SRA with 1000 simulations provide the following probability estimates and

95% confidence intervals: $P_I = 0$ with interval [0, 0.0030], $P_{TS} = 0.099$ with interval [0.0839, 0.1159], and $P_u = 0$ with interval [0, 0.0030]. This is an improvement over the robustness of other controllers A-K listed in Table 2.

Training with modifications was subsequently performed. Two modifications were considered individually: (1) 7 MFs per input and (2) training which also penalizes total control effort. The 1st modification was expected to readily improve fitness at the expense of longer training. The 2nd modification was expected to reduce the jaggedness in the control surface and thereby improve robustness. Controller L in Table 2 corresponds to the 7 MF per input controller. Sample control surfaces for the 7 MF system and 5 MF 'control penalty' system are shown in Figs. 8 and 9. The control surface of the 7 MF system is less smooth than the initial controller, while slightly improving robustness by decreasing $P_{TS} = 0.081$. Improved robustness over other controllers in Table 2 comes at the cost of increased nominal settling time and control effort. The robustness of the control-penalized system is worse than the 7 MF system but better than the initial 5 MF system, providing $P_{TS} = 0.095$.

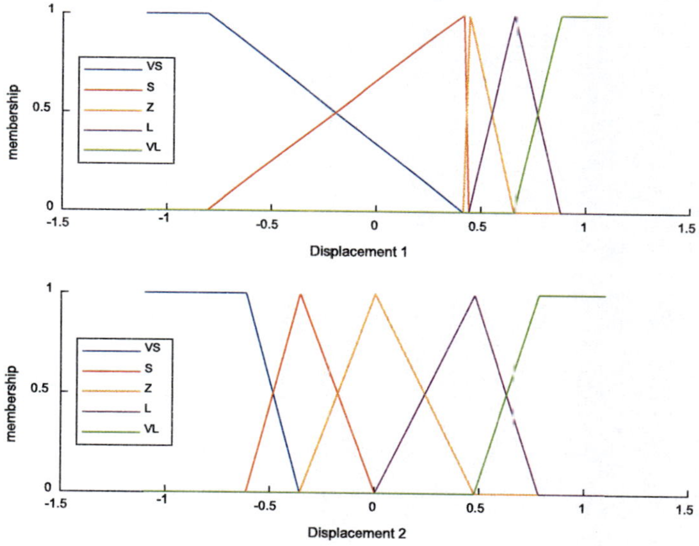

Fig. 5. Best-Fit Membership Functions, 5 MF/input

Table 1. Best-fit Rulebase, 5 MFs/input

Input 2	Input 1				
	VS	S	Z	L	VL
VS	0.1057	1.4013	1.3141	0.5556	1.0755
S	0.9971	1.2009	0.6642	0.9588	1.3324
Z	0.8938	1.2402	1.6988	1.7110	1.0840
L	1.3250	0.7883	1.1588	2.0000	1.4246
VL	1.2881	1.2651	1.2733	1.2427	1.9749

Table 2. Comparison of 7 MF/input controller (L) with other Controllers from Literature (based on references [13, 14])

Controller Description	Design	Nominal Settling Time (s)	Nominal Control Effort	P_I	P_{TS}	P_u
Fixed-order compensators achieving approximate loop-transfer recovery	A	21.0	0.514	0.160	0.971	0.160
Same basic design as A	B	19.5	0.469	0.023	1.000	0.023
Same basic design as A	C	19.7	0.468	0.021	1.000	0.021
H_∞	D	9.9	297.8	0.000	0.000	1.000
Nonlinear constrained optimization	E	18.2	0.884	0.000	1.000	0.000
Structured covariance terms added to linear quadratic Gaussian equations	F	13.7	2.397	0.000	0.633	1.000
Game theoretic controller based on linear exponential Gaussian and H_∞ concepts	G	31.3	1.458	0.000	1.000	1.000

(*continued*)

Table 2. (*continued*)

Controller Description	Design	Nominal Settling Time (s)	Nominal Control Effort	P_I	P_{TS}	P_u
$\boldsymbol{H_\infty}$ using internal model principle	H	14.9	0.574	0.000	0.742	0.000
Same basic design as H	I	17.8	0.416	0.000	0.756	0.000
Same basic design as H	J	43.2	1.047	0.039	1.000	0.857
Optimal control-inspired, fuzzy passive observer-based controller	K	8.8	0.53	0.000	0.468	0.042
Genetic fuzzy passive observer-based controller	L	13.0	0.64	0.000	0.081	0.000

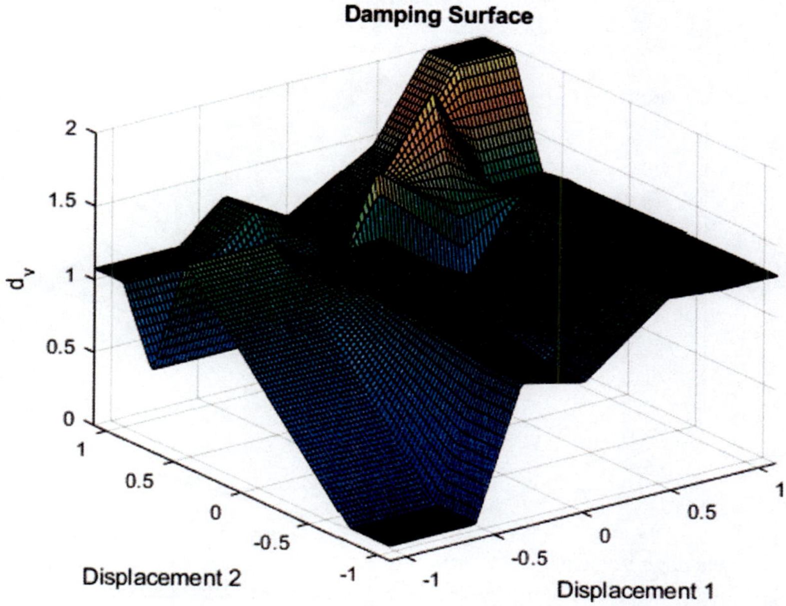

Fig. 6. FIS Control Surface, 5 MF/input

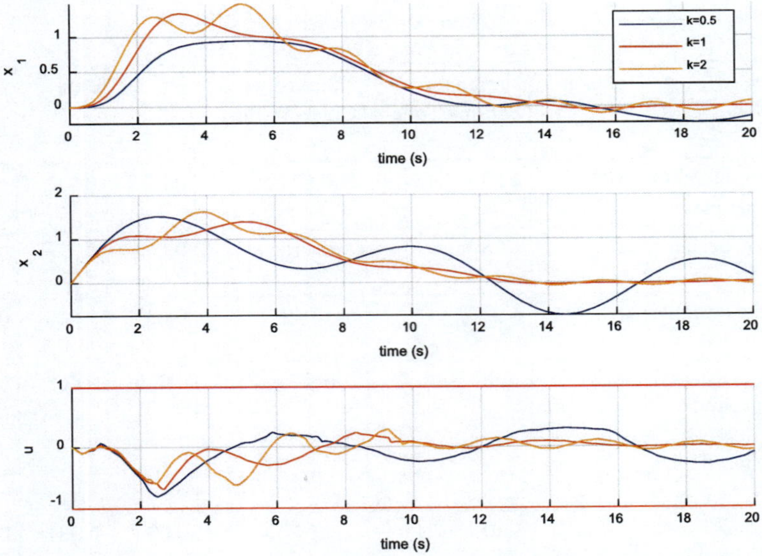

Fig. 7. Displacement and Control Histories due to an Impulse Disturbance, Several Spring Stiffnesses, 5 MF/input

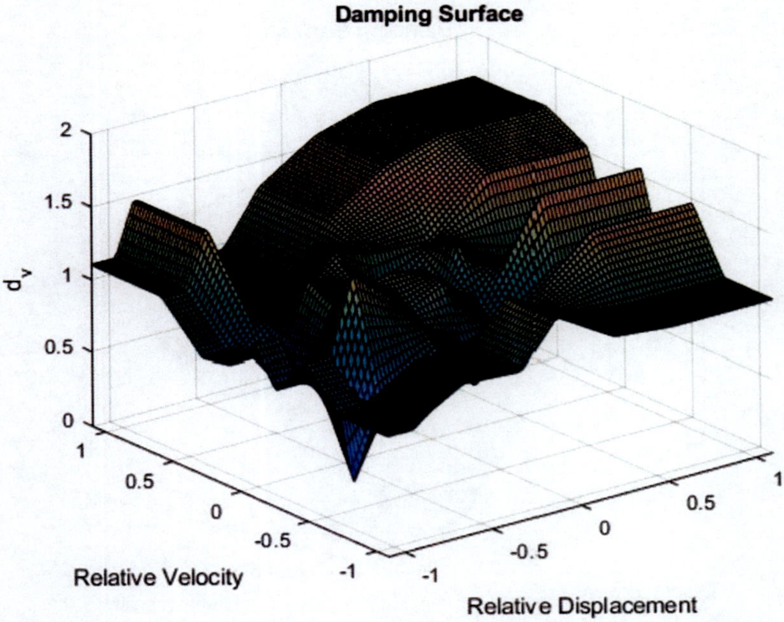

Fig. 8. FIS Control Surface, 7 MF/input

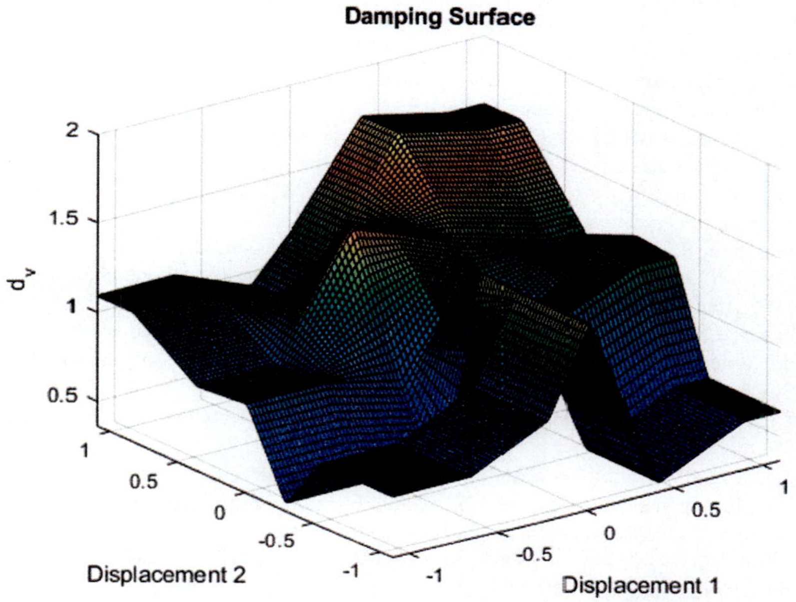

Fig. 9. FIS Control Surface, 5 MF/input with control penalized during training

4 Conclusions

Given a passivity-based system and several performance metrics, a FIS was trained via GA to vary a virtual damping coefficient. This FIS, which takes 2 displacements as input, demonstrated better robustness than the optimal control-inspired FIS of reference [13], which takes a single relative displacement and a single relative velocity as input. Principally, robustness was improved over all other controllers in Table 2. Improved robustness is achieved at the expense of worse nominal settling time and control effort. The arbitrary jaggedness of the produced control surfaces may be due to overfitting and can be partially remedied by penalizing control effort during training. The problem might also be remedied by training on a broader distribution of initial conditions, but this is left for future work. Finally, alternative virtual controllers may be tested for improved performance and robustness. Developing a more general controller which uses the full estimated state of the system to vary virtual damping would combine the controller developed in this work and the optimal control-inspired solution from [13]. This may allow both near-optimal performance and near-optimal robustness at the expense of additional controller complexity. This is left for future work.

Acknowledgements. This work would not have been completed without the guidance and instruction of Dr. Kelly Cohen as well as the support provided by the Jacob D. and Lillian Rindsberg Memorial Fund. All are sincerely appreciated.

References

1. Wang, M., et al.: Application of a new-type damping structure for vibration control in deployment process of satellite antenna component. J. Phys. Conf. Ser. **1748**(4), 042042 (2021). https://doi.org/10.1088/1742-6596/1748/4/042042
2. Durham, M.H., Keller, D., Bennett, R., Wieseman, C.: A status report on a model for benchmark active controls testing. In: 32nd Structures, Structural Dynamics, and Materials Conference (1991). https://doi.org/10.2514/6.1991-1011
3. Bisplinghoff, R.L., et al.: Aeroelasticity. Dover Publications, Mineola (1996)
4. Hoshyari, S., et al.: Vibration-minimizing motion retargeting for robotic characters. ACM Trans. Graph. **38**(4), 1–14 (2019). https://doi.org/10.1145/3306346.3323034
5. Salisbury, J.K.: Active stiffness control of a manipulator in Cartesian coordinates. In: 19th IEEE Conference on Decision and Control Including the Symposium on Adaptive Processes (1980). https://doi.org/10.1109/cdc.1980.272026
6. Den Hartog, J.P.: Mechanical Vibrations. McGraw-Hill Book Company, Inc., New York (1934)
7. Keye, S., et al.: A vibration absorber with variable eigenfrequency for turboprop aircraft. Aerosp. Sci. Technol. **13**(4–5), 165–171 (2009). https://doi.org/10.1016/j.ast.2008.10.001
8. Juang, J.-N., Phan, M.: Robust controller designs for second-order dynamic systems - a virtual passive approach. In: 32nd Structures, Structural Dynamics, and Materials Conference (1991). https://doi.org/10.2514/6.1991-983
9. Morris, K.A, Juang, J.N.: Dissipative controller designs for second-order dynamic systems. Control Flex. Struct. 71–90 (1993). https://doi.org/10.1090/fic/002/03
10. Azadi, M.: Maneuver control and vibration suppression of a smart flexible satellite using robust passivity based control. Adv. Mater. Res. **488–489**, 1803–1807 (2012). https://doi.org/10.4028/www.scientific.net/amr.488-489.1803
11. Cohen, K., et al.: Active control of flexible structures using a fuzzy logic algorithm. Smart Mater. Struct. **11**(4), 541–552 (2002). https://doi.org/10.1088/0964-1726/11/4/309
12. Wie, B., Bernstein, D.S.: A benchmark problem for robust control design. In: American Control Conference (1990). https://doi.org/10.23919/acc.1990.4790876
13. Cohen, K., et al.: Control of linear second-order systems by fuzzy logic-based algorithm. J. Guidance Control Dyn. **24**(3), 494–501 (2001). https://doi.org/10.2514/2.4738
14. Stengel, R.F., Marrison, C.I.: Robustness of solutions to a benchmark control problem. J. Guidance Control Dyn. **15**(5), 1060–1067 (1992). https://doi.org/10.2514/3.20950
15. Sathyan, A., Ma, O.: Collaborative control of multiple robots using genetic fuzzy systems. Robotica **37**(11), 1922–1936 (2019). https://doi.org/10.1017/S0263574719000353
16. Oscar, C.: Genetic Fuzzy Systems: Evolutionary Tuning and Learning of Fuzzy Knowledge Bases, vol. 19. World Scientific, Singapore (2001)
17. Luenberger, D.G.: Observers for multivariable systems. IEEE Trans. Autom. Control **11**(2), 190–197 (1966). https://doi.org/10.1109/TAC.1966.1098323
18. Stengel, R.F., Ryan, L.E.: Stochastic robustness of linear time-invariant control systems. IEEE Trans. Autom. Control **36**(1), 82–87 (1991). https://doi.org/10.1109/9.62270
19. Stengel, R.F., et al.: Probabilistic evaluation of control system robustness. Int. J. Syst. Sci. **26**(7), 1363–1382 (1995). https://doi.org/10.1080/00207729508929105
20. Takagi, T., Sugeno, M.: Fuzzy identification of systems and its applications to moeling and control. IEEE Trans. Syst. Man Cybern. SMC-15(1), 116–132 (1985). https://doi.org/10.1109/TSMC.1985.6313399
21. Wang, Q., Stengel, R.F: Robust nonlinear flight control of a high-performance aircraft. IEEE Trans. Control Syst. Technol. **13**(1), 15–26 (2005). https://doi.org/10.1109/TCST.2004.833651. Accessed 21 Mar 2023

Everything is a Matter of Degree: The Main Idea Behind Fuzzy Logic is Useful in Geosciences and in Authorship

Christian Servin[1], Aaron Velasco[2], Edgar Daniel Rodriguez Velasquez[3,4], and Vladik Kreinovich[5](✉)

[1] Information Technology Systems Department, El Paso Community College (EPCC), 919 Hunter Dr., El Paso, TX 79915-1908, USA
cservin1@epcc.edu
[2] Department of Earth, Environmental, and Resource Sciences, University of Texas at El Paso, 500 W. University, El Paso, TX 79968, USA
aavelasco@utep.edu
[3] Department of Civil Engineering, Universidad de Piura in Peru (UDEP), Av. Ramón Mugica 131, Piura, Peru
edgar.rodriguez@udep.pe, edrodriguezvelasquez@miners.utep.edu
[4] Department of Civil Engineering, University of Texas at El Paso, 500 W. University, El Paso, TX 79968, USA
[5] Department of Computer Science, University of Texas at El Paso, 500 W. University, El Paso, TX 79968, USA
vladik@utep.edu

Abstract. This paper presents two applications of the general principle – the everything is a matter of degree – the principle that underlies fuzzy techniques. The first – qualitative – application helps explain the fact that while most earthquakes occur close to faults (borders between tectonic plates or terranes), earthquakes have also been observed in areas which are far away from the known faults. The second – more quantitative – application is to the problem of which of the collaborators should be listed as authors and which should be simply thanked in the paper. We argue that the best answer to this question is to explicitly state the degree of authorship – in contrast to the usual yes-no approach. We also show how to take into account that this degree can be estimated only with some uncertainty – i.e., that we need to deal with interval-valued degrees.

1 Formulation of the Problem

One of the main ideas behind fuzzy logic is Zadeh's idea that everything is a matter of degree. This idea has been very fruitful in many application areas; see, e.g., [1,4,9,12,13,16].

In this paper, we show that there are still many new areas where this idea can be successfully applied. Specifically, we show that it can help to solve puzzling questions in such diverse applications areas as geosciences and authorship.

K. Cohen et al. (Eds.): NAFIPS 2023, LNNS 751, pp. 219–227, 2023.
https://doi.org/10.1007/978-3-031-46778-3_20

2 Possible Application to Geosciences

Importance of earthquake studies. Earthquakes are among the most devastating events. Their effect depends on our preparedness. If we know that a certain area is prone to have earthquakes of certain magnitude, then we can strengthen all the buildings and structures, so as the minimize the earthquake's damaging effect. This strengthening is reasonably expensive, so it is only used when we are reasonably confident that earthquakes of such magnitude are possible.

Because of this, predicting the magnitudes and location of possible future earthquakes is one of the main objectives of geosciences.

Seismogenic zones: traditional approach to earthquake study. Up to the 19th century, scientists believed that continents remain the same. Then it turned out that continents – or, to be more precise, plates containing continents or parts of the continents – drift with time. This knowledge formed what is now called *plate tectonics*. It was noticed that most strong earthquakes – as well as most volcanos – occur in the borders between these plates.

Later on, it was found that plates themselves are not unchangeable – they consist of smaller pieces called *terranes* that can also drift with respect to each other. The vast majority of earthquakes occurs close to the *faults* – boundaries of terranes.

This is still the main approach to predicting earthquakes: researchers usually assume that the earthquakes occur only at the faults. So at the faults, they recommend engineering measures to mitigate the effect of possible earthquakes, while in the areas inside the terranes no such measures are recommended.

Recent discovery. A recent statistical analysis of earthquake records has shown that, contrary to the above-described traditional beliefs, earthquakes are not limited to the faults; see, e.g., [2]. Most earthquakes do occur at the faults, but there have been earthquakes in other areas as well: as we move away from the fault, the probability of an earthquake decreases but it never goes to 0.

What does this mean in terms of seismogenic zones? In the traditional approach, there was a clear (crisp) distinction between seismogenic zones where earthquakes are possible and other areas where earthquakes are not possible.

The recent discovery shows that earthquakes are possible literally everywhere. At first glance, this implies that the whole Earth is a seismogenic zone, but this would be a useless conclusion. There should be a difference between zones where earthquakes are frequent – e.g., near the major faults – and zones where earthquakes are so rare that it took several decades to notice them. In other words, some areas are clearly seismogenic zones, while other are barely seismogenic.

In other words, being a seismogenic zone is not a crisp property, it is a matter of degree: some areas are more seismogenic, some are less seismogenic. This is perfectly in line with the main idea that Zadeh placed in the foundation of fuzzy logic (see, e.g., [1,4,9,12,13,16]) – that everything is a matter or degree.

What is the physical meaning of this phenomenon. The traditional approach implicitly assumes that a fault is a line. In such a description, we can easily separate regions close to the line from regions which are far away from the line. A similar description was thought to hold when we describe visible cracks: e.g., cracks in rocks, cracks in pavement, etc. A more detailed analysis has shown that visible cracks actually have a fractal structure (see, e.g., [7,15]):

- there is a main crack line, along which the stress is high,
- at several points in the main line, it branches into then smaller-size crack lines, along which the stress is somewhat smaller;
- at several points in each of these "second-order" lines, the line itself branches into even smaller-size crack lines, with even smaller stress, etc.

Because of this structure, there is, in effect, no area completely without cracks and without stress:

- there are areas around the main fault line, in which the crack is the most visible and the stress is the highest;
- there are areas around the second-order fault lines, where the crack is less visible and the stress is somewhat lower;
- there are areas around the "third-order" fault lines, where the crack is even less visible and the stress is even lower, etc.,
- all the way to areas where cracks are microscopic and the stress is barely measurable.

This is what we directly observe in rock cracks, in the pavement cracks, and this is what we indirectly observe for earthquakes: earthquakes can appear everywhere, just in some areas they are more frequent and stronger, while in other areas they are less frequent and weaker. This means, in effect, that faults are everywhere, just in some areas they are larger and correspond to larger stress, while in other areas, they are weaker and the corresponding stress is smaller.

In other words, for this phenomenon, physics is in perfect agreement with Zadeh's principle – at least on the qualitative level.

3 Possible Application to Authorship

Formulation of the problem. In the past, most researchers worked on their own, and most papers had just one author. Nowadays, research is often performed by big groups of researchers: some of them make significant contributions to the research results, while contributions of others are not as significant. So, a question naturally appears: when the results of this joint research are formulated in a paper, who should be included in the list of the paper's authors? This is a subject of many serious discussions, see, e.g., [14].

Why is this a difficult problem? In our opinion, the problem is difficult because there is no crisp, discrete separation between authors on the one hand and contributors who end up being thanked (but who are not listed as authors) on the other hand. In each group, there is an implicit threshold, so that participants whose contribution level is above this threshold are listed as authors, while those whose level of contribution is below this threshold are not. This threshold level varies between different research communities, between different research groups: e.g., many experimental papers have dozens of authors, while most theoretical papers usually have much fewer ones. And within each group, there is a certain level of subjectivity.

Being an author in several papers is critically important for students to defend their dissertations, important for job search, for promotion. As a result, the degree of subjectivity in deciding who is listed as an author (and who is not) often causes conflicts within the research groups – and these conflicts hinder possible collaboration and thus, in the long run, slow down the progress of science. How can we avoid this subjectivity?

What we propose. We propose to take into account that being an author is, in effect, not a crisp notion. In many cases, it is a matter of degree. So instead of listing some collaborators as authors and others as non-authors, why not list everyone who contributed something intellectual to the result as authors – but with the corresponding degree of authorship?

To some extent, this is already done in some journals – where for each submitted paper, the authors have to agree on percentages of their contributions. But at present, this is only done with respect to participants who have already been declared authors. We propose to extend this idea to all the participants, including those who are usually not included in the authors' list.

Of course, for this idea to work, we need to take into account this degree of authorship when evaluating the quality of a student's dissertation work, or the quality of the researcher. We believe that this – yet another – example of using the above-mentioned Zadeh's principle will help resolve this issue.

How to assign the degree of authorship: main idea. For this idea to work, we need to have an acceptable way to assign degrees of authorship. In some cases, the authoring group includes a leader whose opinion everyone respects. In such cases, we can simply ask this trusted leader to provide the degrees of authorship.

However, the very fact that often conflicts appear around this issue means that in many cases, people's opinions differ. In such cases, a natural idea is to ask different participants of the research group to provide such estimates – and then we need to come up with combined estimates that take into account all the opinions.

How to assign the degree of authorship: first approximation. For every two participants i and j, let us denote the degree assigned to the participant i by the participant j by d_{ij}.

In the beginning, we do not know a priori who contributed more – and thus, whose opinion is more informed and more valuable. In the first approximation, we can therefore simply take the average of all the assigned degrees. In other words, to compute the first approximation to the authorship degree assigned to each participant i, we can do the following:

- first, we compute the average values

$$a_i^{(1)} = \frac{1}{n-1} \cdot \sum_{j \neq i} d_{ij}, \tag{1}$$

where n is the number of contributors;
- then, we normalize these values, to make sure the resulting degrees $d_i^{(1)}$ add up to 1:

$$d_i^{(1)} = \frac{a_i^{(1)}}{\sum\limits_{j=1}^{n} a_j^{(1)}}. \tag{2}$$

A natural example shows that we need to go beyond the first approximation. Let us consider a simple case when three people worked on a project:

- Professor Einstein ($i = 1$) came up with the great design idea,
- Engineer Edison ($i = 2$) transformed this idea into the actual step-by-step design, and
- a skilled worker Mr. Dexterous ($i = 3$) actually design this device – which worked exactly as Professor Einstein expected.

How should we allocate authorship of the resulting paper?

- Professor Einstein understands that while his was the main idea, this idea would not have been implemented without the ingenuity of the engineer and the skills of the worker. In his opinion, the engineer's task was clearly more creative, so he assigns, to the engineer, the weight $d_{12} = 0.2$ and to the worker the weight $d_{13} = 0.1$ – thus implicitly assuming that his own contribution was 70%.
- Engineer Edison largely agrees with this assessment, so he assigns $d_{21} = 0.7$ and $d_{23} = 0.1$.
- On the other hand, Mr. Dexterous did not communicate with Professor Einstein at all, all he saw was a great design given to him by the engineer. While the engineer have probably praised Professor Einstein's contribution, Mr. Dexterous attributes this praise to engineers' modesty. As many other people, Mr. Dexterous believes that academicians are talking a lot and getting too much praise for their mostly impractical (thus, largely useless) ideas, while engineers (and, to some extent, workers) are the ones who contribute to the society's progress. So, he assigns, to Professor Einstein, the same small degree as to himself $d_{31} = 0.1$, while he assigns the rest of the degree to the engineer: $d_{32} = 0.8$.

As a result of taking the average, we get

$$a_1^{(1)} = \frac{0.7 + 0.1}{2} = 0.4, \quad a_2^{(1)} = \frac{0.2 + 0.8}{2} = 0.5, \quad a_3^{(1)} = \frac{0.1 + 0.1}{2} = 0.1.$$

These averages happen to add up to 1, so after normalization, we get the exact same degrees: $d_1^{(1)} = 0.4$, $d_2^{(1)} = 0.5$, and $d_3^{(1)} = 0.1$.

So now it looks like the engineer was the major contributor to the project. This is not right.

How can we get more adequate estimates: idea. The problem with the above first-approximation estimate is that in this estimate, the opinion of someone whose contribution to the paper was very small was given the same weight as the opinion of the major contributors. We should give more weight to the opinions of major contributors and less weight to the opinions of minor contributors.

A natural idea is to use the degree of authorship as this weight. Of course, we do not yet know this degree – the whole purpose of this procedure is to come up with such a degree. However, we do know approximate values of these degrees, so let us use them as weight. This way, we can get adjusted – hopefully more adequate – degrees. Thus, we arrive at the following procedure.

How can we get more adequate estimates: algorithm. Formulas (1) and (2) show how to compute the degrees $d_i^{(1)}$ corresponding to the first approximation. Based on these degrees, we can compute the next approximation values $d_i^{(2)}$ as follows:

- first, we compute the weighted averages

$$a_i^{(2)} = \frac{\sum\limits_{j \neq i} d_i^{(1)} \cdot d_{ij}}{\sum\limits_{j \neq i} d_i^{(1)}}, \tag{3}$$

- then, we normalize these values, to make sure the resulting degrees $d_i^{(2)}$ add up to 1:

$$d_i^{(2)} = \frac{a_i^{(2)}}{\sum\limits_{j=1}^{n} a_j^{(2)}}. \tag{4}$$

Example. In the above example, we get:

$$a_1^{(2)} = \frac{0.5 \cdot 0.7 + 0.1 \cdot 0.1}{0.5 + 0.1} = \frac{0.36}{0.6} = 0.6,$$

$$a_2^{(2)} = \frac{0.4 \cdot 0.2 + 0.1 \cdot 0.8}{0.4 + 0.1} = \frac{0.16}{0.5} = 0.32, \text{ and}$$

$$a_3^{(2)} = \frac{0.4 \cdot 0.1 + 0.5 \cdot 0.1}{0.4 + 0.5} = \frac{0.09}{0.9} = 0.1.$$

These degrees add up to 1.02, so normalization leads to

$$d_1^{(2)} \approx 0.59, \quad d_2^{(2)} \approx 0.31, \quad d_3^{(2)} \approx 0.10.$$

Good news is that:

- we now recognize Professor Einstein as the main author, and
- the corresponding degrees are very close to the values $d_1 = 0.7$, $d_2 = 0.2$, and $d_3 = 0.1$ agreed upon by the two major contributors.

Algorithmic Comment. If needed, we can iterate further: once we know the degrees $d_i^{(k)}$ corresponding to the k-th approximation, we can compute the next approximation values $d_i^{(k+1)}$ as follows:

- first, we compute the weighted averages

$$a_i^{(k+1)} = \frac{\sum\limits_{j \neq i} d_i^{(k)} \cdot d_{ij}}{\sum\limits_{j \neq i} d_i^{(k)}}, \tag{5}$$

- then, we normalize these values, to make sure the resulting degrees $d_i^{(k+1)}$ add up to 1:

$$d_i^{(k+1)} = \frac{a_i^{(k+1)}}{\sum\limits_{j=1}^{n} a_j^{(k+1)}}, \text{ etc.} \tag{6}$$

General Comment. What we propose is similar to the iterative process that leads to PageRank – a numerical criterion that Google search uses to rank possible answers to queries; see, e.g., [6]. Crudely speaking, the PageRank algorithm boils down to the following.

In the first approximation, we view the importance $a_i^{(i)}$ of a webpage by the number of other pages j that link to it ($j \to i$):

$$a_i^{(1)} = \sum_{j:j \to i} 1.$$

In the next approximation, we take into account that the linking pages have, in general, different importance, so we use this importance as a weight. We do not yet know the actual importance, so we use approximate importance values obtained on the previous step:

$$a_i^{(2)} = \sum_{j:j \to i} a_j^{(1)}.$$

If needed, we can continue this procedure: once we know the k-th approximation, we can compute the next approximation values as

$$a_i^{(k+1)} = \sum_{j:j \to i} a_j^{(k)}.$$

What if we take into account the uncertainty of the degrees? It is difficult for people to come up with exact numbers d_{ij} describing contributions of others: this is very subjective, and we do not think that it is possible to distinguish between, e.g., 50% and 51%. People are much more comfortable providing a range $[\underline{d}_{ij}, \overline{d}_{ij}]$ of possible values, such as $[0.6, 0.7]$.

In this case, in the first approximation, we come up with interval of possible values of $a_i^{(1)}$:

$$[\underline{a}_i^{(1)}, \overline{a}_i^{(1)}] = \frac{1}{n-1} \cdot \sum_{j \neq i} [\underline{d}_{ij}, \overline{d}_{ij}],$$

where by sum – or any other operation $\oplus$ – between intervals we mean the range of the values $a \oplus b$ when a and b lie in the corresponding intervals $[\underline{a}, \overline{a}]$ and $[\underline{b}, \overline{b}]$ (see, e.g., [3,5,8,10]):

$$[\underline{a}, \overline{a}] \oplus [\underline{b}, \overline{b}] \stackrel{\text{def}}{=} \{a \oplus b : a \in [\underline{a}, \overline{a}], b \in [\underline{b}, \overline{b}]\}.$$

Once we have an approximation $[\underline{a}_i^{(1)}, \overline{a}_i^{(1)}]$, we need to compute the intervals of possible values of the normalized degrees. Each normalized degree (2) is a fraction of two expressions which are linear in $a_i^{(1)}$. There exists an efficient algorithm for computing this range – see, e.g., [11]. This algorithm is, in effect, what is used when we extend centroid defuzzification to the interval-valued fuzzy case; see, e.g., [9].

Once we have the interval-valued degrees $d_i^{(1)}$, we can take into account that the expression (3) is monotonic in d_{ij}. Thus:

- to find the largest possible value of $a_i^{(2)}$, it is sufficient to consider the upper bound $\overline{d}_{ij}$, while
- to find the smallest possible value of $a_i^{(2)}$, it is sufficient to consider the lower bound $\underline{d}_{ij}$.

Once we fix the values d_{ij} this way, the formula (3) also becomes fractionally linear, so we can use the same algorithm to compute the interval of possible values of $a_i^{(2)}$, etc.

Acknowledgments. This work was supported in part by the National Science Foundation grants 1623190 (A Model of Change for Preparing a New Generation for Professional Practice in Computer Science), HRD-1834620 and HRD-2034030 (CAHSI Includes), and by the AT&T Fellowship in Information Technology.

It was also supported by the program of the development of the Scientific-Educational Mathematical Center of Volga Federal District No. 075-02-2020-1478, and by a grant from the Hungarian National Research, Development and Innovation Office (NRDI).

The authors are greatly thankful to the anonymous referees for valuable suggestions.

References

1. Belohlavek, R., Dauben, J.W., Klir, G.J.: Fuzzy Logic and Mathematics: A Historical Perspective. Oxford University Press, New York (2017)
2. Fan, W., et al.: Very low frequency earthquakes in between the seismogenic and tremor zones in Cascadia?. AGU Adv. **3**(2), e2021AV000607 (2022)
3. Jaulin, L., Kiefer, M., Didrit, O., Walter, E.: Applied Interval Analysis, with Examples in Parameter and State Estimation, Robust Control, and Robotics. Springer, London (2001). https://doi.org/10.1007/978-1-4471-0249-6
4. Klir, G., Yuan, B.: Fuzzy Sets and Fuzzy Logic. Prentice Hall, Upper Saddle River (1995)
5. Kubica, B.J.: Interval Methods for Solving Nonlinear Contraint Satisfaction, Optimization, and Similar Problems: from Inequalities Systems to Game Solutions. Springer, Cham (2019). https://doi.org/10.1007/978-3-030-13795-3
6. Langville, A.N., Meyer, C.D.: Google's PageRank and Beyond: The Science of Search Engine Rankings, Princeton University Press. Princeton, New Jersey, and Oxford, UK (2012)
7. Mandelbrot, B.B.: Fractals: Form, Chance and Dimension. Freeman Co., New York (2020)
8. Mayer, G.: Interval Analysis and Automatic Result Verification. de Gruyter, Berlin (2017)
9. Mendel, J.M.: Uncertain Rule-Based Fuzzy Systems: Introduction and New Directions. Springer, Cham (2017). https://doi.org/10.1007/978-3-319-51370-6
10. Moore, R.E., Kearfott, R.B., Cloud, M.J.: Introduction to Interval Analysis. SIAM, Philadelphia (2009)
11. Nguyen, H.T., Kreinovich, V., Wu, B., Xiang, G.: Computing Statistics under Interval and Fuzzy Uncertainty. Springer, Berlin (2012)
12. Nguyen, H.T., Walker, C.L., Walker, E.A.: A First Course in Fuzzy Logic. Chapman and Hall/CRC, Boca Raton (2019)
13. Novák, V., Perfilieva, I., Močkoř, J.: Mathematical Principles of Fuzzy Logic. Kluwer, Boston (1999)
14. Parsons, M.A., Katz, D.S., Langseth, M., Ramapriyan, H., Ramdeen, S.: Credit where credit is due. EOS: Science News from the American Geophysical Union, 103(11), 20–23 (2022)
15. Vallejo, L.E.: Fractal analysis of the cracking and failure of asphalt pavements. In: Proceedings of the 2016 Geotechnical and Structural Engineering Congress, Phoenix, Arizona (2016)
16. Zadeh, L.A.: Fuzzy sets. Inf. Control **8**, 338–353 (1965)

Comparison of Explanation Methods for Genetic Fuzzy Trees for Wine Quality Predictions

Timothy Arnett(✉), Nicholas Ernest, and Zachariah Phillips

Thales Avionics, Inc., Cincinnati, OH 45242, USA
{tim.arnett,nick.ernest,zach.phillips}@defense.us.thalesgroup.com

Abstract. As advanced AI and Machine Learning models are more widely adopted, explanations about what the models learned from the data are desirable in order for human users to be able to interpret, verify, and trust the outputs. There are many different types of Machine Learning models, ranging in human interpretability from easy-to-understand *if-then* statements to large Neural Network-based models with complicated architectures that are effectively black boxes. A large focus in Explainable AI has been to develop methods to generate explanations about what these systems learned *ex post facto*, as many popular models and techniques lack inherent interpretability. Fuzzy-based Systems such as Genetic Fuzzy Trees, can offer inherent explanatory power due to their interpretable nature, while still retaining powerful approximation capabilities. In this work, we train a Genetic Fuzzy Tree on a regression dataset and then compare its own inherent explanations against explanations generated by the Local Interpretable Model-Agnostic (LIME) technique. We show that the inherent explanations given by the Fuzzy Tree are richer and more easily interpreted by human users.

1 Introduction

Explainable AI has been a large focus in recent years [1–4] due to the rapid development and deployment of AI systems trained with Machine Learning (ML). Oftenly these models are opaque to a human designer due to their internal operations and characteristics. Several popular ML architectures and techniques, such as Neural Networks, are effectively black boxes that cannot be investigated to ensure correct functionality or explain errant or interesting outputs that are unexpected. Although several methods exist to try to examine the effect of inputs on the output values, this is typically the extent of *ex post facto* explanations and does not fully explain each step in the model. On the other hand, there are ML models that are inherently interpretable such as Decision Trees. However, most inherently interpretable models lack rich approximation capabilities, and sometimes cannot be effectively applied when it comes to complex problems such as Reinforcement Learning (RL). Genetic Fuzzy Trees (GFTs) [5] were introduced in 2015 and are a happy middle ground in their ability to approximate highly

K. Cohen et al. (Eds.): NAFIPS 2023, LNNS 751, pp. 228–240, 2023.
https://doi.org/10.1007/978-3-031-46778-3_21

nonlinear functions, be applicable and scalable in complex Reinforcement [6] and Supervised Learning [7] problems, and be transparent, interpretable, and explainable to a human operator/designer. GFTs combine the approximation capabilities of Fuzzy Inference Systems with a network or tree-like structure that vastly reduces the number of parameters needed, along with Genetic Algorithms (GA) that can be applied to a wide range of problems. This work uses Thales' GFT toolkit [8] which combines Fuzzy Trees along with a state-of-the-art, GA-based optimizer named EVE that represent the best current software implementation of GFTs.

In this work, a GFT is created and trained on a dataset to predict red wine quality based on 11 input features. The dataset is part of the UCI repository [9]. The GFT is then compared to other ML models trained on the same dataset as a baseline. The trained GFT model is then queried for particular inputs to get explanations for why it predicts particular wine quality outputs. This is then compared to explanations given by the open source package LIME [10] which can give explanations about outputs for a given input to any arbitrary ML model. The results are compared and discussed in the context of quality of explanations and ease of interpretation by a human operator.

The remainder of the paper is organized as follows. Section 2 gives details on the GFT used in the work and soem details of the LIME package. Section 3 shows the results of the GFT learning process as well as comparisons to other common ML models on the data. Section 4 shows the explanations given for selected data instances by the GFT model and LIME. Finally, Sect. 5 concludes with a discussion of the results and opportunities for future work.

2 GFT Creation and Initialization

This section details the architecture, operations, and training methods of the Genetic Fuzzy Tree used to make wine quality predictions.

2.1 GFT Structure

The two primary methods, among others, to determine the structure of Thales GFTs are self-organization by optimization methods and manual definintion by human designers. There are pros and cons to each method. In the case of self-orgranization, the major benefit is that no *a priori* expert/domain knowledge is needed in order to create the structure, and the placement of individual FISs and their inputs are left to an optimization routine. This allows for maximum flexibility to learn what the structure and input ordering should be from the machine learning process, but the downside is that the connections between the FISs do not necessarily have a semantic meaning. This means that although the outputs of individual FISs can be explained by their rules and membership functions, the explanation likely cannot be traced back through the structure to the top most level such that it is interpretable and meaningful to humans.

In constrast to this, a structure can also be defined by a human designer with even partial knowledge of the domain in question. In this work, a GFT is created manually by associating inputs in FISs together based on their contributions to human understandable terms related to wine quality. This was sourced from articles detailing wine experts opinions on the contributions of each to overall wine quality. An example of this is flavor. The flavor of wine, although famously complex, in the context of this problem is mainly due to the saltiness, sweetness, and acidity of the wine. Therefore, grouping together relevant inputs to generate a latent variable *flavor* makes both semantic sense as well as capturing likely correlations in the inputs. This latent variable can then be used to propagate explanations through the tree such that the output is understood in terms of any set of input variables desired, including latent variables. The resulting GFT architecture chosen for this problem is shown in Fig. 1.

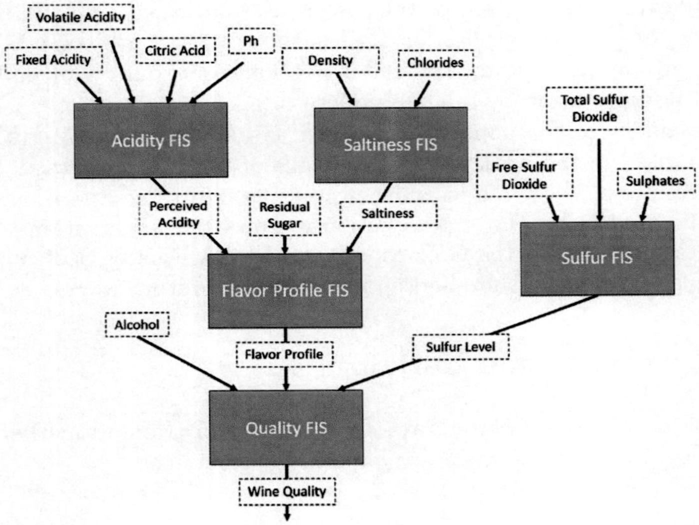

Fig. 1. Genetic Fuzzy Tree architecture for wine quality prediction

This architecture was chosen due to its simplicity, its human understandability, and its grouping of inputs based on expert knowledge of wine characteristics. Each FIS in the GFT is a zero-order Takagi Sugeno type, using Ruspini partitioning and triangular membership functions. Each input has three membership functions. The defuzzification method is weighted average (WAV) on all FISs except the Quality FIS, which uses Mean-of-Maxima (MOM). This allows for regression/continuous values in the input space while somewhat discretizing the output space. Note that the output set of membership functions for the Quality FIS is a function of the input membership functions and is not limited to a fixed set of membership functions *a priori*. WAV defuzzification, normally the default option for regression, was implemented as well during development with similar results, but MOM performed slightly better and allows for marginally more interpretable results. Note that different configurations and numbers of membership functions were evaluated, and this configuration had the best combination of simplicity, performance, and explainability.

2.2 Explanation Methods

2.2.1 LIME

LIME is a package of techniques for generating explanations for any given black box ML model. It works by creating a local, linearized model using a data instance and perturbing it to get different model outputs. It then uses this local, linearized model in order to explain the impact of each input value to the output. This is visualized as a bar graph plot as shown in the example in Fig. 2.

This explanation shows the relative contributions of the relevant input variables to the current output value of the model. However, a downside of LIME is that it only works on specific data instanes (although other variants exist) and creates linear approximations of the model that may not fully accurately capture local behavior in highly complex ML models.

2.2.2 GFT Explanations

In order to generate explanations, the input membership functions and rules are used in order to create linguistic sentences for each FIS based on which rule had the most impact on the output. This is performed by using the activation level of the input membership functions along with membership function labels to construct the sentence. Thales GFTs can output multiple levels of depth of explanations, in this work they are limited to the activated membership functions

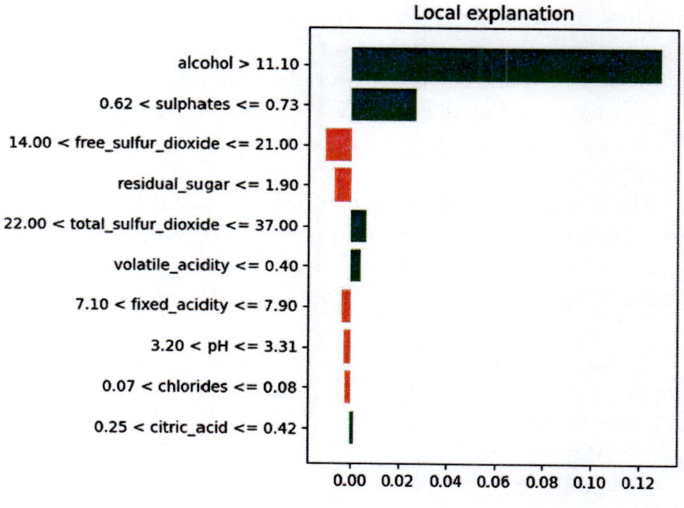

Fig. 2. LIME explanation example showing output impact values

and in some cases their activation levels. An example explanation for the Quality FIS as shown in the structure shown in Fig. 1 is as follows.

- quality_fis: wine_quality=0.6 because alcohol is 82% "Medium" and flavor_profile is 97% "Ok" and sulfur_level is 60% "Medium"

Note that the other FISs are included in the explanations such that every variable (including latent variables) are given. This is shown later in Sect. 4.

3 Supervised Learning Results

In order to train the GFT, the data was split into train and test sets. This was done such that the target variable was somewhat balanced between the train and test sets for each value. A table showing the count of each value is show in Table 6. Note that the train test ratio was 60:40. This was done to ensure better generalization (Table 1).

Table 1. Instances for each target variable value, y_t

y_t Values	3.0	4.0	5.0	6.0	7.0	8.0
Total Value Counts	10	53	681	638	199	18
Train Value Counts	6	32	408	383	119	11
Test Value Counts	4	21	273	255	80	7

The GFT was then trained over the training set using mean absolute error (MAE) as the loss function with early stopping if the test MAE didn't improve for a number of steps. The population was set to 1000 and ran for 47 generations. If early stopping occured, the parameters found during the best test set evaluation were saved. A visualization of the MAE vs. generations is shown in Fig. 3 where early stopping occured and the best GFT configuration was found at approximately generation 25.

The final MAE was 0.440302 on the training set and 0.4328125 on the test set. In order to establish a baseline for performance, a number of ML models from SKLearn were evaluated on the same train and test sets. Their MAE loss values are shown in Table 2 and the GFT had a lower loss than all other models evaluated.

Due to the fact that there are so few values at the extremes, the models had difficulty fitting to those and as such the errors were larger towards the boundaries of the output values. The distribution of MAE per y_t value for the GFT is shown in Fig. 4.

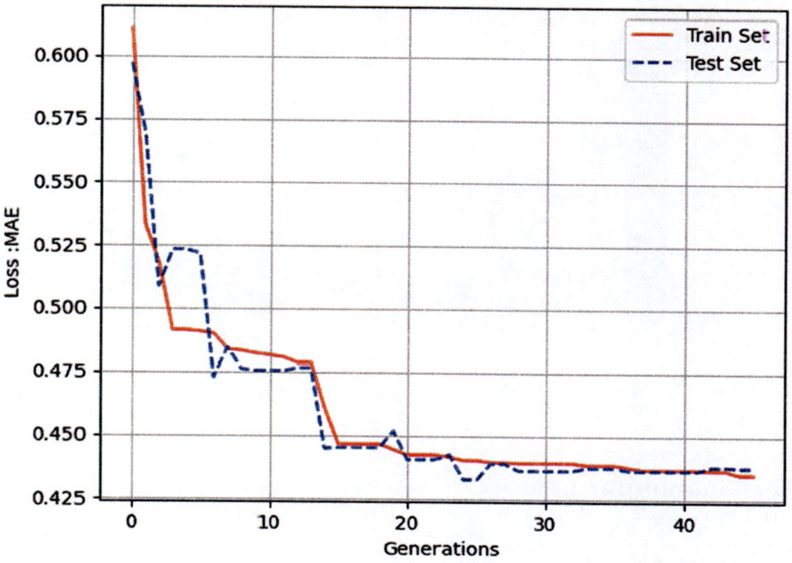

Fig. 3. MAE loss vs. number of generations during GFT training

Table 2. MAE on test set for SK-Learn ML models

ML Model	Test MAE
Linear SVR	0.551682
Radial SVR	0.624015
Polynomial SVR	0.594766
Linear Lasso	0.557889
Linear Bayesian Ridge	0.496017
Decision Tree	0.485938
NN MLPR	0.529352
Random Forest	0.450976

4 Explanations

Typically humans want explanations from ML models when they differ from what humans expect, or when they predict something of particular interest. However, due to the issues mentioned in the previous section with lack of data at the extremes, random selections from each of the non-extremem y_t values were selected (Fig. 5 and Table 3).

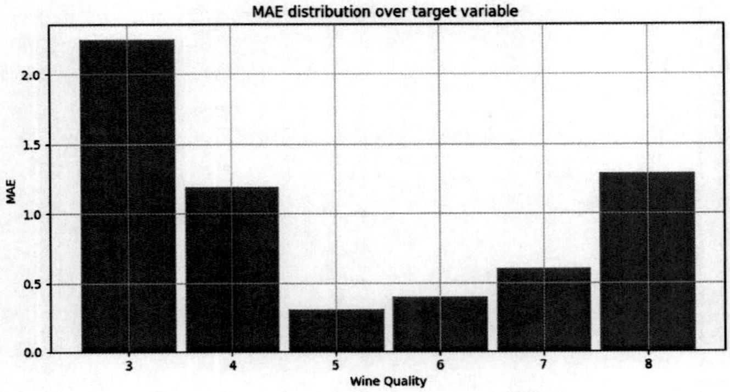

Fig. 4. MAE loss vs. target variable values

4.1 Explanation for $y_t = 4$

Table 3. Input values for $y_t = 4$ explanation

fixed acidity	volatile acidity	citric acid	pH	residual sugar	chlorides	alcohol	free sulfur dioxide	total sulfur dioxide	sulphates	density
6.9	1.09	0.06	3.51	2.1	0.061	11.4	12	31	0.43	0.9948

GFT Explanation

- **acidity_fis**: perceived_acidity=0.375033 because fixed_acidity is 64% "Low" and volatile_acidity is 65% "High" and citric_acid is 92% "Low" and ph is 75% "Medium"
- **saltiness_fis**: saltiness=0.524301 because density is 87% "Low" and chlorides is 51% "Low"
- **flavor_profile_fis**: flavor_profile=0.519593 because perceived_acidity is 66% "Medium" and residual_sugar is 91% "Low" and saltiness is 59% "Medium"
- **sulfur_fis**: sulfur_level=0.409858 because free_sulfur_dioxide is 70% "Low" and total_sulfur_dioxide is 85% "Low" and sulphates is 69% "Low"
- **quality_fis**: wine_quality=5.0 because alcohol is 50% "Low" and flavor_profile is 65% "Ok" and sulfur_level is 78% "Medium"

The LIME model predicted a value of 5.96 and the GFT predicted a value of 5 which highlights one of the issues with LIME. As it creates a local linear model to approximate black box functions, it may have difficulty capturing regions with highly nonlinear behavior. However, this predicted value is off by a single unit and is not particularly surprising due to the large errors for this y_t value. It appears that LIME believes that the wine should be rated at $\tilde{6}$ though because of *alcohol* being greather than 11.10. It does not seem apparent to the authors that the best explanation for rating the wine this value would be from alcohol.

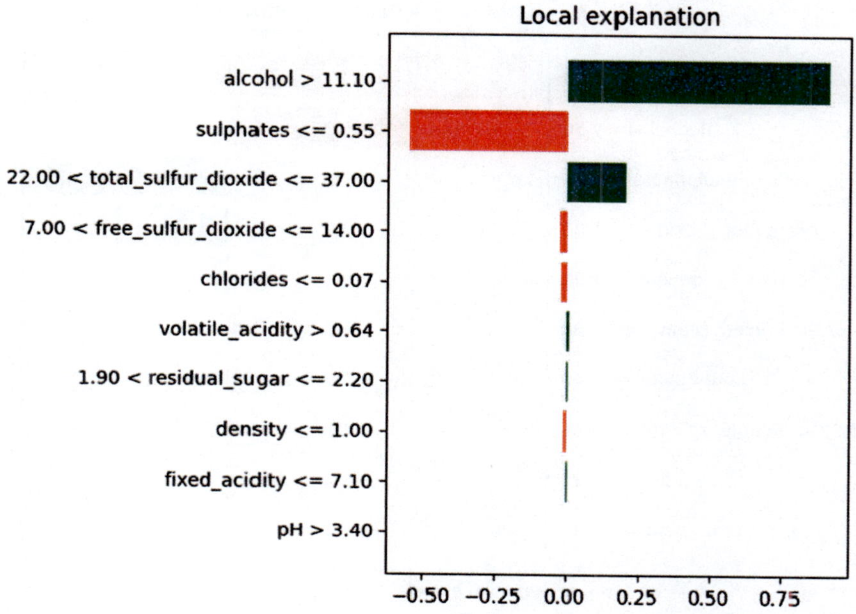

Fig. 5. LIME explanation plot for $y_t = 4$

In contrast, the GFT explanation seems more reasonable in that it rates it at 5.0 because of *flavor* having majority membership in the "Ok" membership set the and *sulfur_level* being "Medium". Interestingly, *alcohol* had equal membership in the "Low" and "Medium" sets (due to Ruspini partitioning) which is somewhat in conflict with the LIME explanation. Of course it still has a full rating error (5.0 instead of 4.0) but the explanation with the discrepancy may lead a human to interpret the model as giving an acceptable output (Fig. 6 and Table 4).

4.2 Explanation for $y_t = 5$

Table 4. Input values for $y_t = 5$ explanation

fixed acidity	volatile acidity	citric acid	pH	residual sugar	chlorides	alcohol	free sulfur dioxide	total sulfur dioxide	sulphates	density
8.9	0.62	0.18	3.16	3.8	0.176	9.2	52	145	0.88	0.9986

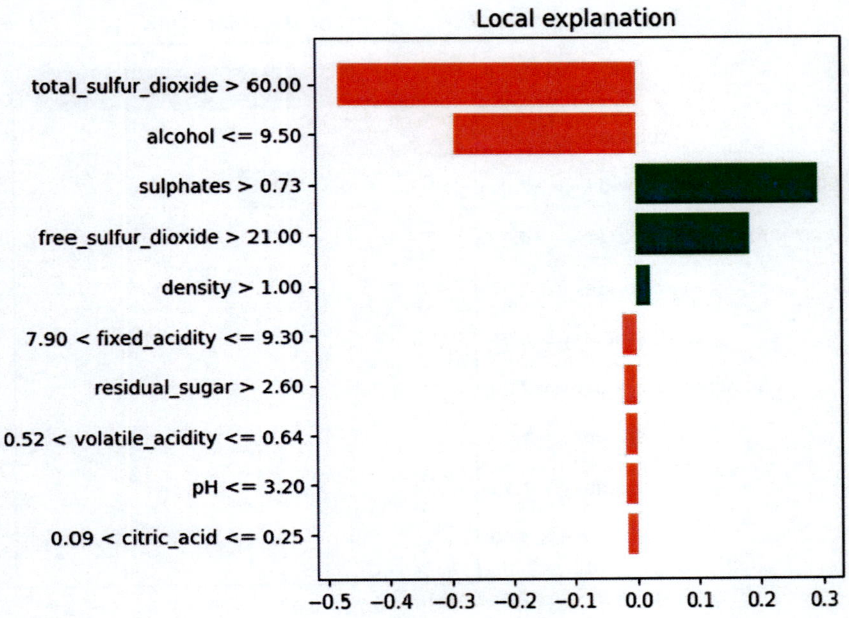

Fig. 6. LIME explanation plot for $y_t = 5$

GFT Explanation

- **acidity_fis**: perceived_acidity=0.280277 because fixed_acidity is 67% "Medium" and volatile_acidity is 69% "Medium" and citric_acid is 76% "Low" ph is 59% "Low"
- **saltiness_fis**: saltiness=0.394885 because density is 56% "Low" and chlorides is 88% "Medium"
- **flavor_profile_fis**: flavor_profile=0.607385 because perceived_acidity is 76% "Medium" and residual_sugar is 78% "Low" saltiness is 75% "Medium"
- **sulfur_fis**: sulfur_level=0.557008 because free_sulfur_dioxide is 59% "Medium" and total_sulfur_dioxide is 86% "Medium" sulphates is 83% "Medium"
- **quality_fis**: wine_quality=5.0 because alcohol is 87% "Low" and flavor_profile is 75% "Ok" sulfur_level is 59% "Medium"

In this case the GFT predicted the correct output value of $y_t = 5.0$. The LIME model predicted 5.36 so fairly accurate. This time it seems that the reason LIME wants to attribute a lower score to this wine is the high amount of sulfur dioxide and low alcohol amount - contrasting with the previous example. The GFT explanation howerever shows that although that is the case, it perceives the latent variable *flavor_profile* to be 75% "Ok" and perceives the sulfur levels to be 59% "Medium" (Fig. 7 and Table 5).

4.3 Explanation for $y_t = 6$

Table 5. Input values for $y_t = 6$ explanation

fixed acidity	volatile acidity	citric acid	pH	residual sugar	chlorides	alcohol	free sulfur dioxide	total sulfur dioxide	sulphates	density
7.7	0.18	0.34	3.37	2.7	0.066	11.8	15	58	0.78	0.9947

GFT explanation

- **acidity_fis**: perceived_acidity=0.248791 because fixed_acidity is 52% "Low" and volatile_acidity is 86% "Medium" and citric_acid is 55% "Low" and ph is 61% "Medium"
- **saltiness_fis**: saltiness=0.522277 because density is 85% "Low" and chlorides is 52% "Low"
- **flavor_profile_fis**: flavor_profile=0.528730 because perceived_acidity is 79% "Medium" and residual_sugar is 86% "Low" and saltiness is 59% "Medium"
- **sulfur_fis**: sulfur_level=0.622573 because free_sulfur_dioxide is 62% "Low" and total_sulfur_dioxide is 68% "Low" and sulphates is 90% "Medium"
- **quality_fis**: wine_quality=7.0 because alcohol is 56% "Medium" and flavor_profile is 66% "Ok" and sulfur_level is 50% "High"

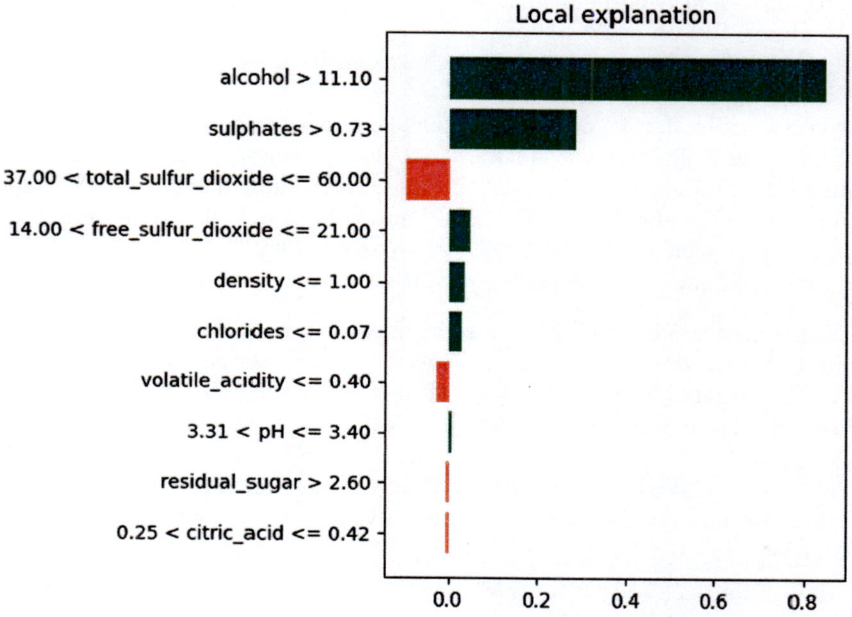

Fig. 7. LIME explanation plot for $y_t = 6$

In this case the GFT predicted even higher, a 7.0 compared to the 6.0 actual and the LIME model was fairly accurate again at 6.46. LIME heavily attributes this to *alcohol* being above 11.10 and sulphates being above 0.73. LIME outputs a seemingly reasonable explanation, although the GFT explanation says that the high rating is due to the *alcohol* level being mostly "Medium", but is bordering on "High", and the *flavor profile* being on the high end of "Ok". The GFT also agrees that the *sulfur level* is "High", and one thing of note is that one of the main effects of sulfur levels is to prevent oxidation and preserve color. So it makes sense that high sulfur levels may lead to a higher perceived wine quality, along with decent flavor and acceptable alcohol levels (Fig. 8).

4.4 Explanation for $y_t = 7$

Table 6. Input values for $y_t = 7$ explanation

fixed acidity	volatile acidity	citric acid	pH	residual sugar	chlorides	alcohol	free sulfur dioxide	total sulfur dioxide	sulphates	density
12.8	0.615	0.66	3.07	5.8	0.083	0	7	42	0.73	1.0022

GFT Explanation

- **acidity_fis**: perceived_acidity=0.537539 because fixed_acidity is 64% "Medium" and volatile_acidity is 69% "Medium" and citric_acid is 87% "Medium" and ph is 68% "Low"
- **saltiness_fis**: saltiness=0.498358 because density is 81% "Low" and chlorides is 62% "High"
- **flavor_profile_fis**: flavor_profile=0.596968 because perceived_acidity is 51% "High" and residual_sugar is 63% "Low" and saltiness is 62% "Medium"
- **sulfur_fis**: sulfur_level=0.630333 because free_sulfur_dioxide is 84% "Low" and total_sulfur_dioxide is 78% "Low" and sulphates is 94% "Medium"
- **quality_fis**: wine_quality=6.0 because alcohol is 73% "Low" and flavor_profile is 74% "Ok" and sulfur_level is 51% "High"

In this case we have another disagreement, and due to the relatively low error for this range of y_t, it is interesting to see why the model predicted this. The GFT predicted a wine quality of 6.0 instead of the actual 7.0. LIME is yet again predicting a lower value of 5.5 due to the low to medium alcohol amount and sulfur dioxide which is congruent with the GFT. There is also almost full agreement on *sulphates* being "Medium" level. However, full transparency into the latent variables and other input contributions allow for a much more robust explanation from the GFT.

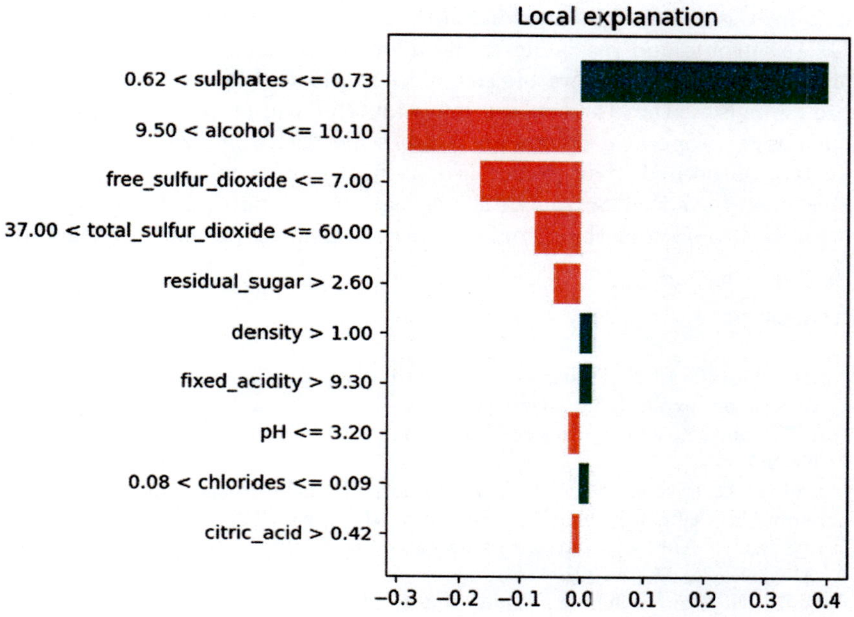

Fig. 8. LIME explanation plot for $y_i = 7$

5 Conclusion

In conclusion, the GFT was relatively high performing on the dataset considering the limitations of unbalanced values. The GFT outperformed all other ML models evaluated over the test set and a GFT could potentially have higher performance with a different structure and more parameters as well.

The explanations from the GFT and LIME both highlighted similar things, although the latent variables that were created by the manually-made GFT structure allowed for a richer, more intuitive explanation. If desired, it can also output the contributions of the input values directly in the form of membership activation and rule firiring weight, but often the amount of activation in the membership functions in the highest weighted rule is sufficient.

Overall, the fact that GFTs can perform approximation on noisy datasets like this one this successfully while also giving usable, interpretable explanations is potentially very helpful for human designers.

5.1 Future Work

There are a number of avenues for future work. Most notably an analysis could be done with other XAI tools such as SHAP, based on Shapley values, which can evaluate more than a single instance at a time. Another possible extension would be to create a much larger GFT for a more balanced dataset with more features for regression/classification. In that case, having full transparency into the GFTs outputs and reasoning would allow for a much richer understanding by human designers and operators as opposed to black-box methods. Something to note here also is that the explanations of a GFT can be traced as deep as the human designer/operator desires, and can scale with the size and complexity of the tree as needed. This could also be investigated with automatic latent variable creation based on semantic association of inputs. There may also be opportunities to do something similar in Reinforcement Learning problems.

References

1. Alonso Moral, J.M., Castiello, C., Magdalena, L., Mencar, C.: Remarks and prospects on explainable fuzzy systems. In: Explainable Fuzzy Systems. SCI, vol. 970, pp. 219–225. Springer, Cham (2021). https://doi.org/10.1007/978-3-030-71098-9_7
2. Gunning, D., Stefik, M., Choi, J., Miller, T., Stumpf, S., Yang, G.Z.: XAI-explainable artificial intelligence. Sci. Rob. **4**(37), eaay7120 (2019)
3. Gunning, D., Aha, D.: Darpa's explainable artificial intelligence (XAI) program. AI Mag. **40**(2), 44–58 (2019)
4. Andreas, H.: From machine learning to explainable AI. In: 2018 World Symposium on Digital Intelligence for Systems and Machines (DISA), pp. 55–66. IEEE (2018)
5. Ernest, N.D.: Genetic Fuzzy Trees for Intelligent Control of Unmanned Combat Aerial Vehicles. University of Cincinnati (2015)
6. Ernest, N., Carroll, D., Schumacher, C., Clark, M., Cohen, K., Lee, G.: Genetic fuzzy based artificial intelligence for unmanned combat aerial vehicle control in simulated air combat missions. J. Defense Manage. **6**(1), 2167–2374 (2016)
7. Fleck, D.E., et al.: Prediction of lithium response in first-episode mania using the lithium intelligent agent (LIThia): pilot data and proof-of-concept. Bipolar Disord. **19**(4), 259–272 (2017)
8. Thales: Thales GFT AI toolkit (2022)
9. Cortez, P.: UCI red wine quality dataset (2009). https://archive.ics.uci.edu/ml/datasets/Wine+Quality. Accessed 01 Mar 2023
10. Ribeiro, M.T., Singh, S., Guestrin, C.: why should i trust you? explaining the predictions of any classifier. In: Proceedings of the 22nd ACM SIGKDD International Conference on Knowledge Discovery and Data Mining, pp. 1135–1144 (2016)

Review of a Fuzzy Logic Based Airport Passenger Flow Prediction System

Javier Viaña[1,2](✉), Kelly Cohen[2], Stephen Saunders[3], Naashom Marx[3], Brian Cobb[3], Hannah Meredith[3], and Madison Bourbon[2]

[1] Massachusetts Institute of Technology, Cambridge, MA 02139, USA
vianajr@mit.edu
[2] University of Cincinnati, Cincinnati, OH 45219, USA
[3] Cincinnati/Northern Kentucky International Airport (CVG), Hebron, KY 41048, USA

Abstract. We have developed a system that predicts accurately the flow of passengers arriving at the airport a week in advance. The algorithm used by the system leverages Fuzzy Logic for the data processing. The system has been integrated in the Cincinnati/Northern Kentucky International Airport (CVG). This paper is a review of this technology and discusses some of the implicit benefits of its usage.

Keywords: Airports · Airport Security · Forecast · Fuzzy Logic · Artificial Intelligence · Explainable AI · Passenger Flow

1 Introduction

Predicting passenger flow at an airport can be challenging due to many factors such as weather, holidays, community events, or unexpected congestions. Being able to accurately estimate the passenger flow a week in advance can help airport officials plan and allocate resources, improving overall operational efficiency, and providing a better passenger experience.

The prediction of the passenger flow at the airport security checkpoint has been an important task for improving airport operations. Other researchers have also utilized the airport security checkpoint as the reference location for performing this prediction [1]. To tackle this problem, neural network models have proven to be very useful [2, 3]. Nevertheless, other techniques such as SARIMA time series models [4] have also been utilized. The flow of passengers within the airport is also a critical factor that has been studied [5]. In fact, these two problems are quite connected since the outputs of the passenger flow prediction at the security checkpoint are often the inputs for the simulation of the flow of passengers inside the airport.

Overall, the use of advanced prediction techniques has become increasingly important for improving the efficiency and effectiveness of airport operations. To that end, researchers continue to explore new and innovative methods for tackling this problem.

K. Cohen et al. (Eds.): NAFIPS 2023, LNNS 751, pp. 241–245, 2023.
https://doi.org/10.1007/978-3-031-46778-3_22

2 The System

Our team has created a system that can accurately predict the arrival flow of airport passengers a week ahead of time.

The system was trained and tested with real data from the Cincinnati/Northern Kentucky International Airport (CVG). This project was developed in the framework of a partnership between the University of Cincinnati and the Innovation Office of CVG.

We used multiple data sources for our passenger flow prediction system:

- Publicly available data sources include the flight schedules at the airport, which provide information on the number of flights, their capacity, carrier, destination, scheduled departing times, etc. We also used the total number of passengers per day throughout the year at the airport to extract seasonal patterns.
- The data that is not publicly available includes the scans of boarding passes at the security checkpoint, which provide real-time information on the number of passengers who have already arrived at the airport, and the average waiting time at the security checkpoint line. The boarding passes provide multiple features that we also leverage, these include the airline, destination, number of checked bags, if the passenger is a fast-track user or not, and other relevant demographics.

Our models use historical data to identify patterns and trends that may be relevant to make the desired predictions. In more updated versions of the system we have also incorporated to the model external factors to the airport, e.g., weather forecast, traffic congestions in the nearby highways, and holidays events.

The hardware developed for this project were the scanners of the boarding passes located at the security checkpoint. The development was done by CVG in a partnership with Differential (Ohio-based company in software development) and DESKO (company that provides airport barcode reading solutions). The device is capable of reading non-identifiable data from passengers' boarding passes (Fig. 1).

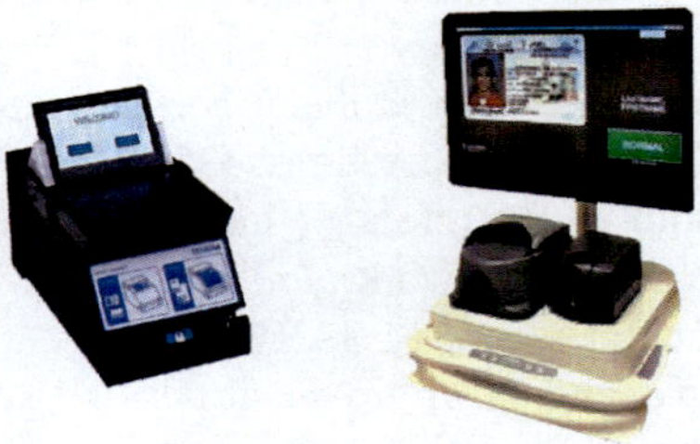

Fig. 1. Scanners developed by DESKO for airport barcode reading of non-personal identifiable data. Courtesy of CVG Airport Innovation Office.

A detailed explanation of the algorithm used for the prediction of the flow of passengers can be found in [6, 7]. This algorithm utilizes Fuzzy Logic in combination with flight-similarity metrics and bio-inspired optimization techniques. First, it converts the discrete nature of the raw scans of the boarding passes into continuous density functions that represent each flight. For the conversion, the use of a fuzzy sine membership function proved to be the most optimal [8]. Then, it performs a hyperparameter optimization of the similarity formulas that compare different flights based on the available features (airline, departure time, capacity, destination, etc.). The optimization process is done by means of a Genetic Algorithm. Using the similarity formulas defined in [6, 7], the next step is to perform a pattern search to obtain the past most similar flights to each of the scheduled flights. Then, the algorithm generates the predicted passenger flow distribution for all the future flights leveraging the data of the past flights. Finally, all the predictions of the individual flights are aggregated to obtain the real flow of passengers arriving at the security checkpoint. This process is also described in Fig. 2.

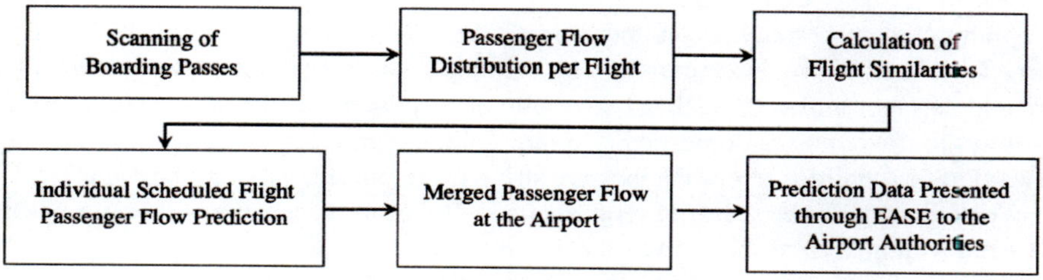

Fig. 2. Block diagram of the data handling process.

The algorithm was integrated as part of the Enterprise Awareness & Situational Exceptions (EASE) platform [9]. EASE is a tool for airport authorities that integrates data enabling operational awareness in real-time, gathers essential data insights, and forecasts customer behavior. An example of the prediction made by the algorithm as part of the EASE interface is shown in Fig. 3.

Fig. 3. EASE Interface and prediction example for March 15th, 2023. Courtesy of EASE and CVG Airport Innovation Office.

3 Reflections on the Success of the Project

In addition to the success within CVG, which was measured using the figures of merit defined in [6] (results and discussion shown in [6]), this project has generated several external tangible outcomes that further demonstrate the capabilities of the system. Namely, two publications [6, 7], one US patent [10], and the recognition of the Airport Cooperative Research Program Graduate Research Award, sponsored by the Federal Aviation Administration, administered by the Transportation Research Board and The National Academy of Sciences, and managed by the Virginia Space Grant Consortium. In addition, there is intent to commercialize the system developed, which is a clear indicator of the value that brings to the airport community.

Quoting Brian Cobb, Chief Innovation Officer of CVG airport: "At CVG, we're redefining predictive analytics and our operational outcomes with Dr. Viaña's Explainable AI. Aviation's long attempt to rely on certainties of peak travel days or seasonality to schedule labor and systems has been obliterated post pandemic where there are no absolutes to travel norms domestically or around the world. Explainable AI is the way forward to the quality and consistency we strive for in our industry and beyond."

We also include the testimony of Madison Bourbon, Licensing Associate of the 1819 Innovation Hub at the University of Cincinnati: "We expect this passenger flow prediction technology to not only improve operations, but also to save time and money for the airline industry."

Gaining an accurate quantitative understanding of the airport's passenger flow is a critical step to understand and control the airport operations. Having the ability to make necessary adjustments to the algorithm's inputs based on specific events that significantly impact travel volumes (i.e., time of year, special events, weather, etc.) further insulates its accuracy and trustworthiness, which is exactly what airport operators need to make confident and informed decisions. Overall, this predictive algorithm has tremendous potential for the aviation industry.

References

1. Monmousseau, P., Jarry, G., Bertosio, F., Delahaye, D., Houalla, M.: Predicting passenger flow at Charles de Gaulle airport security checkpoints. In: 2020 International Conference on Artificial Intelligence and Data Analytics for Air Transportation (AIDA-AT), Singapore, 2020, pp. 1–9. IEEE (2020)
2. Chen, J., Li, J.: Airport passenger flow forecast based on the wavelet neural network model. In: Proceedings of the 2018 2nd International Conference on Deep Learning Technologies (ICDLT 2018). Association for Computing Machinery, New York, NY, USA (2018)
3. Liu, L., Chen, R.-C.: A novel passenger flow prediction model using deep learning methods. Transp. Res. Part C: Emerg. Technol. **84**, 74–91 (2017)
4. Li, Z., Bil, J., Li, Z.: Passenger flow forecasting research for airport terminal based on SARIMA time series model. In: IOP Conference Series: Earth and Environmental Science, vol. 100, 1st International Global on Renewable Energy and Development (IGRED 2017). IOP Publishing Ltd., Singapore (2017)
5. Guo, X., Grushka-Cockayne, Y., De Reyck, B.: Forecasting airport transfer passenger flow using real-time data and machine learning. Manuf. Serv. Oper. Manag. **24**(6), 3193–3214 (2021)

6. Viaña, J., Cohen, K., Saunders, S., Marx, N., Cobb, B.: Explainable algorithm to predict passenger flow at CVG airport. Transportation Research Record (Accepted, to appear)
7. Viaña, J., Cohen, K., Saunders, S., Marx, N., Cobb, B.: ACRP graduate research award: explainable algorithm for passenger flow prediction at the security checkpoint of CVG Cincinnati/Northern Kentucky International Airport. In: Transportation Research Board 102nd Annual Meeting (TRB 2023), Committee on Airport Terminals and Ground Access (AV050), Washington D.C. (2023)
8. Holguin, S., Viaña, J., Cohen, K., Ralescu, A., Kreinovich, V.: Why sine membership functions. In: Dick, S., Kreinovich, V., Lingras, P. (eds.) Applications of Fuzzy Techniques. NAFIPS 2022. Lecture Notes in Networks and Systems, vol. 500, pp. 83–89. Springer, Cham (2023). https://doi.org/10.1007/978-3-031-16038-7_9
9. EASE Homepage. https://ease.aero. Accessed 19 Mar 2023
10. Viaña, J., Cohen, K.: Systems and methods for predicting airport passenger flow, United States Patent and Trademark Office. International Publication Number: WO 2023/012939 A1. International Publication Date: 16.02.2023

Complex-Valued Interval Computations are NP-Hard Even for Single Use Expressions

Martine Ceberio[1], Vladik Kreinovich[1(✉)], Olga Kosheleva[2], and Günter Mayer[3]

[1] Department of Computer Science, University of Texas at El Paso, 500 W. University, El Paso, TX 79968, USA
{mceberio,vladik}@utep.edu

[2] Department of Teacher Education, University of Texas at El Paso, 500 W. University, El Paso, TX 79968, USA
olgak@utep.edu

[3] Fachbereich Mathematik, Universität Rostock, Universitätsplatz 1, 18051 Rostock, Germany
guenter.mayer@uni-rostock.de

Abstract. In practice, after a measurement, we often only determine the interval containing the actual (unknown) value of the measured quantity. It is known that in such cases, checking whether the measurement results are consistent with a given hypothesis about the relation between quantities is, in general, NP-hard. However, there is an important case when this checking problem is feasible: the case of single-use expressions, i.e., expressions in which each variable occurs only once. Such expressions are ubiquitous in physics, e.g., Ohm's law $V = I \cdot R$, formula for the kinetic energy $E = (1/2) \cdot m \cdot v^2$, formula for the gravitational force $F = G \cdot m_1 \cdot m_2 \cdot r^{-2}$, etc. In some important physical situations, quantities are complex-valued. A natural question is whether for complex-valued quantities, feasible checking algorithms are possible for single-use expressions. We prove that in the complex-valued case, computing the exact range is NP-hard even for single-use expressions. Moreover, it is NP-hard even for such simple expressions as the product $f(z_1, \ldots, z_n) = z_1 \cdot \ldots \cdot z_n$.

1 Introduction

Need for hypothesis testing and for interval computations. In many practical situations, the value y of a physical quantity depends on the values $x_1, \ldots, x_n$ of related quantities, but we do not know the exact form of this dependence. In such cases, based on the available observation, researchers come up with a hypothesis that $y = f(x_1, \ldots, x_n)$ for some specific function f.

How can we test this hypothesis? A natural idea is to consider other situations in which we know both the value y and the values $x_1, \ldots, x_n$. In the ideal situation, in which we know the exact values of y and x_i, this testing is

K. Cohen et al. (Eds.): NAFIPS 2023, LNNS 751, pp. 246–257, 2023.
https://doi.org/10.1007/978-3-031-46778-3_23

simple: we simply check whether for these measurement results, y is equal to $f(x_1, \ldots, x_n)$.

In practice, however, measurements are never absolutely accurate. For each quantity q, the measurement result $\widetilde{q}$ is, in general, different from the actual (unknown) value q of this quantity. In many practical situations, the only information that we have about the difference $\Delta q \stackrel{\text{def}}{=} \widetilde{q} - q$ (known as the *measurement error*) is the upper bound Δ on its absolute value: $|\Delta q| \le \Delta$; see, e.g., [15]. In this case, based on the measurement result $\widetilde{q}$, the only information that we have about the actual value q is that this value is contained in the interval $[\widetilde{q} - \Delta, \widetilde{q} + \Delta]$.

Under such interval uncertainty, after the measurement, we have the interval $[\underline{y}, \overline{y}]$ of possible values of y and the intervals $[\underline{x}_i, \overline{x}_i]$ of possible values of each quantity x_i. In this case, the original hypothesis is confirmed if there exist values within these intervals for which $y = f(x_1, \ldots, x_n)$. In other words, the hypothesis is confirmed if the interval $[\underline{y}, \overline{y}]$ has a common point with the range

$$f([\underline{x}_1, \overline{x}_1], \ldots, [\underline{x}_n, \overline{x}_n]) \stackrel{\text{def}}{=}$$
$$\{f(x_1, \ldots, x_n) : x_1 \in [\underline{x}_1, \overline{x}_1], \ldots, x_n \in [\underline{x}_n, \overline{x}_n]\}. \tag{1}$$

A natural way to check this is to compute the range and then to check whether the intersection of the range and the y-interval is non-empty. Computing the range is known as *interval computations*; see, e.g., [4,8,9,11].

Comments.

- The measurement results $\widetilde{q}$ are usually rational numbers – in modern computer age, they come from a computer and are, thus, binary rational, i.e., number of the type $m/2^n$ for integers m and n. Similarly, the bounds Δ on the absolute value of the measurement error are rational. Because of this, the endpoints of the resulting interval $[\widetilde{q} - \Delta, \widetilde{q} + \Delta]$ are usually also rational numbers.
- In the case when we have *fuzzy* information about y and x_i (see, e.g., [1,5,10, 12,13,16]), we have *fuzzy* sets corresponding to y and $f(x_1, \ldots, x_n)$. In this case, a natural idea it to look for the largest α for which the α-cuts of these two sets have a common point – this is the degree to which the given fuzzy data is consistent with the proposed hypothesis. It is known that the α-cut of the fuzzy set $f(x_1, \ldots, x_n)$ is equal to the range (1) of the function f on the α-cuts of x_i. Thus, from the computational viewpoint, the fuzzy version of this problem can be reduced to its interval version.

Computational complexity of interval computations. It is known (see, e.g., [7]) that the problem of computing the range (1) is, in general, NP-hard. It is NP-hard already for quadratic functions $f(x_1, \ldots, x_n)$.

However, there are classes of expressions for which computing the range is feasible. First, these are functions representing elementary arithmetic operations. In these cases, the resulting expressions come from the fact that the corresponding functions are monotonic:

- the range for the sum $f(x_1, x_2) = x_1 + x_2$ is equal to $[\underline{x}_1 + \underline{x}_2, \overline{x}_1 + \overline{x}_2]$;
- the range for the difference $f(x_1, x_2) = x_1 - x_2$ is equal to $[\underline{x}_1 - \overline{x}_2, \overline{x}_1 - \underline{x}_2]$;
- the range for the product $f(x_1, x_2) = x_1 \cdot x_2$ is equal to $[\underline{y}, \overline{y}]$, where

$$\underline{y} = \min(\underline{x}_1 \cdot \underline{x}_2, \underline{x}_1 \cdot \overline{x}_2, \overline{x}_1 \cdot \underline{x}_2, \overline{x}_2 \cdot \overline{x}_2),$$

$$\overline{y} = \max(\underline{x}_1 \cdot \underline{x}_2, \underline{x}_1 \cdot \overline{x}_2, \overline{x}_1 \cdot \underline{x}_2, \overline{x}_2 \cdot \overline{x}_2).$$

- the range for the inverse $f(x_1) = 1/x_1$ is equal to $[1/\overline{x}_1, 1/\underline{x}_1]$ if $0 \notin [\underline{x}_1, \overline{x}_1]$.

Similarly, we can derive explicit formulas for the ranges of elementary functions. For example, for $f(x_1) = x_1^2$, the range is equal:

- to $[\underline{x}_1^2, \overline{x}_1^2]$ if $0 \le \underline{x}_1$;
- to $[\overline{x}_1^2, \underline{x}_1^2]$ if $\overline{x}_1 \le 0$;
- to $[0, \max(\underline{x}_1^2, \overline{x}_1^2)]$ if $\underline{x}_1 \le 0 \le \overline{x}_1$.

These formulas form *interval arithmetic*:

- the range of the sum is called the sum of the intervals,
- the range of the difference is called the difference between the intervals,
- the range of the square is called the square of the interval, etc.

Another example when there is a feasible algorithm for computing the range is the so-called *single use expressions* (SUE, for short), i.e., expressions in which each variable occurs only once. Examples of such expressions include the product $x_1 \cdot \ldots \cdot x_n$, the value $(1/n) \cdot \sum_{i=1}^{n} x_i^2$, etc. Such expressions are ubiquitous in physics: Ohm's law $V = I \cdot R$, formula for the kinetic energy $E = (1/2) \cdot m \cdot v^2$, formula for the gravitational force $F = G \cdot m_1 \cdot m_2 \cdot r^{-2}$, etc. (as opposed to, e.g., expressions of the type $y = x - x^2$ in which the variable x occurs twice). For single-use expressions, the range (1) can obtained if we simply replace each arithmetic operation (that form the computation algorithm) with the corresponding operation of interval arithmetic; see, e.g., [3,4].

Need for complex values and for the corresponding hypothesis testing and interval computations. Many physical quantities are complex-valued, e.g., complex amplitude and impedance in electrical engineering, complex wave function in quantum mechanics, etc. Similarly to the real-valued case, we can have a complex-valued quantity z depend on the complex-values quantities $z_1, \ldots, z_n$, we can have hypotheses $z = f(z_1, \ldots, z_n)$, and we need to check whether the new measurement results are consistent with these hypotheses.

In many such situations, to measure the corresponding complex value $z = x + \mathrm{i} \cdot y$, we separately measure the real part x and the imaginary part y. After the measurement, we get the interval $[\underline{x}, \overline{x}]$ of possible values of x and the interval $[\underline{y}, \overline{y}]$ of possible values of y. It is reasonable to call the set of possible complex numbers

$$\mathbf{z} \stackrel{\text{def}}{=} \{x + \mathrm{i} \cdot y : x \in [\underline{x}, \overline{x}], y \in [\underline{y}, \overline{y}]\}$$

a *complex interval*; see, e.g., [6,14].

Based on these interval-valued measurement results, we need to check whether the complex interval for z has a common point with the range

$$f(\mathbf{z}_1,\ldots,\mathbf{z}_n) \stackrel{\text{def}}{=} \{f(z_1,\ldots,z_n) : z_1 \in \mathbf{z}_1,\ldots,z_n \in \mathbf{z}_n\}. \quad (2)$$

Computational complexity of complex interval computations: what is known and what we want to analyze. Of course, real-valued computations can be viewed as a particular case of complex-values ones – it is sufficient to take all imaginary parts equal to 0. Since real-valued interval computations are NP-hard, this implies that complex-valued interval computations are NP-hard as well.

A natural question is: what about the SUE expressions? In this paper, we show that, in contrast to real-valued case, complex interval arithmetic is NP-hard even for SUE expressions.

2 Main Result

Definition 1. *By a* complex interval, *we mean a set*

$$\mathbf{z} = [\underline{x},\overline{x}] + \mathrm{i}\cdot[\underline{y},\overline{y}] = \{x + \mathrm{i}\cdot y : x \in [\underline{x},\overline{x}] \text{ and } y \in [\underline{y},\overline{y}],$$

where $\underline{x}$, $\overline{x}$, $\underline{y}$, *and* $\overline{y}$ *are rational numbers.*

Definition 2. *By a* problem of complex interval computations *for a complex-valued function* $z = f(z_1,\ldots,z_n)$, *we mean the following problem;*

- *given: complex intervals* $\mathbf{z}$, $\mathbf{z}_1$, $\ldots$, $\mathbf{z}_n$,
- *check whether the complex interval* $\mathbf{z}$ *and the range (2) have a common point.*

Proposition 1. *For each of the following SUE functions, the problem of complex interval computations is NP-hard:*

1. *the scalar (dot) product*

$$f(z_1,\ldots,z_n,t_1,\ldots,t_n) = \sum_{i=1}^{n} z_i \cdot t_i;$$

2. *the second moment*

$$f(z_1,\ldots,z_n) = \frac{1}{n}\cdot\sum_{i=1}^{n} z_i^2;$$

3. *the product* $f(z_1,\ldots,z_n) = z_1 \cdot \ldots \cdot z_n$.

Proof. To prove NP-hardness of all three range computation problems, we will reduce, to this new problem, a known NP-hard *partition problem*; see, e.g., [2,7]. The partition problem is as follows: given n positive integers $s_1,\ldots,s_n$, to check whether there exists values $\varepsilon_i \in \{-1,1\}$ such that $\sum\limits_{i=1}^{m} \varepsilon_i \cdot s_i = 0$.

1. Let us first prove that the first complex interval computations problem is NP-hard.

To prove this, for every instance $s_1, \ldots, s_n$ of the partition problem, we take $\mathbf{z}_i = s_i \cdot (1 + \mathrm{i} \cdot [-1, 1])$, $\mathbf{t}_i = 1 + \mathrm{i} \cdot [-1, 1]$ and $\mathbf{z} = [0, 0]$. Then, possible values $z_i \in \mathbf{z}_i$ have the form $z_i = s_i \cdot (1 + \mathrm{i} \cdot a_i)$ for some $a_i \in [-1, 1]$, and possible values of t_i are of the form $t_i = 1 + \mathrm{i} \cdot b_i$ with $b_i \in [-1, 1]$.

Let us prove that the interval $\mathbf{z}$ has a common point with the range $f(\mathbf{z}_1, \ldots, \mathbf{z}_n, \mathbf{t}_1, \ldots, \mathbf{t}_n)$ – i.e., equivalently, that the point $z = 0$ belongs to the range (2) – if and only if the given instance of the partition problem has a solution.

1.1. If the partition problem has a solution ε_i for which $\sum s_i \cdot \varepsilon_i = 0$, then, as one can easily check, for $z_i = s_i \cdot (1 + \mathrm{i} \cdot \varepsilon_i)$ and $t_i = 1 + \mathrm{i} \cdot \varepsilon_i$, we have

$$z_i \cdot t_i = s_i \cdot (1 + 2\mathrm{i} \cdot \varepsilon_i - 1) = 2\mathrm{i} \cdot s_i \cdot \varepsilon_i,$$

and thus,

$$f(z_1, \ldots, z_n, t_1, \ldots, t_n) = \sum_{i=1}^{n} z_i \cdot t_i = 2\mathrm{i} \cdot \sum_{i=1}^{n} s_i \cdot \varepsilon_i = 0 = z.$$

So, in this case, the interval $\mathbf{z}$ has a common point with the range $f(\mathbf{z}_1, \ldots, \mathbf{z}_n, \mathbf{t}_1, \ldots, \mathbf{t}_n)$.

1.2. Let us now prove that, vice versa, if the interval $\mathbf{z}$ has a common point with the range $f(\mathbf{z}_1, \ldots, \mathbf{z}_n, \mathbf{t}_1, \ldots, \mathbf{t}_n)$, i.e., if $0 = f(z_1, \ldots, z_n, t_1, \ldots, t_n)$ for some values $z_i \in \mathbf{z}_i$ and $t_i \in \mathbf{t}_i$, then the corresponding instance of the partition problem has a solution.

Indeed, for each i, the product $z_i \cdot t_i$ is equal to

$$s_i \cdot (1 + \mathrm{i} \cdot a_i) \cdot (1 + \mathrm{i} \cdot b_i) = s_i \cdot ((1 - a_i \cdot b_i) + \mathrm{i} \cdot (a_i + b_i)).$$

Since $|a_i| \le 1$ and $|b_i| \le 1$, we have $|a_i \cdot b_i| \le 1$, and therefore, $1 - a_i \cdot b_i \ge 0$. Thus, the real part of the sum $\sum\limits_{i=1}^{n} z_i \cdot t_i$ is equal to the sum of n non-negative numbers $s_i \cdot (1 - a_i \cdot b_i)$. The only possibility for this sum to be equal to 0 if when all n non-negative terms are equal to 0, i.e., when $a_i \cdot b_i = 1$.

Since $|a_i| \le 1$ and $|b_i| \le 1$, the absolute value of the product $|a_i \cdot b_i|$ cannot exceed 1, and the only possibility for this product to be equal to 1 is when both absolute values are equal to 1, i.e., when $a_i = \pm 1$ and $b_i = \pm 1$. Since $a_i \cdot b_i = 1$, the signs must coincide, i.e., we must have $a_i = b_i \in \{-1, 1\}$. Let us denote the common value of a_i and b_i by ε_i. For these values $a_i = b_i = \varepsilon_i$, the imaginary part of $z_i \cdot t_i$ is equal to $2 \cdot \varepsilon_i \cdot s_i$, so the fact that the imaginary part of the sum $\sum\limits_{i=1}^{n} z_i \cdot t_i$ is equal to 0 is equivalent to $2 \cdot \sum\limits_{i=1}^{n} \varepsilon_i \cdot s_i = 0$ – i.e., to the fact that the original instance of the partition problem has a solution.

The statement is proven.

2. Let us now prove that complex interval computation problem is NP-hard for the second function.

2.1. If all the values s_i are squares of integers, then we can take $\mathbf{z}_i = \sqrt{s_i}\cdot(1+\mathrm{i}\cdot[-1,1])$ and $\mathbf{z} = [0,0]$. In this case, all possible values of z_i have the form $z_i = \sqrt{s_i}\cdot(1+\mathrm{i}\cdot a_i)$ for some $a_i \in [-1,1]$. Then, we have $z_i^2 = s_i\cdot((1-a_i^2)+\mathrm{i}\cdot(2a_i))$.

If the corresponding instance of the partition problem has a solution, then, as one can easily check, we have $0 = f(z_1,\ldots,z_n)$ for $z_i = \sqrt{s_i}\cdot(1+\mathrm{i}\cdot\varepsilon_i)$.

Vice versa, let us assume that $0 = f(z_1,\ldots,z_n)$ for some values

$$z_i = \sqrt{s_i}\cdot(1+\mathrm{i}\cdot a_i) \in \mathbf{z}_i.$$

In this case, since $|a_i| \le 1$, we have $1 - a_i^2 \ge 0$, and the only possibility for the expression $\frac{1}{n}\cdot\sum\limits_{i=1}^{n} z_i^2$ to have a zero real part is to have $1 - a_i^2 = 0$ for all i, i.e., to have $a_i = \pm 1$ for every i. In other words, we have $a_i = \varepsilon_i \in \{-1,1\}$. For these a_i, the imaginary part of the expression $\frac{1}{n}\cdot\sum\limits_{i=1}^{n} z_i^2$ is equal to $\frac{2}{n}\cdot\sum\limits_{i=1}^{n} s_i\cdot\varepsilon_i$. Thus, the fact that the imaginary part is equal to 0 is equivalent to $\sum\limits_{i=1}^{n}\varepsilon_i\cdot s_i = 0$, i.e., to the existence of the solution to the original instance of the partition problem.

2.2. When the values s_i are not full squares, in defining $\mathbf{z}_i$, instead of $\sqrt{s_i}$, we can take rational numbers r_i for which r_i^2 is ε-close to $\sqrt{s_i}$ for some small $\varepsilon > 0$. Instead of $\mathbf{z} = [0,0]$, we take $\mathbf{z} = \mathrm{i}\cdot[-\delta,\delta]$ for some small $\delta > 0$. Let us prove that for appropriately chosen ε and δ, the original instance of the partition problem has a solution if and only if the sets $\mathbf{z}$ and $f(\mathbf{z}_1,\ldots,\mathbf{z}_n)$ have a common point.

2.2.1. If the corresponding instance of the partition problem has a solution, then we can take $z_i = r_i\cdot(1+\mathrm{i}\cdot\varepsilon_i)$, in which case

$$\frac{1}{n}\cdot\sum_{i=1}^{n} z_i^2 = \mathrm{i}\cdot\frac{2}{n}\cdot\sum_{i=1}^{n} r_i^2.$$

Since each value r_i^2 is ε-close to s_i, we conclude that

$$\left|\frac{2}{n}\cdot\sum_{i=1}^{n} r_i^2 - \frac{2}{n}\cdot\sum_{i=1}^{n} s_i\right| \le \frac{2}{n}\cdot\sum_{i=1}^{n}\varepsilon = \frac{2}{n}\cdot n\cdot\varepsilon = 2\varepsilon.$$

So, for $2\varepsilon \le \delta$, we get $f(z_1,\ldots,z_n) \in \mathbf{z}$, and thus, the sets $\mathbf{z}$ and $f(\mathbf{z}_1,\ldots,\mathbf{z}_n)$ have a common point.

2.2.2. Vice versa, let us assume that the sets $\mathbf{z}$ and $f(\mathbf{z}_1,\ldots,\mathbf{z}_n)$ have a common point. Let us prove that in this case, the original instance of the partition problem has a solution.

Indeed, in this case, similarly to Part 2.1 of this proof, from the fact that there is a common point, we can still conclude that for the corresponding common point, we have $a_i = \pm 1$. For these a_i, the imaginary part I of the expression $\frac{1}{n}\cdot\sum\limits_{i=1}^{n} z_i^2$ is equal to $\frac{2}{n}\cdot\sum\limits_{i=1}^{n} r_i^2\cdot\varepsilon_i$.

Thus, the fact that absolute value of the imaginary part is bounded by δ is equivalent to

$$|I| = \left|\frac{1}{n}\cdot\sum_{i=1}^{n}\varepsilon_i\cdot r_i^2\right| \le \delta.$$

Since each value r_i^2 is ε-close to s_i, we conclude that

$$\left|I - \frac{2}{n}\cdot\sum_{i=1}^{n} s_i\cdot\varepsilon_i\right| \le 2\varepsilon.$$

Thus,

$$\left|\frac{2}{n}\cdot\sum_{i=1}^{n} s_i\cdot\varepsilon_i\right| \le |I| + \left|I - \frac{2}{n}\cdot\sum_{i=1}^{n} s_i\cdot\varepsilon_i\right| \le 2\varepsilon + \delta,$$

hence

$$\left|\sum_{i=1}^{n} s_i\cdot\varepsilon_i\right| \le \frac{n}{2}\cdot(2\varepsilon+\delta).$$

The sum $\sum s_i\cdot\varepsilon_i$ is an integer, so if

$$\frac{n}{2}\cdot(2\varepsilon+\delta) < 1,$$

this implies that $\sum s_i\cdot\varepsilon_i = 0$, i.e., that the values $\varepsilon_i\in\{-1,1\}$ provide a solution to the original instance of the partition problem.

2.2.3. By combining the results from Parts 2.2.1 and 2.2.2 of this proof, we conclude that the desired equivalence can be proven if we have $2\varepsilon \le \delta$ and $(n/2)\cdot(2\varepsilon+\delta) < 1$. These two inequalities are satisfies for sufficiently small ε and δ, e.g., for $\delta = 1/(2n)$ and $\varepsilon = 1/(4n)$. The statement is proven.

3. Let us now prove that for the third function, the complex interval computation problem is NP-hard.

For every instance of the partition problem, we compute $k = 1/\left(\sum\limits_{j=1}^{n} s_j\right)$, $\theta_i = k\cdot s_i$, and $\tan(\theta_i)$.

3.1. Let us first consider the case when all the values $\tan(\theta_i)$ and the product $\prod\limits_{i=1}^{n}\sqrt{1+\tan^2(\theta_i)}$ are rational numbers. In this case, we can take $\mathbf{z}_i = 1 + \mathrm{i}\cdot[-t_i, t_i]$ for $t_i = \tan(\theta_i)$ and $z = \prod\limits_{i=1}^{n}\sqrt{1+t_i^2}$.

Let us prove that the selected number z belongs to the range of the product if and only if the original instance of the partition problem has a solution.

In this proof, we will use the known fact that every complex number $z = x + \mathrm{i} \cdot y$ can be represented in a polar form $z = \rho \cdot e^{\mathrm{i} \cdot \alpha}$, where $\rho = \sqrt{x^2 + y^2}$ is the absolute value (magnitude) of z, and the "phase" θ is the angle between the direction from 0 to z and the positive real semi-axis. When we multiply complex numbers, their magnitudes multiply and their phases add.

3.1.1. Let us first prove that if the original instance has a solution ε_i, then z is equal to the product of n values $z_i = 1 + \mathrm{i} \cdot \varepsilon_i \cdot t_i \in \mathbf{z}_i$.

Indeed, since $|z_i| = \sqrt{1 + t_i^2}$, the product of the magnitudes is the desired value z. The angle α_i corresponding to each z_i is equal to $\alpha_i = \varepsilon_i \cdot \theta_i$, so the sum α of these angles is equal to $\sum\limits_{i=1}^{n} \varepsilon_i \cdot \theta_i$. Since $\theta_i = k \cdot s_i$, we conclude that $\alpha = k \cdot \sum\limits_{i=1}^{n} \varepsilon_i \cdot s_i$, i.e., $\alpha = 0$. So, this product $z_1 \cdot \ldots \cdot z_n$ has the right magnitude and the correct angle and is, thus, equal to z.

3.1.2. Vice versa, let us assume that z belongs to the range, i.e., that z can be represented as the product $z_1 \cdot \ldots \cdot z_n$ for some $z_i \in \mathbf{z}_i$. In other words, for this product, the magnitude is equal to z, and the phase α is 0. For each value $z_i = 1 + \mathrm{i} \cdot y_i \in \mathbf{z}_i$, its magnitude is equal to $\sqrt{1 + y_i^2}$. Since $|y_i| \le t_i$, this magnitude cannot exceed $\sqrt{1 + t_i^2}$, and it is equal to $\sqrt{1 + t_i^2}$ only for the two endpoints $y_i = \pm t_i$.

If for some i, we have $y_i \in (-t_i, t_i)$, then the resulting magnitude is the product of several numbers all of which are $\le \sqrt{1 + t_i^2}$ and some are smaller – thus, the magnitude of the product will be smaller than z. Since the magnitude of the product is equal to z, then, for each i, we have $z_i = 1 + \mathrm{i} \cdot \varepsilon_i \cdot t_i$ for some $\varepsilon_i \in \{-1, 1\}$. For each of these numbers z_i, the phase α_i is equal to $\varepsilon_i \cdot \theta_i$. Thus, from the fact that the overall angle $\alpha = \sum\limits_{i=1}^{n} \alpha_i$ is equal to 0, we conclude that $\sum\limits_{i=1}^{n} \varepsilon \cdot \theta_i = 0$, and, since $\theta_i = k \cdot s_i$, that $\sum\limits_{i=1}^{n} \varepsilon_i \cdot s_i = 0$ – i.e., the original instance of the partition problem indeed has a solution.

3.2. In the general case, when the values $\tan(\theta_i)$ and the product $\prod\limits_{i=1}^{n} \sqrt{1 + \tan^2(\theta_i)}$ are not rational, we can take, as t_i, rational values which are ε-close to t_i for some small $\varepsilon > 0$, take $\mathbf{z}_i = 1 + \mathrm{i} \cdot [-t_i, t_i]$, and take the interval $\mathbf{z} = [z_0 - \delta, z_0 + \delta]$ for some small δ, where z_0 is a rational number which is δ-close to the value $\prod\limits_{i=1}^{n} \sqrt{1 + \tan^2(\theta_i)}$. The proof then follows from the arguments presented in Part 2.2 of this proof.

The proposition is proven.

3 First Auxiliary Result

Situation. What if we can only measure the real part of the quantity y? In this case, checking the hypothesis means comparing the interval of all possible real values of $f(z_1, \ldots, z_n)$ with the given interval. Being able to solve this checking problem means being able to compute the range of real values and/or the range of imaginary values. It turns out that this problem is also NP-hard already for the product.

Definition 3. *By a* problem of real-valued complex interval computations *for a complex-valued function* $z = f(z_1, \ldots, z_n)$*, we mean the following problem;*

- *given: a real-valued interval* $\mathbf{x}$ *and complex intervals* $\mathbf{z}_1, \ldots, \mathbf{z}_n$*,*
- *check whether there exist values* $z_i \in \mathbf{z}_i$ *for which the real part of* $f(z_1, \ldots, z_n)$ *belongs to the interval* $\mathbf{x}$*.*

Proposition 2. *For the product* $f(z_1, \ldots, z_n) = z_1 \cdot \ldots \cdot z_n$*, the problem of real-valued complex interval computations is NP-hard.*

Discussion. In the real-valued case, for SUE expressions, we can compute the smallest interval containing the actual range by using straightforward interval computations, i.e., by replacing each operation with numbers in the original formula f by the corresponding operation with intervals. For the product of complex numbers, not only we do not get the smallest box at the end, but the result of the corresponding operation-by-operation interval operations may actually depend on the order of multiplications – because for complex intervals, multiplication is, in general, not associative. For example, for $(1 - \mathrm{i}) \cdot (1 + \mathrm{i}) \cdot ([0, 1] - \mathrm{i})$, we get:

$$(1 + \mathrm{i}) \cdot ([0, 1] - \mathrm{i}) = ([0, 1] + 1) + \mathrm{i} \cdot ([0, 1] - 1) = [1, 2] + \mathrm{i} \cdot [-1, 0];$$

$$(1 - \mathrm{i}) \cdot ([1, 2] + \mathrm{i} \cdot [-1, 0]) = ([1, 2] + [-1, 0]) + \mathrm{i} \cdot ([-1, 0] - [1, 2]) = [0, 2] + \mathrm{i} \cdot [-3, -1],$$

while $(1 - \mathrm{i}) \cdot (1 + \mathrm{i}) = 2$ hence

$$2 \cdot ([0, 1] - \mathrm{i}) = [0, 2] - 2\mathrm{i} \neq [0, 2] + \mathrm{i} \cdot [-3, -1].$$

Proof of Proposition 2. We will prove that computing the largest possible real part $\overline{x}$ of the product is NP-hard.

In this proof, we use the same reduction as in Part 3 of Proposition 1. In that proof, we showed that when $z_i \in \mathbf{z}_i$, the product $z = z_1 \cdot \ldots \cdot z_n$ has a magnitude which cannot exceed $z \stackrel{\text{def}}{=} \prod_{i=1}^{n} \sqrt{1 + t_i^2}$. Since the real part of a complex number cannot exceed its magnitude, the largest possible value $\overline{x}$ of the real part cannot exceed z. The only possibility for $\overline{x}$ to be equal to z is when there is a point with real value z; since the magnitude cannot exceed z, the imaginary part of this point must be 0. Thus, the only way for $\overline{x}$ to be equal to z is to have z itself represented as a product of values $z_i \in \mathbf{z}_i$, and we already know that checking this condition is NP-hard. Thus, computing $\overline{x}$ is also NP-hard.

Comment. To take care of the general case, when the tangents and the product are not all rational, we can use of appropriate ε-close and δ-close rational numbers.

4 Second Auxiliary Result

Situation. In some practical situations, we can directly measure the corresponding complex value z. In this case, as a result of the measurement, we get a rational-valued complex number $\widetilde{z}$ for which $|z-\widetilde{z}| \le \Delta$ for some known bound Δ. After this measurement, all we know about the actual (unknown) value z of this quantity is that this value belongs to the set $(\widetilde{z}, \Delta) \stackrel{\text{def}}{=} \{z : |z-\widetilde{z}| \le \Delta\}$. This is known as *circular complex interval uncertainty.*

Definition 4. *By a* circular complex interval, *we mean a set*

$$\mathbf{z} = (\widetilde{z}, \Delta) \stackrel{\text{def}}{=} \{z : |z-\widetilde{z}| \le \Delta\},$$

where $\widetilde{z} = \widetilde{x} + \mathrm{i} \cdot \widetilde{y}$, and the values $\widetilde{x}$, $\widetilde{y}$, and r are rational numbers.

Definition 5. *By a* problem of circular complex interval computations *for a complex-valued function $z = f(z_1, \ldots, z_n)$, we mean the following problem;*

- *given: circular complex intervals $\mathbf{z}$, $\mathbf{z}_1$, ..., $\mathbf{z}_n$,*
- *check whether the circular complex interval $\mathbf{z}$ and the range (2) have a common point.*

Proposition 3. *For the scalar (dot) product and for the second moment, the problem of circular complex interval computations is NP-hard.*

Proof. Let us first consider the scalar (dot) product. The main idea is that similarly to the proof of Part 1 of Proposition 1, let us take $\mathbf{z}_i = \mathbf{t}_i = \sqrt{s_i} \cdot \left(1, \dfrac{\sqrt{2}}{2}\right)$ and $z = 0$. One can easily check that for complex numbers $z_i \in \mathbf{z}_i$, the phase takes the values from $-45°$ to $45°$. The phase is equal to $45°$ only at a point $\sqrt{s_i} \cdot (0.5 + \mathrm{i} \cdot 0.5)$, and the phase is equal to $-45°$ only at a point $\sqrt{s_i} \cdot (0.5 - \mathrm{i} \cdot 0.5)$.

When we multiply complex numbers, their phases add up. Thus, for the product $z_i \cdot t_i$, the angle is always between $-90°$ and $90°$, i.e., the real part of the product is always non-negative. So, the real part of the sum $\sum\limits_{i=1}^{n} z_i \cdot t_i$ is also always non-negative. The only possibility for this real part to be 0 is when the real parts of all the terms in the sum are equal to 0, i.e., when for each i, the phase of the product $z_i \cdot t_i$ is equal to either $90°$ or to $-90°$. This, in turn, is possible only if either both z_i and t_i have phases $45°$ or both z_i and t_i have phases $-45°$.

- In the first case, we have $z_i = t_i = \sqrt{s_i} \cdot (0.5 + \mathrm{i} \cdot 0.5)$, hence $z_i \cdot t_i = 0.5 \cdot s_i \cdot \mathrm{i}$.
- In the second case, we have $z_i = t_i = \sqrt{s_i} \cdot (0.5 - \mathrm{i} \cdot 0.5)$, hence $z_i \cdot t_i = -0.5 \cdot s_i \cdot \mathrm{i}$.

In both cases, we have $z_i \cdot t_i = 0.5 \cdot \varepsilon_i \cdot s_i \cdot \mathrm{i}$ for some $\varepsilon_i \in \{-1, 1\}$.

Thus, the imaginary part of the sum $\sum\limits_{i=1}^{n} z_i \cdot t_i$ is equal to $0.5 \cdot \sum\limits_{i=1}^{n} \varepsilon_i \cdot s_i$. This imaginary part is equal to 0 if and only if $\sum\limits_{i=1}^{n} \varepsilon_i \cdot s_i = 0$, i.e., if and only if the original instance of the partition problem has a solution.

Comments.

- To take care of the fact that the square roots are, in general, not rational, we can use appropriate ε-close and δ-close rational numbers.
- The proof for the second moment is similar.

Acknowledgments. This work was supported in part by the National Science Foundation grants 1623190 (A Model of Change for Preparing a New Generation for Professional Practice in Computer Science), HRD-1834620 and HRD-2034030 (CAHSI Includes), and by the AT&T Fellowship in Information Technology.

It was also supported by the program of the development of the Scientific-Educational Mathematical Center of Volga Federal District No. 075-02-2020-1478, and by a grant from the Hungarian National Research, Development and Innovation Office (NRDI).

The authors are greatly thankful to the anonymous referees for valuable suggestions.

References

1. Belohlavek, R., Dauben, J.W., Klir, G.J.: Fuzzy Logic and Mathematics: A Historical Perspective. Oxford University Press, New York (2017)
2. Garey, M.R., Johnson, D.S.: Computers and Intractability, a Guide to the Theory of NP-Completeness. W. H. Freeman, San Francisco (1979)
3. Hansen, E.: Sharpness in interval computations. Reliable Comput. **3**, 7–29 (1997)
4. Jaulin, L., Kieffer, M., Didrit, O., Walter, E.: Applied Interval Analysis: With Examples in Parameter and State Estimation, Robust Control and Robotics. Springer, London (2001)
5. Klir, G., Yuan, B.: Fuzzy Sets and Fuzzy Logic. Prentice Hall, Upper Saddle River (1995)
6. Kolev, L.V.: Interval Methods for Circuit Analysis. World Scientific, Singapore (1993)
7. Kreinovich, V., Lakeyev, A., Rohn, J., Kahl, P.: Computational Complexity and Feasibility of Data Processing and Interval Computations. Kluwer, Dordrecht (1997)

8. Kubica, B.J.: Interval Methods for Solving Nonlinear Constraint Satisfaction, Optimization, and Similar Problems: from Inequalities Systems to Game Solutions. Springer, Cham (2019)
9. Mayer, G.: Interval Analysis and Automatic Result Verification. de Gruyter, Berlin (2017)
10. Mendel, J.M.: Uncertain Rule-Based Fuzzy Systems: Introduction and New Directions. Springer, Cham (2017)
11. Moore, R.E., Kearfott, R.B., Cloud, M.J.: Introduction to Interval Analysis. SIAM, Philadelphia (2009)
12. Nguyen, H.T., Walker, C.L., Walker, E.A.: A First Course in Fuzzy Logic. Chapman and Hall/CRC, Boca Raton (2019)
13. Novák, V., Perfilieva, I., Močkoř, J.: Mathematical Principles of Fuzzy Logic. Kluwer, Boston (1999)
14. Petkovic, M.S., Petkovic, L.D.: Complex Interval Arithmetic and Its Applications. Wiley, New York (1998)
15. Rabinovich, S.G.: Measurement Errors and Uncertainty: Theory and Practice. Springer, New York (2005)
16. Zadeh, L.A.: Fuzzy sets. Inf. Control **8**, 338–353 (1965)

Check for updates

On Truncating Fuzzy Numbers with α-Levels

Juan-Carlos Figueroa-García[1(✉)], Roman Neruda[2], and Carlos Franco[3]

[1] Universidad Distrital Francisco José de Caldas, Bogotá, Colombia
jcfigueroag@udistrital.edu.co
[2] Department of Informatics, Charles University, Prague, Czech Republic
roman@cs.cas.cz
[3] Universidad del Rosario, Bogotá, Colombia
carlosa.franco@urosario.edu.co

Abstract. Unbounded fuzzy sets (in particular fuzzy numbers) are popular in different applications, but the implementation of a unbounded support is inconvenient since it is hard to do. This chapter presents a method for truncating a unbounded fuzzy number based on α-levels and a method to re-scale it in order to obtain a closed support fuzzy set i.e. bounded. An application to compute a nonlinear equation is presented and the proposed method is tested over different fuzzy sets in order to see its properties.

1 Introduction and Motivation

Some fuzzy applications are based on α-cuts and by consequence use the support of the fuzzy sets a.k.a $\alpha = 0$ level, such applications include fuzzy optimization models like fuzzy linear programming, fuzzy decision making techniques like TOPSIS, fuzzy differential equations among others. The most common membership functions used in those applications have bounded support since they use the $\alpha = 0$ level to obtain the widest interval of solutions and by consequence the use of unbounded fuzzy sets/numbers is restricted.

Truncating a fuzzy set/number is a common practice in control applications where the analyst divides a selected interval into uniform subintervals to perform fuzzy inference. This method is based on the expertise of the analyst and it depends highly on an arbitrary selection of the support of each fuzzy set. An interesting analysis of the effect of truncating fuzzy sets into fuzzy control problems is provided by Moon, Moon & Lee [9] and its extension to a Type-2 fuzzy sets environment was given by Greenfield & Chiclana [6].

To overcome the issue of having unbounded support fuzzy sets in some applications, we propose to define a β-level set in order to re-scale the membership function of a unbounded fuzzy set using the support of such β-level. The proposed method obtains a re-scaled fuzzy set which can be easily represented by α-levels and used wherever bounded support sets are required.

K. Cohen et al. (Eds.): NAFIPS 2023, LNNS 751, pp. 258–267, 2023.
https://doi.org/10.1007/978-3-031-46778-3_24

The chapter is organized as follows: Sect. 1 shows the introduction. Section 2 presents definitions on fuzzy sets/numbers; Sect. 3 presents a method for truncating unbounded fuzzy numbers using a β-level; Sect. 4 presents three examples and Sect. 5 presents some concluding remarks.

2 Basics on Fuzzy Numbers

Firstly, we establish basic notations. $\mathcal{F}(X)$ is the class of all fuzzy sets defined over a universe of discourse X. A fuzzy set $A : X \to [0, 1]$ is the set of ordered pairs between $x \in X$ and its membership function $\mu_A(x) \in [0, 1]$ i.e.,

$$A = \{(x, \mu_A(x)) \,|\, x \in X\}. \tag{1}$$

A fuzzy number *FN* (see Bede [1], Diamond & Kloeden [2] and Klir & Yuan [8]) is as follows:

Definition 1. *Let $A : \mathbb{R} \to [0, 1]$ be a fuzzy subset of the reals, then $A \in \mathcal{F}_1(\mathbb{R})$ its called fuzzy number if there exists a closed interval $[x_l, x_r] \neq \emptyset$ with membership degree $\mu_A(x)$ such that:*

$$\mu_A(x) = \begin{cases} 1 & \text{for } x \in [\check{x}_c, \hat{x}_c], \\ l(x) & \text{for } x \in [-\infty, x_l], \\ r(x) & \text{for } x \in [x_r, \infty], \end{cases} \tag{2}$$

where $\mu_A(x) = 1$ for the subset $x \in [\check{x}_c, \hat{x}_c]$, $l : (-\infty, x_l) \to [0, 1]$ is monotonic non–increasing, continuous from the right i.e. $l(x) = 0$ for $x < x_l$; $l : (x_r, \infty) \to [0, 1]$ is monotonic non–decreasing continuous from the left i.e. $r(x) = 0$ for $x > x_r$.

A fuzzy number A satisfies the following properties (see Bede [1]):

i) A is normal, i.e. $\exists\, x' \in \mathbb{R}$ such that $A(x') = 1$;
ii) A is α-convex (i.e. $A(\alpha x + (1 - \alpha)y) \geqslant \min\{A(x), A(y)\}$, $\forall\, \alpha \in [0, 1]$);
iii) A is upper semicontinuous on $\mathbb{R}$, i.e. $\forall \epsilon > 0\, \exists \delta > 0$ such that

$$A(x) - A(x') < \epsilon,\ |x - x'| < \delta;$$

The α*-cut* of a fuzzy number $A \in \mathcal{F}_1(\mathbb{R})$ namely ${}^{\alpha}A$ is the set of values with a membership degree equal or greatest than α i.e.

$${}^{\alpha}A = \{x \,|\, \mu_A(x) \geqslant \alpha\}\ \forall\, x \in X, \tag{3}$$

$${}^{\alpha}A = \left[\inf_x {}^{\alpha}\mu_A(x),\ \sup_x {}^{\alpha}\mu_A(x)\right] = \left[\check{x}_\alpha,\ \hat{x}_\alpha\right]. \tag{4}$$

The *support* of $\tilde{A} \in \mathcal{F}_1(X)$ namely $S(\tilde{A})$ is the set of all values $x \in X$ with positive membership i.e.

$$S(A) = \{x \,|\, \mu_A(x) > 0\} = [\check{x}, \hat{x}]. \tag{5}$$

The *core* of $A \in \mathcal{F}_1(X)$ namely $K(A)$ is the set of all values $x \in X$ with maximum membership i.e.

$$K(A) = \{x \mid \sup_{\forall x} \mu_{\tilde{A}}(x)\} = [\check{x}_c, \hat{x}_c]. \tag{6}$$

The cardinality $|A|$ is

$$|A| = \int_X \mu_A \, dx \tag{7}$$

its *expected value* $E_A[X]$ namely centroid $C(A)$ is defined as follows.

Definition 2. *Let μ_A be integrable over $X \subseteq \mathbb{R}$, the expected value of A namely centroid of A i.e. $C(A)$ is defined as:*

$$E_A[X] = C(A) = \frac{1}{|A|} \int_X x \, \mu_A \, dx. \tag{8}$$

and its *variance* $V(A)$ is as follows (see Figueroa-García et al. [4,5]):

$$V(A) = E_A[(X - E_A[X])^2] \tag{9}$$

$$= E_A[X^2] - C(A)^2 \tag{10}$$

$$= \frac{1}{|A|} \int_X (x - C(A))^2 \mu_A \, dx. \tag{11}$$

3 A β-Level Method for Truncating Fuzzy Numbers

Nonlinear fuzzy numbers (Gaussian, exponential, logistic membership functions, etc.) are among the most popular shapes in mathematical analysis and applications, but they are unbounded i.e. $S(A) \in [-\infty, \infty]$ which leads to have non-compact support with unbounded/unfeasible results. To overcome this issue we propose a method to truncate and re-scale unbounded fuzzy sets via β-levels similar to the *tolerance value* approach proposed by Bede [1] (see Bede [1], Section 4.3), as follows.

Definition 3 (β truncated method). *Let $A \in \mathcal{F}_1(\mathbb{R})$ be a unbounded fuzzy number and $\beta \in [0,1]$ be an α-cut, then a β-truncated fuzzy number A^β is an ordered pair $\{(A, \mu_A^\beta(x)) | x \in X\}$ such that:*

$$\mu_A^\beta(x) = \begin{cases} \dfrac{\mu_A(x) - \beta}{1 - \beta}, & x \in [\inf \mu_A^{-1}(\beta), \sup \mu_A^{-1}(\beta)] \\ 0, & x \notin [\inf \mu_A^{-1}(\beta), \sup \mu_A^{-1}(\beta)] \end{cases} \tag{12}$$

where $\beta \in (0,1)$ is a scale factor and its support is the α-level given by $\beta \in [0,1]$:

$$S(A^\beta) = [\check{x}^\beta, \hat{x}^\beta]. \tag{13}$$

As expected, there is some bias induced by β since it affects the shape of μ_A, so the centroid and variance of A^β can be expressed as follows.

Definition 4 (Some order statistics for A^β). *Let A^β be a β-truncated fuzzy number. The cardinality $|A^\beta|$ is*

$$|A^\beta| = \frac{1}{1-\beta}\int_{\check{x}_\beta}^{\hat{x}_\beta} (\mu_A(x) - \beta)\, dx = \frac{1}{1-\beta}\left(\int_{\check{x}_\beta}^{\hat{x}_\beta} \mu_A(x)\, dx - \beta(\hat{x}_\beta - \check{x}_\beta)\right) \quad (14)$$

its centroid $C(A^\beta)$ is defined as follows.

$$C(A^\beta) = \frac{1}{|A^\beta|}\int_{\check{x}_\beta}^{\hat{x}_\beta} \frac{x(\mu_A(x) - \beta)}{1-\beta}\, dx. \quad (15)$$

and its variance *$V(A^\beta)$ is as follows:*

$$V(A^\beta) = \frac{1}{|A^\beta|}\int_{\check{x}_\beta}^{\hat{x}_\beta} (x - C(A^\beta))^2 \frac{(\mu_A(x) - \beta)}{1-\beta}\, dx. \quad (16)$$

The α-level for the re-scaled set A^β is then renamed as follows:

$$^\alpha A^\beta = \left[\inf_x {}^\alpha\mu_{A^\beta}(x),\ \sup_x {}^\alpha\mu_{A^\beta}(x)\right] = [\check{x}_\beta,\ \hat{x}_\beta]. \quad (17)$$

where $[\check{x},\ \hat{x}] \supseteq [\check{x}^\beta,\ \hat{x}^\beta]$ and $[\check{x}_\alpha,\ \hat{x}_\alpha] \supseteq [\check{x}_\beta,\ \hat{x}_\beta]\, \forall \beta > 0$.

It is clear that the bigger β the bigger the bias on $C(A^\beta)$ and $V(A^\beta)$ so the selection of β is a degree of freedom to be kept in mind. Large values of β i.e. $\beta > 0.01$ lead to modify $C(A^\beta)$ and $V(A^\beta)$ in a considerable amount while very small values of β i.e. $\beta < 0.00001$ can lead to have a large support which is what the analyst is trying to avoid when truncating a fuzzy set. In order to see the behavior of β and how it affects $\mu_A, C(A)$ and $V(A)$ some examples are introduced in next section.

4 Examples

Two of the most popular unbounded membership functions used in fuzzy logic and fuzzy sets theory are Gaussian and exponential fuzzy sets. However, in some applications (control, optimization, modelling) is important to have support–bounded fuzzy sets, so its β-truncated form can be helpful in the analysis. Exponential FNs namely $E(c, \lambda_l, \lambda_r)$ are defined as follows:

$$A(x) = \begin{cases} exp\,(-\lambda_l(c - x))\,; & x \leqslant c \\ exp\,(-\lambda_r(x - c))\,; & x > c \end{cases}$$

and Gaussian FNs namely $G(c, \delta_l, \delta_r)$ are defined as follows:

$$A(x) = \begin{cases} exp\left(-\frac{1}{2}\left(\frac{x-c}{\delta_l}\right)^2\right); & x \leqslant c \\ exp\left(-\frac{1}{2}\left(\frac{x-c}{\delta_r}\right)^2\right); & x > c \end{cases}$$

whose β-truncated representations are described in Definitions 3 and 4.

4.1 Symmetric Examples

This example presents the β-truncated forms for symmetric Exponential i.e. $\lambda_l = \lambda_r$ and Gaussian FNs i.e. $\delta_l = \delta_r$ for different values of β. The obtained results are shown in Table 1.

Table 1. Results of the symmetric examples

Exponential set $c = 5, \lambda_l = \lambda_r = 0.5$						Gaussian set $c = 50, \delta_l = \delta_r = 3$					
β	$\check{x}^\beta$	$\hat{x}^\beta$	$C(A)$	$Var(A)$	$\|A\|$	β	$\check{x}^\beta$	$\hat{x}^\beta$	$C(A)$	$Var(A)$	$\|A\|$
0	$-\infty$	∞	5	0,5	1	0	$-\infty$	∞	50	9	7,520
0,0001	0,395	9,605	5	0,491	0,999	0,0001	37,124	62,876	50	8,981	7,518
0,001	1,546	8,454	5	0,460	0,993	0,001	38,849	61,151	50	8,877	7,504
0,01	2,697	7,303	5	0,358	0,954	0,01	40,895	59,105	50	8,312	7,394
0,02	3,044	6,956	5	0,304	0,920	0,02	41,609	58,391	50	7,897	7,291
0,03	3,247	6,753	5	0,269	0,892	0,03	42,055	57,945	50	7,564	7,198
0,04	3,391	6,609	5	0,242	0,866	0,04	42,388	57,612	50	7,277	7,111
0,05	3,502	6,498	5	0,220	0,842	0,05	42,657	57,343	50	7,024	7,029
0,06	3,593	6,407	5	0,202	0,820	0,06	42,884	57,116	50	6,793	6,950
0,07	3,670	6,330	5	0,186	0,800	0,07	43,081	56,919	50	6,582	6,874
0,08	3,737	6,263	5	0,173	0,780	0,08	43,257	56,743	50	6,387	6,800
0,09	3,796	6,204	5	0,161	0,762	0,09	43,416	56,584	50	6,204	6,728
0,1	3,849	6,151	5	0,150	0,744	0,1	43,562	56,438	50	6,032	6,659

The results for the shapes, variances and cardinalities of the exponential shape are displayed in Fig. 1.

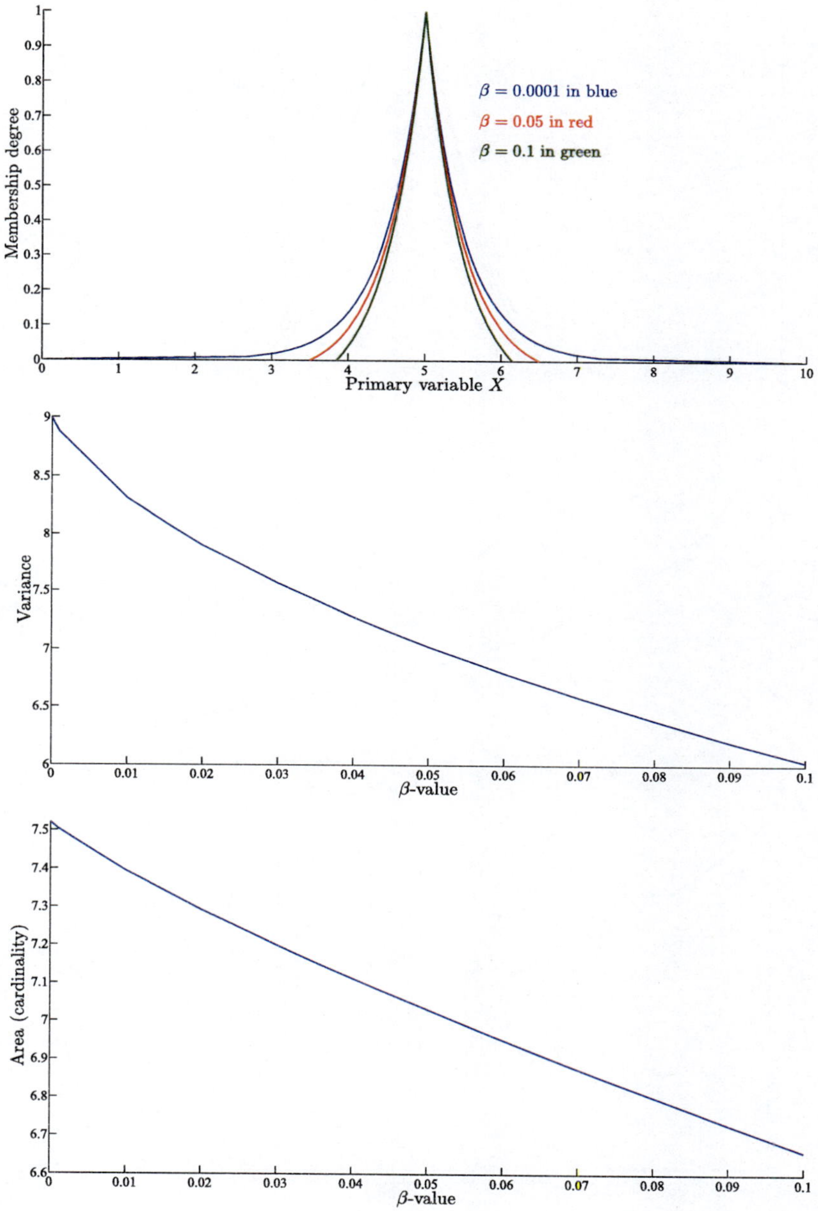

Fig. 1. Symmetric exponential example (β-truncated, variance and cardinality)

And the results for thei shapes, variances and cardinalities of the Gaussian shape are displayed in Fig. 2.

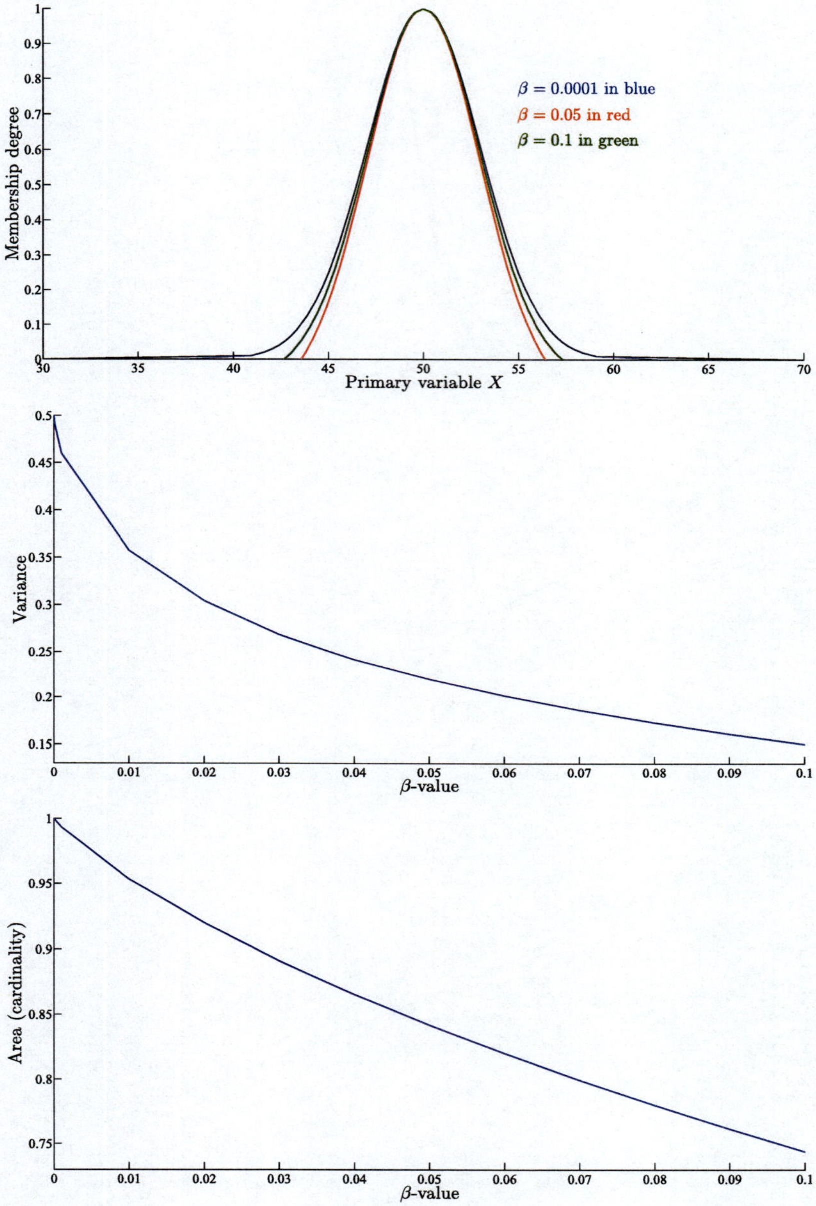

Fig. 2. Symmetric Gaussian example (β-truncated, variance and cardinality)

Note that the shape for $\beta = 0.0001$ in Figs. 1 and 2 is very close to the original μ_A while $\beta = 0.01$ and $\beta = 0.1$ (in red and green respectively) exhibit bigger differences when compared to the original μ_A.

As expected, there is a bias in $|A|$ and $V(A)$ of both exponential and Gaussian shapes induced by the value of β. However, $C(A)$ has no change since μ_{A^β} is symmetric. It is also interesting to see that $|A|$ and $V(A)$ decrease as β increases in a nonlinear way (specially $V(A)$ for small values of β).

4.2 Asymmetric Examples

This example presents the β-truncated forms for asymmetric Exponential i.e. $\lambda_l \neq \lambda_r$ and Gaussian FNs i.e. $\delta_l \neq \delta_r$ for different values of β. The obtained results are shown in Table 2.

Table 2. Results of the asymmetric examples

Exponential set $c = 12, \lambda_l = 1, \lambda_r = 2$						Gaussian set $c = 12, \delta_l = 2.2, \delta_r = 1.1$					
β	$\check{x}^\beta$	$\hat{x}^\beta$	$C(A)$	$Var(A)$	$\lvert A\rvert$	β	$\check{x}^\beta$	$\hat{x}^\beta$	$C(A)$	$Var(A)$	$\lvert A\rvert$
0	$-\infty$	$-\infty$	11.5	1.25	1.5	0	$-\infty$	$-\infty$	11.122	2.865	4.136
0,0001	2.790	16.605	11.502	1.226	1.499	0,0001	2.558	16.721	11.123	2.853	4.135
0,001	5.092	15.454	11.512	1.143	1.490	0,001	3.823	16.089	11.127	2.817	4.127
0,01	7.395	14.303	11.556	0.876	1.430	0,01	5.323	15.338	11.149	2.628	4.066
0,02	8.088	13.956	11.585	0.741	1.380	0,02	5.846	15.077	11.167	2.492	4.010
0,03	8.493	13.753	11.607	0.651	1.337	0,03	6.174	14.913	11.183	2.382	3.959
0,04	8.781	13.609	11.625	0.584	1.299	0,04	6.418	14.791	11.196	2.289	3.911
0,05	9.004	13.498	11.640	0.530	1.264	0,05	6.615	14.693	11.209	2.207	3.866
0,06	9.187	13.407	11.654	0.485	1.231	0,06	6.781	14.609	11.221	2.133	3.822
0,07	9.341	13.330	11.666	0.447	1.200	0,07	6.926	14.537	11.232	2.065	3.781
0,08	9.474	13.263	11.678	0.414	1.171	0,08	7.055	14.472	11.243	2.002	3.740
0,09	9.592	13.204	11.688	0.385	1.143	0,09	7.172	14.414	11.253	1.944	3.701
0,1	9.697	13.151	11.698	0.359	1.116	0,1	7.279	14.361	11.262	1.889	3.662

In this example, bias appears not only in $|A|$ and $V(A)$ but in $C(A)$ as well. It is also interesting to see that $|A|$ and $V(A)$ decrease as β increases and $C(A)$ seems to go in the opposite direction of β. We point out that these experiments do not proof that $C(A^\beta)$ increases as β decreases and vice versa.

4.3 Computation of the Economic Order Quantity (EOQ)

The EOQ model was introduced by Ford W. Harris in 1913 (see Harris [7] and Erlenkotter [3]) and became one of the most influential models in management and operations research. It is composed by three deterministic parameters: Cost of order a batch Co, unitary inventory cost Ci and the demand of the product D whose solution is the optimal quantity-to-order Q^* (by batch):

$$Q^* = \sqrt{Co \cdot D/Ci}$$

If we consider all parameters as fuzzy numbers then we can compute the fuzzy set of optimal quantities-to-order namely μ_{Q^*} as follows:

$$Q^* \equiv \bigcup_{\alpha \in L} {}_{\alpha}Q^* = \bigcup_{\alpha \in L} \{[\check{Q}^*_{\alpha}, \hat{Q}^*_{\alpha}] \,|\, L \subseteq [0,1]\},$$

$$\check{Q}^*_{\alpha} = \sqrt{\check{Co}_{\alpha} \cdot \check{D}_{\alpha} / \hat{Ci}_{\alpha}};\ \hat{Q}^*_{\alpha} = \sqrt{\hat{Co}_{\alpha} \cdot \hat{D}_{\alpha} / \check{Ci}_{\alpha}}$$

where $L \subseteq [0,1]$ is a set of α–cuts and the shapes of $\mu_{Co}, \mu_{Ci}, \mu_D$ are defined as Gaussian $G(50,3,3)$, Gaussian $G(3,0.2,0.2)$, and exponential $E(200,0.25,0.25)$. The obtained results for 200 α-cuts and two β-levels: $\beta = 0.0001$ (in blue) and $\beta = 0.1$ (in red) are displayed in Fig. 3.

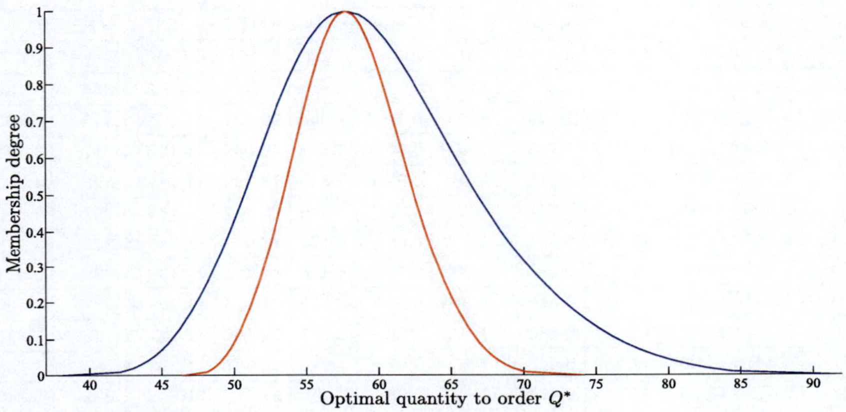

Fig. 3. Fuzzy set μ_{Q^*} of optimal quantities to order for $\beta = 0.0001$ and $\beta = 0.1$

The original support of μ_{Q^*} is unbounded i.e. $S(Q^*) = [-\infty, \infty]$ which heads up to misleading results and interpretability. When applying the β truncated method we obtain the following results:

$$\beta = 0.0001 \rightarrow |Q^*| = 17.37, C(Q^*) = 59.8, V(Q^*) = 51.74$$
$$\beta = 0.1 \rightarrow |Q^*| = 9.58, C(Q^*) = 58.17, V(Q^*) = 14.48$$

By using β truncated sets, the optimal quantity-to-order has a bounded support i.e. $S(Q^*) = [37.87, 91.57]$ for $\beta = 0.0001$ and $S(Q^*) = [46.41, 74.25]$ for $\beta = 0.1$ and still represents the nonlinear nature of Q^* so the analyst can see the set of possible values of Q^* in order to plan operations based on this information. It is also clear that $\beta = 0.0001$ produces a closer representation than $\beta = 0.1$. However, it depends on the selection of the analyst and how restrictive he/she is, so the selection of β is an additional degree of freedom.

5 Concluding Remarks

The proposed β-truncated method allows to transform unbounded into bounded fuzzy sets which is required in many fuzzy optimization, fuzzy decision making and fuzzy control applications. The proposed method is sensitive to the selection of β, so it is recommended to select small values i.e. $\beta = \{0.0001, 0.001, 0.01\}$ in order to modify the μ_A the less. Figures 1 and 2 show exponential and Gaussian fuzzy sets with their truncated transformations which can be used into any application as shown in Example 4.3.

There is a bias effect over $|A|, C(A)$ and $V(A)$ (except on $C(A)$ for symmetric fuzzy sets) which increases as β does, so the analyst should be careful when selecting β. The bigger β the smaller $|A|$ and $V(A)$ while $C(A)$ seems to go in the opposite direction than its symmetry.

Further experiments and applications to the proposed method including simulation (continuous/discrete), optimization (linear/nonlinear), decision making, etc. are required. Also its extension to other fuzzy sets representations such as Type-2 fuzzy sets, intuitionistic fuzzy sets, neutrosophic fuzzy sets, etc. is an interesting topic for future applications.

References

1. Bede, B.: Mathematics of Fuzzy Sets and Fuzzy Logic. Springer, Heidelberg (2015)
2. Diamond, P., Kloeden, P.: Metric topology of fuzzy numbers and fuzzy analysis. In: Dubois, D., Prade, H. (eds.) Fundamentals of Fuzzy Sets. FSHS, vol. 7, pp. 583–641. Springer, Boston (2000). https://doi.org/10.1007/978-1-4615-4429-6_12
3. Erlenkotter, D.: Ford Whitman Harris and the economic order quantity model. Oper. Res. **38**(6), 937–946 (1990). https://doi.org/10.1287/opre.38.6.937
4. Figueroa-García, J.C., Melgarejo-Rey, M.A., Soriano-Mendez, J.J.: On computing the variance of a fuzzy number. In: Orjuela-Cañón, A.D., Figueroa-García, J.C., Arias-Londoño, J.D. (eds.) ColCACI 2018. CCIS, vol. 833, pp. 89–98. Springer, Cham (2018). https://doi.org/10.1007/978-3-030-03023-0_8
5. Figueroa-García, J.C., Soriano-Mendez, J.J., Melgarejo-Rey, M.A.: On the variance of a fuzzy number based on the yager index. In: Proceedings of ColCaCi 2018, pp. 1–6. IEEE (2018)
6. Greenfield, S., Chiclana, F.: Type-reduced set structure and the truncated type-2 fuzzy set. Fuzzy Sets Syst. **352**, 119–141 (2018). https://doi.org/10.1016/j.fss.2018.02.012
7. Harris, F.: How many parts to make at once. Factory Mag. Manag. **10**, 135–136 (1913)
8. Klir, G.J., Yuan, B.: Fuzzy Sets and Fuzzy Logic: Theory and Applications. Prentice Hall, Hoboken (1995)
9. Moon, B.S., Moon, J.S., Lee, J.: Truncation effects of the fuzzy logic controllers. J. Fuzzy Logic Intell. Syst. **6**(1), 35–40 (1994)

A Fuzzy Inference System for an Optimal Spacecraft Attitude State Trajectory

Alex R. Walker(✉)

Sierra Lobo, Inc., Milan, OH, USA
walkerax@mail.uc.edu

Abstract. Attitude is the rotational orientation of an object, such as a rigid body spacecraft, in three-dimensional space. A spacecraft's attitude affects everything from a its ability to generate electrical power (via solar panels) and communicate with operators on the ground to its ability to dissipate heat and keep its internal electronics within their operating temperature ranges. Spacecraft mission design requires development of a concept of operations which includes defining the fixed attitude(s) or attitude state trajectories required to meet mission objectives. Usually, mission designers use their intuition to select from a small set of widely used attitudes and verify their choices satisfy power, communications, thermal, and other mission objectives and constraints. However, some missions may have objectives that are not easily met using one or more typical attitudes. In this case, a constrained optimization problem can be used to solve for an optimal attitude state trajectory. These trajectory optimization problems are usually transcribed into finite-dimensional nonlinear parameter optimization problems and solved using numerical methods. In this work, the optimal attitude of a satellite with a cryogenic fluid management experiment is considered. The "best" attitude minimizes experiment temperature subject to control hardware constraints. A fuzzy inference system is used to encode the attitude state trajectory as a function of sun vector parameters, and a genetic algorithm is used to tune the fuzzy inference system, achieving an optimal fuzzy attitude state trajectory.

Keywords: fuzzy · satellite · attitude · trajectory · genetic · optimization · quaternion

1 Introduction

When designing a mission for a satellite, it is important to develop a concept for what attitudes the spacecraft must attain in order to meet its mission objectives. For many missions, the decision of how to orient the satellite is obvious to experienced mission designers [1]. For a fixed hardware design, the attitude of a rigid body spacecraft affects its ability to generate power, communicate over radio frequencies, and manage temperature of its internal components [1, 2]. The attitude of a spacecraft is also typically important for accomplishing specific mission objectives. For instance, earth-observing satellites carry instruments which must point toward some point(s) of interest on or near the earth,

K. Cohen et al. (Eds.): NAFIPS 2023, LNNS 751, pp. 268–278, 2023.
https://doi.org/10.1007/978-3-031-46778-3_25

communications satellites carry antennas which must point toward ground stations on the earth or other satellites in orbit, and space telescopes carry instruments which must point toward some point(s) of interest in space [2].

Though many missions have obvious attitude concepts of operation, CryoCube, the 3U cubesat mission which motivated this work, is one such example of a mission with requirements that did not have an attitude solution which clearly met its mission objectives. The CryoCube mission was a cryogenic fluid management technology demonstration mission which carried a small tank of gaseous xenon and a sun shield intended to passively cool the tank to cryogenic temperatures, thereby liquefying the xenon in the tank [3]. This passive thermal control strategy required active attitude control to point the open end of the sun shield toward deep space, away from radiation heat sources like the earth and sun. The research team sought to minimize tank temperature for this demonstration mission [3].

A fuzzy inference system (FIS) was chosen to provide the ideal attitude state of the spacecraft, effectively acting as the guidance law for spacecraft attitude. A genetic algorithm was used to find the optimal state trajectory FIS. Recently, genetic fuzzy methods have been increasingly applied to aerospace applications. Such systems have been successfully applied to critical decision-making and control tasks for unmanned combat aerial vehicles (UCAVs) [4, 5], separation assurance and collision avoidance of heterogeneous networks of unmanned aircraft systems (UAS) [6, 7], position estimation of UAS [8], attitude control of small satellites [9, 10], and collaborative control of distributed space robotic systems [11].

2 Satellite Mathematical Model

2.1 Kinematics

The spacecraft attitude is represented using a unit quaternion. Quaternion parameterization of attitude offers many advantages and is the "preferred parameterization for spacecraft attitude control systems" [12]. The quaternion attitude representation is governed by the following ordinary differential equation:

$$\dot{q} = \begin{bmatrix} \dot{q}_0 \\ \dot{q}_1 \end{bmatrix} = \begin{bmatrix} -\omega^T q_1 \\ q_0\omega + q_1 \times \omega \end{bmatrix} \tag{1}$$

Here, q is the four-parameter quaternion, q_0 is the scalar real part of the quaternion, q_1 is the three-vector imaginary part of the quaternion, ω is the three-vector angular velocity of the spacecraft, and over-dots represent time derivatives. Generally, quaternions describe the orientation of any coordinate system with respect to any other coordinate system. The quaternion parameters returned by the FISs in this work express the ideal attitude with respect to a local-level coordinate system.

2.2 Dynamics

Rotational dynamics of a rigid body satellite with reaction wheels are derived from the angular momentum, h, of the system with mass moment of inertia, I:

$$h = I_b\omega_b + I_w\omega_w \tag{2}$$

Here subscripts b indicate quantities for the main spacecraft body and subscripts w indicate quantities for reaction wheels. The rate of change of angular momentum is equal to external disturbance (*dist*) plus control (*cntrl*) torques, τ, applied to the spacecraft. Expressed in spacecraft body-fixed coordinates, this is:

$$\dot{h} = \omega_b \times (I_b\omega_b + I_w\omega_w) + I_b\dot{\omega}_b + I_w\dot{\omega}_w = \tau = \tau_{dist} + \tau_{cntrl} \tag{3}$$

Rearranging to a form for use by an ordinary differential equation solver yields:

$$\dot{\omega}_b = -I_b^{-1}(\omega_b \times (I_b\omega_b)) - I_b^{-1}(I_w\dot{\omega}_w + \omega_b \times (I_w\omega_w)) + I_b^{-1}\tau_{dist} + I_b^{-1}\tau_{cntrl} \tag{4}$$

On the right-hand side, the first term is the typical rigid-body quadratic off-axis coupling, the second term is the contribution of the reaction wheels, the third term is the contribution of external disturbance torques, and the fourth term is the contribution of external (i.e. momentum dumping) control torques. The second term within the parenthesis of the reaction wheel contribution term is a small internally generated disturbance which is not directly controlled by the reaction wheels, whereas the first term of the reaction wheel contribution term is directly controlled with the demanded reaction wheel acceleration. Models of gravity gradient and aerodynamic disturbance torques and a model of the magnetic control torque were used in simulation, but not explicitly presented in this paper.

2.3 Temperature

The temperature state of any spacecraft is based on a complex history of internal and external temperatures and heat fluxes. External fluxes come from known, predictable objects, like the sun and earth. To calculate temperature, a finite element heat transfer model, Fig. 1, was developed in COMSOL Multiphysics. A time-dependent solution of the finite element model takes several minutes to solve, whereas a steady-state solution takes several seconds to solve. Because many evaluations of the model are required for optimization, steady-state temperature of the experiment tank is used as a proxy for actual time-dependent experiment tank temperature for optimization. Steady-state solutions are solved for a number of sun vector and earth vector positions prior to optimization, and spacecraft experiment tank temperatures are interpolated from these pre-computed solutions using sun and earth vectors needed during optimization. Though attitude dynamics could drive dynamics of the thermal model to a state significantly different from steady state, this is assumed to be extremely unlikely.

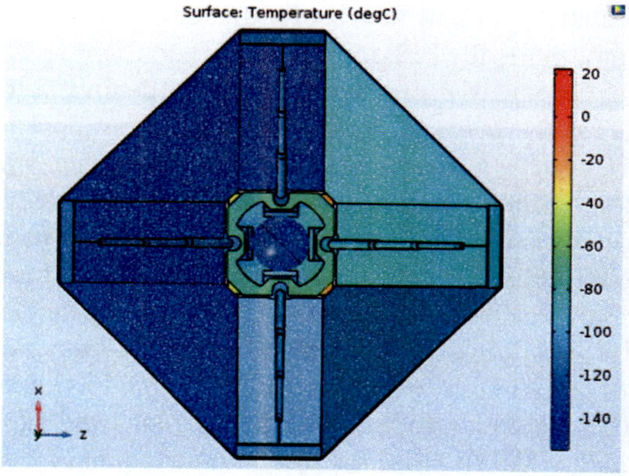

Fig. 1. COMSOL Multiphysics thermal model of CryoCube-1.

3 Fuzzy Inference System

The FIS encoding the optimal attitude state trajectory acts as a function outputting the best attitude state, given some inputs. General trajectory optimization yields a time-dependent series of states from a known start to a target end. However, these open-loop trajectories are often not very robust to errors in initial conditions or errors in modeling, so it is desirable that FIS input is not just time since an initial condition, but that the trajectories are valid for a variety of orbit geometries and do not have to be computed periodically for specific epochs. To determine variables most relevant to producing a desired attitude, it is helpful to think of a simple strategy one might use to accomplish the given mission objective. For minimizing experiment temperature, an obvious strategy might be to point the open end of the sunshield away from the sun and earth. This suggests both sun and earth vector are important for temperature minimization. Careful selection of reference coordinate system reduces required FIS input. For instance, the reference coordinate system is chosen to be a local-level coordinate system in which earth nadir is always along the –X-axis. This choice eliminates earth vector input to the FIS, leaving only sun vector input needed for the FIS. The unit sun vector is represented using a minimum set of parameters: an azimuth (α) and a coelevation (ε).

The output of the attitude trajectory FIS is an axis-angle representation of attitude. The axis-angle attitude representation is used here because it easily relates to the attitude quaternion via Eq. 5, and it can be expressed compactly as: azimuth of the axis (α_q), coelevation (or polar angle) of the axis (ε_q), and angle of rotation about the axis (θ_q).

$$\begin{gathered} q = \left[\cos\left(\frac{\theta_q}{2}\right)\ v_q \sin\left(\frac{\theta_q}{2}\right)\right] \\ v_q = \left[\sin(\varepsilon_q)\cos(\alpha_q)\ \sin(\varepsilon_q)\sin(\alpha_q)\ \cos(\varepsilon_q)\right]^T \end{gathered} \quad (5)$$

A Mamdani-type FIS was chosen. Precedents for each rule are combined using minimization. Inference and aggregation of outputs are accomplished using maximization. And defuzzification uses the centroid method. The membership functions (MFs) are chosen to provide a relatively simple means of evaluation onboard the spacecraft and to provide a means of modifying the FIS during optimization.

All input MFs are triangular, parameterized by three distinct values. Azimuth input linguistic variables span the range [0 360]° and coelevation input linguistic variables span the range [0 180]°. MFs of these linguistic variables are constrained to provide a consistent treatment of the inputs at the apparent singularities of the spherical coordinate representation used. These singularities occur at the boundaries of the input ranges: 0 and 360° azimuth and 0 and 180° coelevation. MFs are placed at these locations so their maximum membership occurs on the boundary. (Note this placement leaves only one degree of freedom for specifying boundary MFs, whereas all three degrees of freedom are available to specify other input MFs.) Because 0° azimuth is equal to 360° azimuth, rules including the 0 degree azimuth boundary MF must match the rules including the 360 degree azimuth boundary MF to prevent a discontinuity. Additionally, at coelevations of 0° and 180°, the direction of the vector parameterized in spherical coordinates is independent of the azimuth, so the rule base collapses to one rule, rather than one rule per azimuth MF, there.

Like input MFs, output MFs are all triangular. The output axis azimuth variable spans the range [0 360]°, the output axis coelevation angle spans the range [0 180]°, and the output rotation angle spans the range [0 180]°. All three degrees of freedom of each output MF can be modified by the optimization algorithm. However, the centroids of the output MFs are constrained to lie within the output range, ensuring the composite output MF centroid lies within the acceptable output range. Additionally, width of output MFs is constrained to be less than or equal to the size of the range of the output linguistic variable. The centroid is calculated over the range [–180 540]° for the azimuth variable, over the range [–90 270]° for the coelevation variable, and over the range [–90 270]° for the rotation angle variable (Fig. 2).

In general, an attitude trajectory is a time-dependent sequence of attitude states which may or may not be continuous. Discontinuous trajectories define a series of waypoints which an attitude control algorithm could track, possibly by performing large angle slew maneuvers between trajectory waypoints. When tracking a discontinuous trajectory, the attitude states a spacecraft attains between waypoints may be decidedly non-optimal. Therefore, the attitude trajectory FIS must produce attitude trajectories that are at least continuous in time. This FIS may be viewed generically as the set of functions, F, in Eq. 6 that take time-continuous azimuth and elevation of the sun vector and output attitude parameters which can be input to Eq. 5 to find the attitude quaternion.

$$\begin{aligned} \alpha_q &= F_\alpha(\alpha_s, \varepsilon_s) \\ \varepsilon_q &= F_\varepsilon(\alpha_s, \varepsilon_s) \\ \theta_q &= F_\theta(\alpha_s, \varepsilon_s) \end{aligned} \tag{6}$$

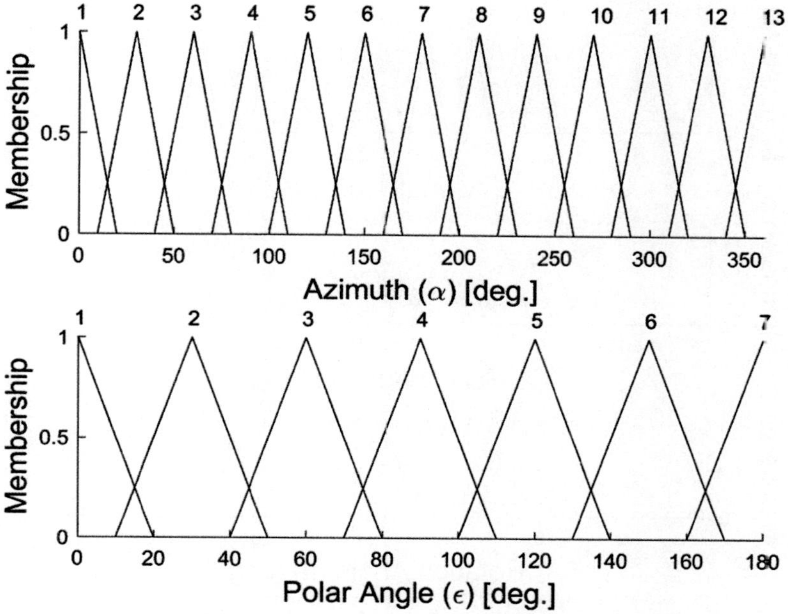

Fig. 2. Input MFs for FIS consistent with constraints of optimal trajectory.

In order for the attitude trajectory FIS to produce a time-continuous attitude output, it is sufficient for the FIS to have the following properties: for every point in the input range, the maximum value in the set of all degree of membership values output by all input MFs used in the rule base is greater than zero, and no input MF has a step change in membership value at any point in the input range. By definition, any point in the input domain which has no mapping to the output domain is undefined, so any point in a FIS's input domain that has zero input linguistic variable membership value for all input MFs has an undefined output; in practice, a FIS outputs some predetermined value, such as the average of the output range, which can yield discontinuities at the points the mapping becomes undefined. If just a single point is undefined, the output value will transition abruptly, a discontinuity, from the centroid of one composite output MF to the centroid of another composite output MF, given the rule base does not map the adjacent input MFs to identical output MFs, though even in this instance, at the transition point, the FIS may output an inconsistent value based on the value it returns when the mapping is undefined. As for the other condition that could cause a FIS to have a discontinuous output, it is clear that a step change in input MF value could result in a step change to the shape of the composite output MF, which could further result in a step change to the composite output MF centroid (Fig. 3).

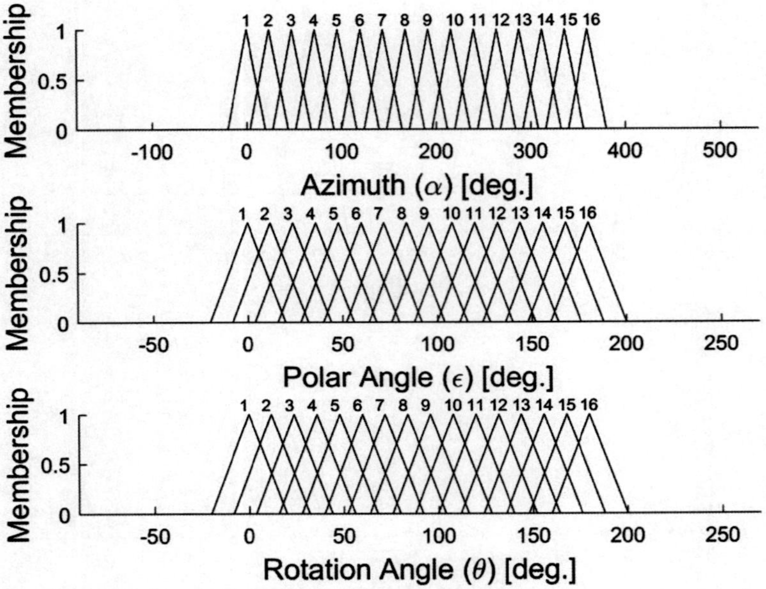

Fig. 3. Output MFs for FIS consistent with constraints of optimal trajectory.

Ideally, the state trajectory would also be consistent with the kinematic and dynamic equations of motion, such that control torques required to achieve the attitude are not undefined. However, inspection of Eqs. 1, 4, 5, and 6 suggest that the FIS must also be twice differentiable so the angular accelerations it outputs are at least continuous, but the optimal attitude FIS is not differentiable; in general, MFs and aggregation and inference operators must be carefully selected to guarantee FIS differentiability. Indeed, it was found that some FIS designs yielded trajectories with unacceptable reaction wheel accelerations, so a filter was implemented at the output of the FIS to sufficiently smooth the trajectory.

4 Optimization Problem

An attitude trajectory FIS which minimizes average experiment temperature, subject to reaction wheel angular velocity and angular acceleration constraints, is sought. Mathematically, the objective of the optimization problem is:

$$\begin{aligned} &\underset{q(t)}{\text{argmin}} \left(\frac{1}{P} \int_{t=0}^{P} T_{payload}(t)dt \right) \\ &\qquad\qquad orbit,\ attitude\ dynamics \\ subject\ to:\ &\qquad \|\omega_w\| \leq 6500\ RPM \\ &\qquad \|\dot{\omega}_w\| \leq 33\ RPM/s \end{aligned} \tag{7}$$

Here, P is the simulation period, t is time, and $T_{payload}$ is the payload temperature. A genetic algorithm (GA) was used to solve the attitude trajectory FIS optimization problem, because GAs are a robust metaheuristic search method which may be used for solving parameter optimization problems, like the FIS optimization problem [13]. However, constraints are not intrinsically handled by GAs, as they just use a fitness function to evaluate performance of a particular solution. One method of dealing with constraints is to transform the constrained optimization problem into an unconstrained optimization problem by modifying the fitness function to penalize solutions which violate constraints [13]. Often, the penalty is incorporated into the fitness function by subtracting some positive-definite function (e.g. the square) of the amount by which the constraint is violated. This approach does not guarantee the GA will find a solution which satisfies all constraints; the constraints may be so restrictive that no point in the solution space satisfies them all. However, this approach does allow the GA to search for a feasible region of the solution space. Modifying Eq. 7 for suitable use with the GA resulted in Eq. 8.

$$
\begin{gathered}
\underset{q(t)}{\operatorname{argmax}}\left(\left(500K - \frac{1}{P}\int_{t=0}^{P} T_{payload}(t)dt\right)\Delta + \Phi_{w,\omega}(P) + \Phi_{w,\alpha}(P)\right) \\
\Delta = \delta_{(1)(\Phi_{w,\omega}(T))}\delta_{(1)(\Phi_{w,\alpha}(T))} \\
\Phi_{w,\omega}(P) = \frac{1}{P}\int_{t=0}^{P} \phi_{w,\omega}(t)dt \\
\phi_{w,\omega}(t) = \begin{cases} 1 & \|\omega_w(t)\| \leq 6500\ RPM \\ 0 & \|\omega_w(t)\| > 6500\ RPM \end{cases} \\
\Phi_{w,\alpha}(P) = \frac{1}{P}\int_{t=0}^{P} \phi_{w,\alpha}(t)dt \\
\phi_{w,\alpha}(t) = \begin{cases} 1 & \|\dot{\omega}_w(t)\| \leq 33\ RPM/s \\ 0 & \|\dot{\omega}_w(t)\| > 33\ RPM/s \end{cases}
\end{gathered}
\tag{8}
$$

Here, 500 K is a constant 500 K, Δ is a function that equals zero until both constraints are met, $\Phi_{w,\omega}$ is the fraction of simulation time wheel velocity is acceptable, and $\Phi_{w,\alpha}$ is the fraction of simulation time wheel acceleration is acceptable. The modified objective function first seeks to maximize the amount of simulated time over which the constraints are not violated, then seeks the optimum of the true objective; the Kronecker delta functions, δ_{ij}, keep the contribution of the true objective equal to zero until all constraints are satisfied. This effectively forces the search into the feasible region, then to an optimum within the feasible region. Also, note that the problem has been reformulated into a maximization problem because GAs solve for maxima of fitness or objective functions.

5 Results

The optimal attitude profile FIS was found by the GA in 314 generations, taking 30 h 4 min, or an average of 5 min 45 s per generation to solve. The optimal FIS has an objective function value of 336.04[1].

Figure 4 shows the time-dependent spacecraft attitude with respect to the reference local-level coordinate system and the local-level-referenced celestial sphere with projected images of the earth, sun (yellow dot), and ground station (magenta dot) location. It clearly shows the open end of the sunshield always points generally away from both the sun and earth. Often, the sunshield open end (spacecraft-body-fixed + Z) points toward the + X-direction of the local-level coordinate system.

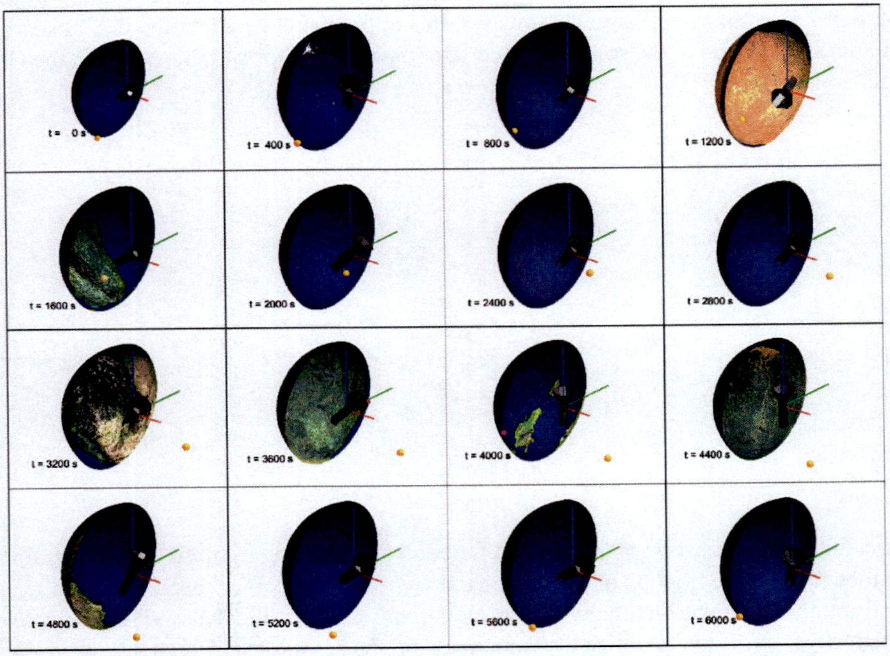

Fig. 4. Representative orbit pass orientation with respect to the reference coordinate system (red X, green Y, blue Z) is shown with the earth, sun (yellow sphere), and ground station (magenta sphere) projected on the celestial sphere.

Figure 5 shows that the experiment temperature drops to about 120 K in eclipse and rises to about 200 K in the sun. Again, it should be noted that these are not true temperatures but are interpolated temperatures calculated using a steady-state heat transfer model which is only a function of earth-sun geometry and vehicle attitude; true temperatures are time-dependent with dynamics that depend on prior temperature state(s)

[1] An estimate for the maximum theoretical value of this objective is 393.37 as the minimum achievable temperature is 108.63 (at only one particular earth-sun geometry and spacecraft orientation) and there are two constraints, each with a maximum value of 1.0.

in addition to sun-earth geometry and vehicle attitude. These steady-state temperature minima are achieved by keeping earth nadir and sun vectors an average of about 120° and no less than 90° from the spacecraft + Z-axis, shading the experiment from direct views of sun and the earth at all times.

Figures 6, 7, 8, 9 and 10, below, show input and output MFs of the optimal attitude profile FIS. All input sun vector azimuth MFs and all output MFs were used for the orbit, or pass, shown in Fig. 4, but only one input sun vector elevation MF was used for this orbit. All but two of the input elevation MFs participated in GA training.

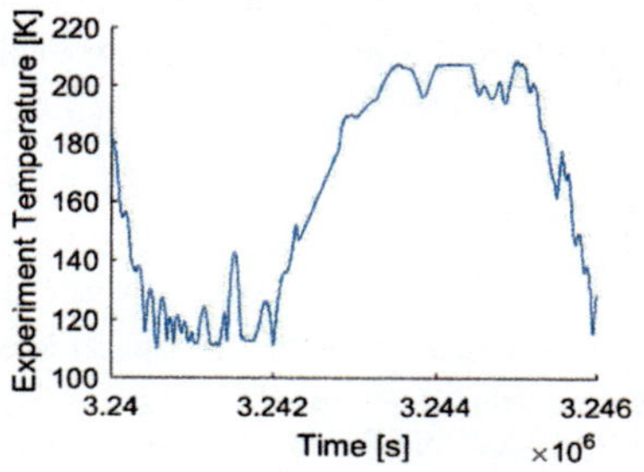

Fig. 5. Experiment temperature for a representative orbit using the optimal attitude FIS.

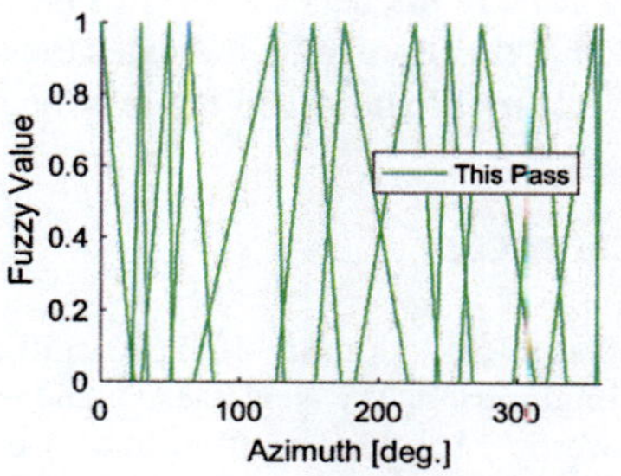

Fig. 6. Sun vector azimuth input MFs.

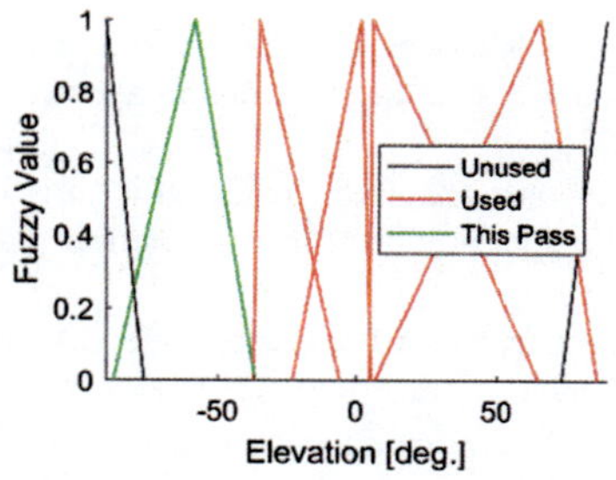

Fig. 7. Sun vector elevation input MFs.

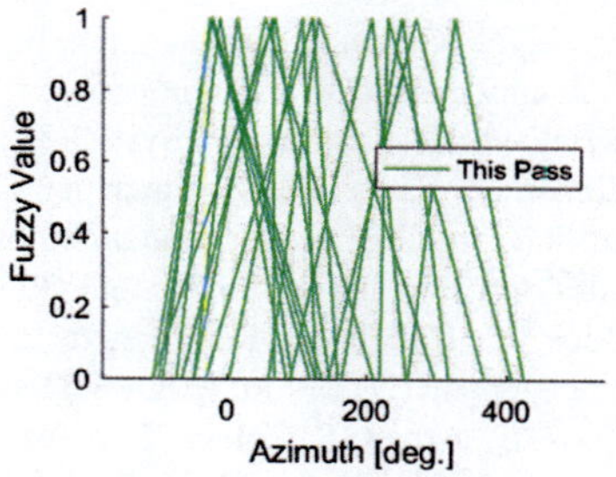

Fig. 8. Axis-angle axis azimuth output MFs.

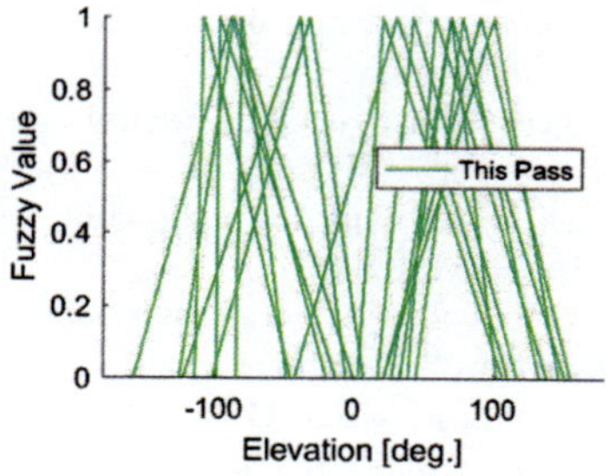

Fig. 9. Axis-angle axis elevation output MFs.

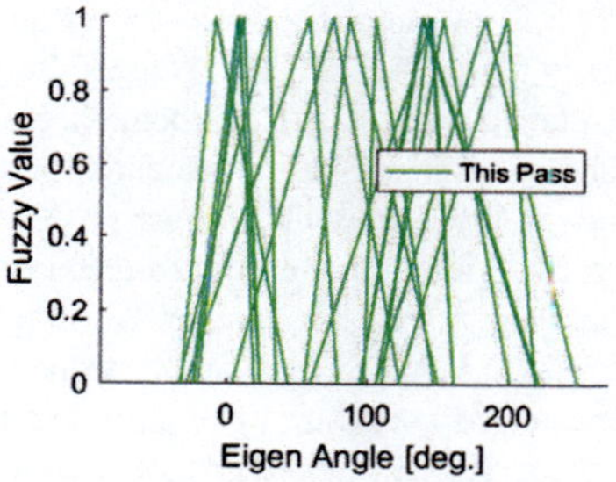

Fig. 10. Axis-angle angle output MFs

6 Conclusion

Solution of a temperature-optimal attitude state trajectory problem was found using a novel genetic-fuzzy approach. The optimal state trajectory was encoded in a FIS as a function of sun vector parameters rather than being expressed strictly as a function of time. This encoding scheme allows the FIS to be used generically for many different time periods rather than having to periodically compute optimal state trajectories and upload them to the spacecraft. A filter was designed to smooth the attitude profile output from the FIS, because the chosen Mamdani-style FIS does not guarantee certain differentiability constraints needed for it to be consistent with spacecraft attitude dynamics. The objective function value indicates the satellite is able to maintain an average experiment temperature of about 165 K, near the cryogenic range ($<$150 K).

References

1. Starin, S.R., Eterno, J.: Spacecraft subsystems II—control systems. In: Space Mission Engineering: The New SMAD, 2nd ed., Microcosm Press, pp. 565–600 (2015)
2. Wertz, J.R.: Spacecraft Attitude Determination and Control. Reidel, Dordrecht, Holland (1978)
3. Berg, J.J., Oliveira, J.M., Congiardo, J.F., Walls, L.K., Putman, P.T., Haberbusch, M.S.: Thermal performance of a cryogenic fluid management cubesat mission. Thermal Fluids Analysis Workshop, KSC-2013–233, Daytona Beach, FL (2013)
4. Ernest, N., Cohen, K., Kivelevitch, E., Schumacher, C., Casbeer, D.: Genetic fuzzy trees and their application towards autonomous training and control of a squadron of unmanned combat aerial vehicles. Unmanned Syst. **3**(3), 185–204 (2015)
5. Ernest, N., Carroll, D., Schumacher, C., Clark, M., Cohen, K., Lee, G.: Genetic fuzzy based artificial intelligence for unmanned combat aerial vehicles control in simulated air combat missions. J. Defense Manage. **6**(1) (2016)
6. Cook, B., Cohen, K., Kivelevitch, E.: A fuzzy logic approach for low altitude UAS traffic management (UTM). In: AIAA SciTech, AIAA Infotech @ Aerospace, San Diego, CA (2016)
7. Cook, B., Arnett, T., Cohen, K.: A fuzzy logic approach for separation assurance and collision avoidance for unmanned aerial systems. Mod. Fuzzy Control Syst. Appl. Intech, 225–256 (2017)
8. Cook, B., Cohen, K.: Multi-source sensor fusion for small unmanned aircraft systems using fuzzy logic. FUZZ-IEEE, Naples, Italy (2017)
9. Walker, A., Putman, P., Cohen, K.: Solely magnetic genetic/fuzzy-attitude-control algorithm for a cubesat. J. Spacecr. Rocket.Spacecr. Rocket. **52**(6), 1627–1639 (2015)
10. Calvo, D., Avilés, T., Lapuerta, V., Laverón-Simavilla, A.: Fuzzy attitude control for a nanosatellite in low earth orbit. Expert Syst. Appl. **58**, 102–118 (2016)
11. Sathyan, A., Ma, O., Cohen, K.: Intelligent approach for collaborative space robot systems. In: 2018 AIAA SPACE and Astronautics Forum and Exposition (2018)
12. Shuster, M.D.: A survey of attitude representations. J. Astronaut. Sci. **41**(4), 439–517 (1993)
13. Goldberg, D.E.: Genetic Algorithms in Search, Optimization, and Machine Learning. Addison Wesley Longman Inc, Crawfordsville (1989)

Causality: Hypergraphs, Matter of Degree, Foundations of Cosmology

Cliff Joslyn[1], Andres Ortiz-Muñoz[2], Edgar Daniel Rodriguez Velasquez[3,4], Olga Kosheleva[5], and Vladik Kreinovich[6](✉)

[1] Mathematics of Data Science, Pacific Northwest National Laboratory, 1100 Dexter Avenue N # 500, Seattle, WA 98109, USA
cliff.joslyn@pnnl.gov

[2] Santa Fe Institute, 1399 Hyde Park Road, Santa Fe, NM 87501, USA
aortiz@santafe.edu

[3] Department of Civil Engineering, Universidad de Piura in Peru (UDEP), Av. Ramón Mugica 131, Piura, Peru
edgar.rodriguez@udep.pe

[4] Department of Civil Engineering, University of Texas at El Paso, 500 W. University, El Paso, TX 79968, USA
edrodriguezvelasquez@miners.utep.edu

[5] Department of Teacher Education, University of Texas at El Paso, 500 W. University, El Paso, TX 79968, USA
olgak@utep.edu

[6] Department of Computer Science, University of Texas at El Paso, 500 W. University, El Paso, TX 79968, USA
vladik@utep.edu

Abstract. The notion of causality is very important in many applications areas. Because of this importance, there are several formalizations of this notion in physics and in AI. Most of these definitions describe causality as a crisp ("yes"-"no") relation between two events or two processes – cause and effect. However, such descriptions do not fully capture the intuitive idea of causality: first, often, several conditions are needed to be present for an effect to occur, and, second, the effect is often a matter of degree. In this paper, we show how to modify the current description of causality so as to take both these phenomena into account – in particular, by extending the notion of directed acyclic graph to hypergraphs. As a somewhat unexpected side effect of our analysis, we get a natural explanation of why, in contrast to space-time of Special Relativity – in which division into space and time depends on the observer, in cosmological solutions there is a clear absolute separation between space and time.

1 Formulation of the Problem

Causality is an important notion in many areas of research, in many practical situations. To be able to solve the corresponding practical problems, it is important to have an adequate description of what is causality.

Because this notion is so important, there are many formal descriptions of causality, in physics (see, e.g., [2,4,5,10,11]), in Artificial Intelligence (AI) – see, e.g., [9] – and

K. Cohen et al. (Eds.): NAFIPS 2023, LNNS 751, pp. 279–289, 2023.
https://doi.org/10.1007/978-3-031-46778-3_26

in many other areas. The very fact that there are many different descriptions is a good indication that the current definitions are not perfect – and most authors writing about causality explicitly state that their definitions, while useful in many practical cases, do not fully capture the intuitive idea of causality.

One of the important aspects of causality that is not properly captured by the current definitions is that in some situations, causality is a matter of degree. This has been stated, on many occasions, by Lotfi Zadeh and other researchers working in fuzzy techniques; see, e.g., [1,3,6–8,12]; this has been shown on a physical level; see, e.g., [4,5]. In this paper, we show how we can extend the current descriptions of causality to take this – and other – important aspects of causality into account.

2 What is Causality: A Brief Reminder and the Need to Distinguish Between Two Aspects of Causality

Two Main Areas Where the Notion of Causality is Used. In order to motivate the following definitions, let us recall what we usually mean by causality. This notion is very important both:

- in everyday life – and, as an extension, in Artificial Intelligence (AI), and
- in physics.

Let us briefly describe how this notion is used in both application areas. Once we come up with this description, we will emphasize the fact that is not as widely known as it should be – that while in these two areas, we use the same term "causality", in these two areas, we talk about two *different* aspects of this notion. Thus, a proper formalization of this notion must take this difference into account.

Commonsense and AI Aspect of Causality: An Example. If a road pavement needs repairs while it was supposed to still last, we need to know what caused it:

- it could be that there were mistakes in design,
- it could be that the design was correct, but the wrong materials were used, or, more generally, that some mistakes were made when this road segment was built;
- it could be that there were unusual extreme weather conditions that were not predicted and that caused the premature pavement deterioration;
- it could be that the load on this road segment was much higher than anticipated, and this caused the pavement deterioration;
- it could be that the formulas describing how pavement deteriorates with time – formulas that were used in the original design – were incorrect, etc.

In this sense, when we say that a caused b, this means that if – everything else being equal – we would not have a, then we would not see b either. For example, if we conclude that the use of a wrong asphalt mix caused the pavement deterioration, this means that if proper material was used, the road would not have deteriorated – and that in a similar situation, if we use the proper material, the road will not deteriorate before its expected lifetime.

Causality in Physics: An Example. The notion of causality is very important in physics too, but there, it means something else. Specifically, in physics, when we say that an event e causally affect an event e', this means that a change in e *can* make some change in e'.

What is the Difference Between These Two Aspects. If a soccer player kicks a ball and it goes into the opponent's gate, then, both from the commonsense viewpoint and from the physics viewpoint, this kick was the cause of the resulting goal. So far so good, it may look like these two notions are the same, but here is a counterexample. In real life, there is usually some wind, and that wind slightly affects the trajectory of the ball.

- From the viewpoint of common sense, the wind was clearly *not* the cause of the goal – unless it was such a strong gust of wind that the ball (that was originally supposed to miss) was blown into the gate.
- However, from the viewpoint of theoretical physics, the wind *did* have a causal effect on the ball's trajectory – in contrast, e.g., to the wind 1000 km away at this same moment that did not have any way to causally affect the ball's trajectory.

In short, in the first approximation:

- in common sense, a causes b if whenever we have a, we have b as well, i.e., when the presence of a determines the presence of b;
- in theoretical physics, a causally affects b if a has some effect on b – but does not necessarily uniquely determines all the aspects of b.

Towards a More Realistic Description. The above first-approximation description implicitly assumed that everything is deterministic. In real life – and in physics – we can rarely fully predict the future events. At best, we can predict the *probabilities* of different future events; see, e.g., [2, 11]. With this in mind, the proper description of the two aspects of causality takes the following form:

- in common sense, a causes b if a *uniquely determines* the probability distribution of possible values of the quantities characterizing the event b, and
- in theoretical physics, a causally effects b if a has *some effect* on the probability distribution of possible values of the quantities characterizing the event b.

Comment. Of course, these explanations are still approximate: in reality, in common sense, we take into account that some minor changes may still happen, so we can have slightly different probability distributions of b – but, from the practical viewpoint, this description provides a good description of real-life situations; see, e.g., [9].

3 How Causality is Usually Described – and What is Missing in this Usual Description

How Causality is Usually Described. The usual description of causality is based on the following two natural properties of causality:

- if a causally influences b, then b cannot casually influence a – unless, of course, a and b are the same (and we extend the definition of causality to this trivial case);
- if a causally influences b and b causally influences c, then a casually influences c.

It makes sense, by the way, to formally extend the notion "a causally influences b" to the case when $a = b$ – since, as we have mentioned earlier, this notion means that a change in a can cause a change in b – and, of course, a change in a causes a change in a.

The same two properties are true if instead of "causally influences", we use "causes". In both cases, if we denote "a causally influences b" or "a causes b" by $a \leq b$, and if we extend the definition to allow the event to casually influence itself, we get the following properties for all a, b, and c:

- $a \leq a$
- if $a \leq b$ and $b \leq a$, then $a = b$; and
- if $a \leq b$ and $b \leq c$, then $a \leq c$ (*transitivity*).

In mathematics, a binary relation with these properties is known as an *order relation* – sometimes called *partial order*, to distinguish it from the case of *linear* (*total*) order, when for every a and b, we have either $a \leq b$ or $b \leq a$.

Because of this, causality is usually described as a *partial* order.

The First (Known) Limitation of the Usual Description. In practice, sometimes, several phenomena have to be present to cause the observed effect. For example, a building may be designed so as to withstand an earthquake, and it can be designed to withstand a tsunami, but if both unexpectedly happen at about the same time – as happened in Japan in 2011 – the building may collapse. In this case, the cause of the collapse is not just the earthquake – an earthquake itself would not have caused the collapse, and not the tsunami, but the fact that both were present.

A more mundane example is that to open a bank vault (or even to open a safe deposit box stored in a bank), you need two keys – none of them would open the vault by itself. It is difficult to describe this phenomenon by an order relation between the events.

A similar phenomenon can be imagined in physics, where it is important to describe cases in which an event a cannot affect an event e' unless some other event e'' was present.

The Second (Known) Limitation of the Usual Description. The second known limitation of the current definition of causality as a partial order is that it does not take into account that causality is a matter of degree: in some cases, a strongly influences b, in other cases, it only affects b with some degree. As a result, when we have a long causal chain in which a_1 affects a_2, a_2 affects a_3, etc., until a_{n-1} affects a_n, we can formally say that a_1 affects a_n – it is usually expressed by saying that the flapping of a butterfly's wings affects a future storm on another continent – but in reality, the degree of this effect decreases at every link of this chain, so that at the very end, the effect is not practically noticeable.

What We Do in the Following Sections. In the following sections, we show how these two limitations can be overcome, i.e., how we can generalize the usual definition to take these two features of causality into account.

4 How to Take into Account that We May Need Several Events to Cause the Observed Phenomenon

Towards a Natural Generalization of the Order Relation. In the deterministic approximation to the traditional commonsense description, the relation $a \le b$ means that if we know the state of a, we can uniquely predict the state of b. In the case that we want to describe, knowing the state of a is not enough to uniquely predict the state of b – to make this prediction, we need to known the state of several events $a_1, \ldots, a_n$. In other words, here, instead of knowing the state of a single event a, we need to know the state of all the events from some set A. It is therefore reasonable to consider the corresponding relation $A \le b$ that means that once we know the states of all the events from the set A, then we can uniquely determine the state of the event b.

Of course, the way we described it means that if we add other elements to the set A, we would still get this relation $A \le b$. So, a natural idea to distinguish "true causality" – which we will denote by $A < b$ – from the above relation is to define it as $A \le b$ (i.e., A influences b) but b is not an element of the set A and no proper subset of A influences the event b. So, if $A < b$, then we cannot have $A' < b$ for any proper subset $A' \subset A$, $A' \ne A$.

Comment. In particular, when A is the empty set, we can have $\emptyset < b$, meaning that all the properties of the event b are pre-determined, nothing can affect them.

What is the Natural Analogue of Avoiding the Casuality Loop, When a Causes b and b Causes a. The original definition of causality avoids non-intuitive loops by requiring that if $a \le b$ and $b \le a$, then $a = b$. In our case, a proper description is that if $A < b$, then we cannot have $B < a$ for any element $a \in A$ and for any set B including b.

What is the Natural Analogue of Transitivity. In the traditional transitivity, if b causes c, and a causes b, then a causes c. What if, in our case, we have $B < c$? Then, to be able to conclude that $A < c$, we need to make sure that all events $b \in B$ can be predicted by A, i.e., that $A < b$ for all $b \in B$. In this case, the set A uniquely determine c – but maybe not all elements of A are needed to predict c, i.e., all we can conclude is that $A \le c$, i.e., that some subset of the set A causes c.

Resulting Definition. Thus, we arrive at the following definition.

Definition 1. *Let U be a set. By a* causality relation *on U, we mean a relation $< \subseteq 2^U \times U$ between subsets of U and elements of U for which the following properties are satisfied:*

- *if $A < b$, then $b \notin A$;*
- *if $A < b$, $A' \subset A$ and $A' \ne A$, then $A' \not< b$;*
- *if $A < b$, then for any $a \in A$ and for any set B containing b, we cannot have $B < a$;*
- *if $A < b$ for all $b \in B$ and $B < c$, then we have $A' < c$ for some subset $A' \subseteq A$ of the set A.*

Terminological Comment. Partial order can be described by a directed acyclic graph (DAG), in which objects are vertices, and pairs (a,b) for which $a \le b$ and $a \ne b$ are edges. In this section, we consider a generalization of this notion to the case when the relation $A \le b$ involves, in general, more than one element $a \in A$ and thus, more than two objects. Such a generalization of a graph – in which we consider subsets consisting of more than 2 elements – is known as a *hypergraph*. From this viewpoint, the above definition provides a generalization of the notion of a directed acyclic graph to hypergraphs.

An Alterative Description. We started this section with the "determination" relation $A \le b$ meaning that knowing A uniquely determines b. In terms of the causality relation $A < b$, this means that either $b \in A$ or there exists a subset $A' \subset A$ for which $A' < b$.

A somewhat inconvenient issue with this notion is that, in contrast to the original definition of causality $a \le b$ that relates two objects of the same type, the relation $A \le b$ related objects of different types: sets and elements. It is therefore natural to extend this relation to relation between two objects of the same type, namely, between sets: we can define $A \le B$ if knowing all information about A provides us with the ability to predict all the values of all the elements $b \in B$. This possibility means that we can predict the values of each state $b \in B$, i.e., that we have $A \le b$ for all $b \in B$.

This new relation between sets has the following natural properties:

- if $B \subseteq A$, then $A \le B$;
- if $A \le B$ and $A \subseteq A'$, then $A' \le B$;
- if $A \le B$ and $B' \subseteq B$, then $A \le B'$.

In terms of the new relation, transitivity takes a very straightforward form: if $A \le B$ and $B \le C$, then $A \le C$.

Other properties of the causality relation can also be described in terms of this relation $A \le B$ if we take into account:

- that $A \le b$ simply means $A \le \{b\}$, and
- that, as we have mentioned, $A < b$ means that $b \notin A$ and that for all proper subsets $A' \subset A$, we have $A' \not\le b$.

5 How to Take into Account that Causality is a Matter of Degree

Why it is Necessary to Take this into Account. Another limitation of the traditional approach to causality is that this approach is black-and-white (crisp): in this approach, either an event a can influence event b or it cannot. In practice, at each moment of time, due to imperfection of measuring instruments, we can only detect effects of certain size – i.e., we can only confirm for certain pairs (a,b) that a can influence b. As technology progresses, we can detect weaker and weaker effects – and thus, confirm that other pairs are also causally connected.

A natural way to describe this situation is to assign, to each pair of events (a,b), a degree $d(a,b)$ to which a can influence b: the larger the effect, the larger this degree. At each moment of time, we can measure the effect with a certain accuracy and thus, we know this degree with the similar accuracy. A natural way to describe such uncertainty is by saying that we know the approximate value $\widetilde{d}$ of this degree, and we know the accuracy δ, i.e., that the only thing we know about the actual (unknown) degree d is that this degree is located somewhere in the interval $[\widetilde{d}-\delta,\widetilde{d}+\delta]$.

What are the Consequences of Taking this into Account. When we consider causality as crisp, then transitivity can be used indefinitely: if a_1 causes a_2, and a_2 causes a_2, etc., and a_{n-1} causes a_n, then a_1 causes a_n. In practice, the effect decreases with each transition: a_1 has a strong effect on a_2, a_2 may also have a strong direct effect on a_3, but the resulting indirect effect of a_1 on a_3 is smaller – since every time we have an effect, there is also some random influence that interferes with this effect. After many steps, the original effect becomes practically non-observable – just like if we transmit a message many times though noisy channels, the message will become corrupted more and more and eventually, will become completely unrecognizable.

In precise terms, instead of crisp transitivity, it is natural to expect the degree-based (fuzzy) version, when for each a, b, and c, the degree $d(a,c)$ to which a affects c is bounded by an "and"-operation (t-norm) $f_\&$ applied to the degrees $d(a,b)$ and $d(b,c)$:

$$d(a,c) \geq f_\&(d(a,b),d(b,c)).$$

This leads to a more realistic picture of causality. For example, if we use a t-norm for which we can have $f_\&(a,b)=0$ for some $a>0$ and $b>0$, then eventually, the degree of influence can get to 0. Thus, we can have, for example, a closed time loop, in which going to the future brings us back to the past – provided that the path was sufficiently long; see, e.g., [10]. This possibility makes physical sense: in billions of years, all traces of the original event will disappear.

Interesting Consequence of this Idea. The idea of causality as a matter of degree does not just provide us with a better fit with physical intuition. In this section, we will show that this idea also – somewhat surprisingly – helps us to understand a certain fact from fundamental physics, namely, a somewhat counterintuitive transition between:

- special relativity theory (and local effects of generic relativity) – that describe the local space-time, and
- cosmology that describes the global structure of space-time.

Let us explain what is counterintuitive in this transition.

One of the main principles of special relativity theory is the relativity principle, according to which, by observations inside the system, there is no way to determine whether the system is at rest or moving with a constant speed in the same direction. This principle was first formulated by Galileo who noticed that when a ship is moving in a calm sea, then, if we are inside a cabin with no windows, we cannot tell whether the ship is moving or not. Einstein combined this principle with the empirical fact that the speed of light c_0 has the same value for all observers – and concluded that a priori, there is no fixed time coordinate: time (and corresponding notion of simultaneity) differs for different observers. Neither time interval Δt nor spatial distance ρ are invariant: they change from observer to observer, The only quantity that remains invariant is the so-called *proper time*, which, in distance units, takes the form $\tau=\sqrt{c_0^2\cdot(\Delta t)^2-\rho^2}$. The impossibility to separate time and space is the main reason why in special relativity, we talk about 4-dimensional space-time, and its division into space and time depends on the observer. Transformation between coordinate systems corresponding to different observers is described by so-called Lorenz transformations.

In contrast, in most cosmological models, there is a very clearly determined time coordinate, there is a clear separation into space and time. Why is that? Why cannot we have cosmology that has the same symmetries as local space-time?

When we consider the usual – crisp – causality, the only answer to this question is that observations support cosmological models in which time and space are separated, and support equations of General Relativity that lead to these models. In this description, there is nothing fundamental about these types of cosmologies – from the purely mathematical viewpoint, there is nothing wrong with considering space-time of special relativity.

Interestingly, the situation changes if we take into account that causality is a matter of degree. It turns out that if we make this assumption, then we cannot keep all the symmetries of special relativity – cosmological split between time and space becomes inevitable. Let us describe this result in precise terms.

To come up with this explanation, let us recall that in Special Relativity, the speed of light c is the largest possible speed, so an event $e=(t,x_1,x_2,x_3)$ can influence the event $e'=(t',x_1',x_2',x_3')$ if and only if we can get from e to e' by traveling with a speed not exceeding the speed of light, i.e., if and only if the distance between the corresponding spatial points (x_1,x_2,x_3) and (x_1',x_2',x_3') is smaller than or equal to the distance $c_0\cdot(t'-t)$ that someone traveling with the speed of light can cover during the time interval $t'-t$.

Definition 2

- *Let c be a positive constant; we will call it* speed of light.
- *By* space-time of special relativity, *we mean an ordered set* $(M,\le)$ *in which* $M=\mathbb{R}^4$ *is the set of all 4-D tuples* $e=(t,x_1,x_2,x_3)$ *of real numbers, and the ordering relation has the form*
$$e=(t,x_1,x_2,x_2)\le e'=(t',x_1',x_2',x_3')\Leftrightarrow$$
$$c_0\cdot(t'-t)\ge\sqrt{(x_1-x_1')^2+(x_2-x_2')^2+(x_3-x_3')^2}.$$
- *If* $e=(t,x_1,x_2,x_2)\le e'=(t',x_1',x_2',x_3')$, *then by the* proper time $\tau(e,e')$ *we mean the quantity*
$$\tau(e,e')=\sqrt{c_0^2\cdot(t'-t)^2-(x_1-x_1')^2-(x_2-x_2')^2-(x_3-x_3')^2}.$$
- *We say that a mapping* $f:M\to M$ *is a* symmetry *if it preserves causality relation and preserves proper time, i.e., if* $e\le e'$ *is equivalent to* $f(e)\le f(e')$ *and* $\tau(e,e')=\tau(f(e),f(e'))$ *for all events e and* e' *for which* $e\le e'$.

Comment. It is known that for every two pairs (e,e') and (g,g') for which $\tau(e,e')=\tau(g,g')$, there exists a symmetry f that transforms e into g and e' into g'.

Let us now describe what we mean by taking into account that causality is a matter of degree. We will call the corresponding descriptions *realistic causality functions*. Their definition uses the notion of an "and"-operation, so let us first define this auxiliary notion.

Definition 3. *By an* "and"-operation, *we mean a continuous function* $f_\&: [0,1] \times [0,1] \to [0,1]$ *for which* $f_\&(1,1) = 1$.

Comments.

- Every t-norm – as defined in fuzzy logic – satisfies this condition. However, this definition is much weaker than the usual definition of t-norm in fuzzy logic: it allows many functions which are not t-norms – e.g., they are not necessarily associative. The reason for us formulating such a weak definition is that we want to prove the main result of this section under most general assumptions. For example, our result stands if we consider non-commutative and non-associative operations instead of t-norms – as long as they satisfy the above definition.
- Continuity at the point $(1,1)$ means that for every $\varepsilon > 0$ there exists $\delta > 0$ such that if $x \geq 1 - \delta$ and $y \geq 1 - \delta$, then $f(x,y) \geq 1 - \varepsilon$.

We want to define a function $d(e,e')$ that describes, for each two events e and e' for which $e \leq e'$, the degree to which e can influence e'. What are the natural property of such a function?

- Of course, it is a fact that each event e "causes" itself – in the sense that if we know the event e, we can uniquely describe this same event. So, we must have $d(e,e) = 1$.
- The fact that causality is a matter of degree means that the case when $e = e'$ should be the only case when we have $d(e,e') = 1$. In all other cases, we must have $d(e,e') < 1$.
- In physics, most dependencies are continuous, so it is reasonable to require that the function $d(e,e')$ should also be continuous.
- Finally, for the case when $e \leq e' \leq e''$, we should have

$$d(e,e'') \geq f_\&(d(e,e'), d(e',e'')).$$

Thus, we arrive at the following definition.

Definition 4. *Let* $f_\&$ *be an "and"-operation. By a* realistic causality function *on* $(M, \leq)$, *we mean a continuous function* $d(e,e')$ *with values from the interval* $[0,1]$ *that is defined for all pairs* (e,e') *for which* $e \leq e'$ *and that satisfies the following properties:*

- $d(e,e') = 1$ *if and only if* $e = e'$, *and*
- $d(e,e'') \geq f_\&(d(e,e'), d(e',e''))$ *for all* $e, e', e'' \in M$.

Definition 5. *We say that a realistic causality function is* invariant *with respect to a symmetry* f *if* $d(e,e') = d(f(e), f(e'))$ *for all* $e \leq e'$.

Proposition 1. *No realistic causality function is invariant with respect to all the symmetries.*

Comment. This result explains that, when we take into account that causality is a matter of degree, then we cannot have a space-time that has the same invariance properties as the space-time of Special Relativity: some corresponding symmetries have to be abandoned – and this explains why none of the current cosmological models has all these symmetries.

Proof. Let us prove this statement by contradiction. Let us assume that a realistic causality function d is invariant with respect to all the symmetries. Let us denote, for all $t \geq 0$, $e(t) \stackrel{\text{def}}{=} (t,0,0,0)$. We want to prove that in this case, we will have $d(e(0),e(1)) = 1$. This will contradict to the first part of the definition of the realistic causality function, according to which $d(e,e') = 1$ is only possible when $e = e'$. To prove the equality $d(e(0),e(1)) = 1$, we will prove that for every $\varepsilon > 0$, we have $d(e(0),e(1)) \geq 1-\varepsilon$.

According to the second comment after the definition of an "and"-operation, for this $\varepsilon > 0$, there exists a number $\delta > 0$ for which:

- if $x \geq 1-\delta$ and $y \geq 1-\delta$,
- then $f_{\&}(x,y) \geq 1-\varepsilon$.

Thus:

- if we have an event e for which $d(e(0),e) \geq 1-\delta$ and $d(e,e(1)) \geq \delta$,
- then we have $f_{\&}(d(e(0),e),d(e,e(1))) \geq 1-\varepsilon$
- and hence, since $d(e(0),e(1)) \geq f_{\&}(d(e(0),e),d(e,e(1)))$, we have $d(e(0),e(1)) \geq 1-\varepsilon$.

To find such e, let us recall that, as we have mentioned, every two pairs (e,e') and (g,g') for which the proper time is the same can be transformed into each other by an appropriate symmetry f: $f(e) = g$ and $f(e') = g'$. Since the realistic causality function is invariant with respect to all the symmetries, this means that $d(e,e') = d(f(e),f(e')) = d(g,g')$. In other words:

- if two pairs have the same proper time,
- then these pairs have the same degree of causality.

In mathematical terms, this means that the realistic causality function is a function of proper time, i.e., that $d(e,e') = F(\tau(e,e'))$ for some function $F(x)$.

In particular, for $e(t) = (t,0,0,0)$ with $t \geq 0$, we have $\tau(e(0),e(t)) = c_0 \cdot t$. So, for these pairs, we have $d(e(0),e(t)) = F(c_0 \cdot t)$. The function d is continuous and $e(t) \to e(0)$ as $t \to 0$. In the limit $t = 0$, we get $d(e(0),e(0)) = 1$. Thus, the function $F(x)$ is also continuous, tending to $F(0) = 1$ as $x \to 0$. By definition of the limit, this means that for every $\delta > 0$, there exists an $\nu > 0$ such that:

- if $x \leq \nu$,
- then $F(x) \geq 1-\delta$.

Let us take $e = (t/2, c_0 \cdot t/2 - \alpha, 0, 0)$. Here, as one can easily check, $e(0) \leq e \leq e(1)$, $\tau(e(0),e) = \tau(e,e(1))$ and the common value of proper time tends to 0 as $\alpha \to 0$. Thus, for sufficiently small α, we have $\tau(e(0),e) \leq \nu$. Hence, we have $d(e(0),e) = F(\tau(e,e(0)) \geq 1-\delta$ and similarly $d(e,e(1)) \geq 1-\delta$. We have already shown that these two inequalities imply that $d(e(0),e(1)) \geq 1-\varepsilon$. Since this is true for every $\varepsilon > 0$, this means that $d(e(0),e(1)) = 1$ – which contradicts to the definition of the realistic causality function.

This contradiction proves that our assumption was wrong, and thus, indeed, no realistic causality function can be invariant with respect to all the symmetries.

Acknowledgments. This work was supported in part by the National Science Foundation grants 1623190 (A Model of Change for Preparing a New Generation for Professional Practice in Computer Science), HRD-1834620 and HRD-2034030 (CAHSI Includes), and by the AT&T Fellowship in Information Technology.

It was also supported by the program of the development of the Scientific-Educational Mathematical Center of Volga Federal District No. 075-02-2020-1478, and by a grant from the Hungarian National Research, Development and Innovation Office (NRDI).

The authors are thankful to Art Duval and Razieh Nabi for valuable discussions, and to the anonymous referees for useful suggestions.

References

1. Belohlavek, R., Dauben, J.W., Klir, G.J.: Fuzzy Logic and Mathematics: A Historical Perspective. Oxford University Press, New York (2017)
2. Feynman, R., Leighton, R., Sands, M.: The Feynman Lectures on Physics. Addison Wesley, Boston (2005)
3. Klir, G., Yuan, B.: Fuzzy Sets and Fuzzy Logic. Prentice Hall, Upper Saddle River (1995)
4. Kreinovich, V., Ortiz, A.: Towards a better understanding of space-time causality: Kolmogorov complexity and causality as a matter of degree. In: Proceedings of the Joint World Congress of the International Fuzzy Systems Association and Annual Conference of the North American Fuzzy Information Processing Society IFSA/NAFIPS 2013, Edmonton, Canada, 24–28 June 2013, pp. 1349–1353 (2013)
5. Kreinovich, V., Kosheleva, O., Ortiz-Muñoz, A.: Need for simplicity and everything is a matter of degree: how Zadeh's philosophy is related to Kolmogorov complexity, quantum physics, and deep learning. In: Shahbazova, S.N., Abbasov, A.M., Kreinovich, V., Kacprzyk, J., Batyrshin, I.Z. (eds.) Recent Developments and the New Directions of Research, Foundations, and Applications. STUDFUZZ, vol. 422, pp. 203–216. Springer, Cham (2023). https://doi.org/10.1007/978-3-031-20153-0_16
6. Mendel, J.M.: Uncertain Rule-Based Fuzzy Systems: Introduction and New Directions. Springer, Cham (2017)
7. Nguyen, H.T., Walker, C.L., Walker, E.A.: A First Course in Fuzzy Logic. Chapman and Hall/CRC, Boca Raton (2019)
8. Novák, V., Perfilieva, I., Močkoř, J.: Mathematical Principles of Fuzzy Logic. Kluwer, Boston (1999)
9. Pearl, J.: Causality: Models, Reasoning and Inference. Cambridge University Press, Cambridge (2009)
10. Pimenov, R.I.: Kinematic Spaces: Mathematical Theory of Space-Time. Consultants Bureau, New York (1970)
11. Thorne, K.S., Blandford, R.D.: Modern Classical Physics: Optics, Fluids, Plasmas, Elasticity, Relativity, and Statistical Physics. Princeton University Press, Princeton (2021)
12. Zadeh, L.A.: Fuzzy sets. Inf. Control **8**, 338–353 (1965)

Faster Algorithms for Estimating the Mean of a Quadratic Expression Under Uncertainty

Martine Ceberio[1], Vladik Kreinovich[1](✉), Olga Kosheleva[2], and Lev Ginzburg[3]

[1] Department of Computer Science, University of Texas at El Paso, 500 W. University, El Paso, TX 79968, USA
{mceberio,vladik}@utep.edu

[2] Department of Teacher Education, University of Texas at El Paso, 500 W. University, El Paso, TX 79968, USA
olgak@utep.edu

[3] Department of Ecology and Evolution, Stony Brook University, Stony Brook, NY 11794, USA

Abstract. In many practical situations, we can safely approximate the actual nonlinear dependence by a quadratic expression. For such situations, there exist techniques for estimating the uncertainty of the result of data processing based on the known information about the uncertainty of the input data – for example, for estimating the mean value of the corresponding approximation error. However, many such techniques are somewhat time-consuming. In this paper, we propose faster algorithms for solving this problem.

1 Formulation of the Problem

Need for Uncertainty Propagation. In many practical situations, we are interested in the value of a physical quantity y which is difficult – or even impossible – to measure directly. For example, we may want to estimate tomorrow's temperature, the amount of oil in a given oilfield, or a distance to a faraway star. Since we cannot measure the quantity y directly, a natural idea is to measure this quantity indirectly, i.e., to measure easier-to-measure quantities $x_1, \ldots, x_n$ that are related to y by a known dependence

$$y = f(x_1, \ldots, x_n), \tag{1}$$

and then to use the results $\widetilde{x}_i$ of these measurements to provide the estimate

$$\widetilde{y} = f(\widetilde{x}_1, \ldots, \widetilde{x}_n) \tag{2}$$

for y.

The problem is that measurements are never absolutely accurate, the measurement result $\widetilde{x}$ is, in general, different from the actual (unknown) value of the corresponding quantity. In other words, there is, usually, a non-zero difference $\Delta x \stackrel{\text{def}}{=} \widetilde{x} - x$ known as a *measurement error*. As a result, the estimate (2) is, in general, different from the desired value (1). It is therefore desirable to use available information about measurement errors Δx_i to gauge the difference $\Delta y \stackrel{\text{def}}{=} \widetilde{y} - y$ between the estimate and the actual value. Estimating this difference is known as *uncertainty propagation*.

K. Cohen et al. (Eds.): NAFIPS 2023, LNNS 751, pp. 290–300, 2023.
https://doi.org/10.1007/978-3-031-46778-3_27

Comment. In many practical situations, even if we know the exact values of the measured quantities $x_1, \ldots, x_n$, we can only approximately determine the value y. For example, in weather prediction, even if we know the exact values of the temperature, atmospheric pressure, etc. at all the locations where we placed the sensors, we will still retain some uncertainty, since tomorrow's temperature is also affected by temperature at other locations. In such cases, the difference $\delta y \stackrel{\text{def}}{=} y - f(x_1, \ldots, x_n)$ between the actual value of the quantity y and the expression $f(x_1, \ldots, x_n)$ – known as *approximation error* – is unpredictable, i.e., random. In other words, instead of the Eq. (1), we have a more accurate formula

$$y = f(x_1, \ldots, x_n) + \delta y. \tag{1a}$$

From the previous experience of predicting the value y, we can get a sample of the values of the approximation error. Based in this sample, we can determine the probability distribution of the approximation error.

If it turns out that the mean value m of the approximation error is different from 0, i.e., that the approximate expression (1) has a bias, then practitioners usually compensate for this bias by replacing the original function $f(x_1, \ldots, x_n)$ with a new function $f_{\text{new}}(x_1, \ldots, x_n) = f(x_1, \ldots, x_n) + m$. For this new function, the mean value of approximation error is 0. Thus, without loss of generality, we can safely assume that the mean value of the approximation error is 0.

Possibility of Linear and Quadratic Approximations. By definition of a measurement error, we have $x_i = \widetilde{x}_i - \Delta x_i$. Thus, the estimation error Δy can be represented as

$$\Delta y = f(\widetilde{x}_1, \ldots, \widetilde{x}_n) - f(\widetilde{x}_1 - \Delta x_1, \ldots, \widetilde{x}_n - \Delta x_n). \tag{3}$$

Measurement errors are usually reasonably small. Thus, terms which are quadratic in Δx_i are much smaller than linear terms, and terms which are cubic in Δx_i are even smaller. For example, for a 20% measurement error, its square is 4%, and its cube is 0.8%. In such situations, from the practical viewpoint, cubic (and higher order) terms can be safely ignored in comparison with linear and quadratic ones. So, to simplify computations, we can expand the right-hand side of the expression (3) and only keep linear and quadratic terms in this expansion. In this case, we have the expression

$$\Delta y = \sum_{i=1}^{n} a_i \cdot \Delta x_i + \sum_{i=1}^{n} a_{ii} \cdot (\Delta x_i)^2 + \sum_{i=1}^{n} \sum_{j \neq i} a_{ij} \cdot \Delta x_i \cdot \Delta x_j \tag{4}$$

for some values a_i and a_{ij}.

Probabilistic, Interval, and Fuzzy Uncertainty. In the ideal case, we should know the probability distribution of each measurement error – and, since the measurement errors can be correlated, the joint probability distribution of these measurement errors. However, determining these probability distributions is very time-consuming and thus, rarely done in practice.

In many practical situations, all we know is the upper bound Δ on the absolute value of the measurement error Δx: $|\Delta x| \le \Delta$; see, e.g., [15]. In such situations, after we learn the measurement result $\widetilde{x}$, the only information that we have about the actual (unknown) value of the corresponding quantity x is that this value is contained in the interval $[\widetilde{x}-\Delta, \widetilde{x}+\Delta]$. This situation is known as *interval uncertainty*; see, e.g., [4,8,10,12].

In some cases, all we know is the expert estimates on the possible values of the measurement error, estimates expressed by using imprecise ("fuzzy") words from natural language, such as "small", "approximately 0.1". To describe the resulting knowledge in precise terms, it is reasonable to use fuzzy techniques – techniques specifically designed for such a description; see, e.g., [1,5,11,13,14,17]. In these techniques, in effect, for each level of confidence α, we provide a bound $\Delta(\alpha)$ that is satisfied with this degree of confidence. So, from the computational viewpoint, this is similar to interval uncertainty – with the main difference that for each quantity, we have several different intervals corresponding to different levels of α. In view of this similarity, in the following text, we will only talk about interval uncertainty – but all the results are also applied to the case when we have fuzzy uncertainty (expressed by the corresponding intervals).

In addition to the upper bounds on the measurement errors Δx_i, we often have partial information on the probabilities of different values of $\Delta x \stackrel{\text{def}}{=} (\Delta x_1, \ldots, \Delta x_n)$.

What is the Typical Case of Partial Information About the Probabilities? In addition to the interval range of each variable Δx_i, we often know the mean ΔE_i of Δx_i. We get it, e.g., from the results of the testing the measuring instrument, when the mean is estimated as the average of measurement errors. The more tests we undertake, the more information we get about the probability distribution, and the more characteristics of the probability distribution we can determine. In the ideal situation, we can perform as many tests as necessary to determine the probability distribution of Δx_i. In many real-life situations, however, we can only afford to determine one (or two) characteristics. In such situations, a natural choice is to determine the mean (and, if possible, the standard deviation; see, e.g., [15]).

In measurement terms, the mean value of the measurement error is called the *systematic error* of the measurement procedure. In measurements, it is a common practice to *calibrate* the measuring instrument so that the systematic error (bias) is eliminated. Calibration means that, instead of the original measured value $\widetilde{x}_i$ of the desired property, we return the value $E_i \stackrel{\text{def}}{=} \widetilde{x}_i - \Delta E_i$ for which the mean value of the re-calibrated measurement error $\Delta x_i' \stackrel{\text{def}}{=} E_i - x_i$ is exactly 0.

The original measurement error Δx_i can attain any value from the interval $[-\Delta_i, \Delta_i]$. As a result, the re-calibrated measurement error Δ_i' can take all possible values from the interval $[-\Delta_i^-, \Delta_i^+]$, where $\Delta_i^- \stackrel{\text{def}}{=} \Delta_i + \Delta E_i$ and $\Delta_i^+ \stackrel{\text{def}}{=} \Delta_i - \Delta E_i$.

For example, if we know that $\Delta x_i \in [-0.1, 0.1]$, and its mean is $\Delta E_i = 0.05$, then for $\Delta x_i' = \Delta x_i - \Delta E_i$, the mean is 0, and the interval of possible values is $[-0.15, 0.05]$.

The mean is the only information that we have about *each* measurement error. What do we know about the dependence *between* the corresponding random variables? In many applications, we know that the same source of noise contributes to the errors of different measurements, so these errors Δx_i are not independent. Since we do not have enough statistics to get any information about each distribution except for its mean, we also cannot determine the correlation between Δx_i. So, if we are interested in guaranteed estimates, we must consider all possible n-dimensional distributions, with all possible correlations.

Formulation of the Problem. In the case of interval uncertainty, we know the intervals of possible values of the measurement errors Δx_i, and we want to find the interval of possible values of the desired quantity $y = f(x_1, \ldots, x_n)$.

In the situations when, in addition to the interval of possible values of measurement error, we also know the mean of the measurement errors, in addition to knowing the interval of possible values for the result y of data processing, it is desirable to also know the interval of possible values of the mean value $E \stackrel{\text{def}}{=} E[\Delta y]$ of the quantity Δy.

In this paper, we consider the case when the dependence $f(x_1, \ldots, x_n)$ between the desired quantity y and the directly measurable quantities $x_1, \ldots, x_n$ can be safely described by a quadratic function (4). So, the problem is:

- we know the intervals $[-\Delta_i, \Delta_i]$ of possible values of Δx_i, and we know the means $\Delta E_i = E[\Delta x_i]$;
- based on this information, we want to estimate the mean $E[\Delta y]$ of the expression (4).

Alternatively, if we re-calibrate all the measuring instruments and take the re-calibrated measured values $E_i = \widetilde{x}_i - \Delta E_i$, then we can expand f around these new values E_i:

$$\Delta y = a'_0 + \sum_{i=1}^{n} a'_i \cdot \Delta x'_i + \sum_{i=1}^{n} a'_{ii} \cdot (\Delta x'_i)^2 + \sum_{i=1}^{n} \sum_{j \neq i} a'_{ij} \cdot \Delta x'_i \cdot \Delta x'_j. \quad (5)$$

After this calibration, the problem takes the following form:

- we know the intervals $[-\Delta_i^-, \Delta_i^+]$ of possible values of $\Delta x'_i$, and we know that $E[\Delta x'_i] = 0$ for all i;
- based on this information, we want to find the interval of possible values of $E[\Delta y]$.

Comment. In situations when the dependence (1) is only approximate, we get a formula similar to the formula (5), but with the additional term δy describing the approximation error:

$$\Delta y = a'_0 + \sum_{i=1}^{n} a'_i \cdot \Delta x'_i + \sum_{i=1}^{n} a'_{ii} \cdot (\Delta x'_i)^2 + \sum_{i=1}^{n} \sum_{j \neq i} a'_{ij} \cdot \Delta x'_i \cdot \Delta x'_j + \delta y. \quad (5a)$$

Since, as we have mentioned earlier, the mean value of the approximation error is 0, the expected value of the expression (1) is exactly the same as the expected value of the expression (5) – that ignores the approximation error.

Thus, in the problem of estimating the mean value of a quadratic expression, we can safely ignore the approximation error.

What is Currently Known. The expression (4) (or (5)) is a linear combination of linear and quadratic terms Δx_i and $\Delta x_i \cdot \Delta x_j$ (correspondingly, $\Delta x_i'$ and $\Delta x_i' \cdot \Delta x_j'$). The expected value of a linear combination is equal to the linear combination of the corresponding expected values. The expected values of Δx_i and $\Delta x_i'$ are known. So, to be able to estimate the expected value of y, it is sufficient to be able to estimate the expected value of a product $\Delta x_i \cdot \Delta x_j$.

It is known [6,7] that if we have two random variables v_1 and v_2 with known ranges $[\underline{v}_i, \overline{v}_i]$ and known means E_i, then the interval $[\underline{E}, \overline{E}]$ of possible values of $E \stackrel{\text{def}}{=} E[v_1 \cdot v_2]$ can be computed as follows. First, we compute the auxiliary values $p_i \stackrel{\text{def}}{=} (E_i - \underline{v}_i)/(\overline{v}_i - \underline{v}_i)$, and then compute

$$\underline{E} = \min(p_1 + p_2 - 1, 0) \cdot \overline{v}_1 \cdot \overline{v}_2 + \min(p_1, 1 - p_2) \cdot \overline{v}_1 \cdot \underline{v}_2 + \min(1 - p_1, p_2) \cdot \underline{v}_1 \cdot \overline{v}_2 + \max(1 - p_1 - p_2, 0) \cdot \underline{v}_1 \cdot \underline{v}_2; \quad (6)$$

$$\overline{E} = \min(1 - p_1, 1 - p_2) \cdot \underline{v}_1 \cdot \underline{v}_2 + \max(p_1 - p_2, 0) \cdot \overline{v}_1 \cdot \underline{v}_2 + \max(p_2 - p_1, 0) \cdot \underline{v}_1 \cdot \overline{v}_2 + \min(p_1, p_2) \cdot \overline{v}_1 \cdot \overline{v}_2. \quad (7)$$

In principle, we can use these formulas to estimate $E[\Delta y]$.

Remaining Problem. In [6,7], our objective was to come up with an expression for a *single* exact range of $E[v_1 \cdot v_2]$ for two variables v_1 and v_2. The fact that we found an explicit analytical expression makes it easy to compute the range.

In our new problem, however, we need to estimate *many* ($\sim n^2$) such ranges – because n may be large. Each range computation requires two divisions (to compute p_1 and p_2) and several multiplications – and division is known to take longer to compute. Since we need to repeat these computations $\sim n^2$ times, it is desirable to look for simpler expressions for $\underline{E}$ and $\overline{E}$, expressions that would hopefully avoid division altogether and require fewer multiplications – and thus, will lead to faster computations.

2 Our Result

The main result of this paper is that such faster-to-compute expressions are indeed possible:

Proposition 1. *If we have two random variables v_1 and v_2 with known means E_i and known ranges $[E_i - \Delta_i^-, E_i + \Delta_i^+]$, then the interval $[\underline{E}, \overline{E}]$ of possible values of $E = E[v_1 \cdot v_2]$ is equal to*

$$[E_1 \cdot E_2 - \min(\Delta_1^- \cdot \Delta_2^-, \Delta_1^+ \cdot \Delta_2^+), E_1 \cdot E_2 + \min(\Delta_1^- \cdot \Delta_2^+, \Delta_1^+ \cdot \Delta_2^-)]. \quad (8)$$

For the expression v_i^2, the range can be computed by using the general techniques from [9,16]. For readers' convenience, let us give an explicit derivation.

Proposition 2. *If we have a random variable v_i with a known mean E_i and a known range $[E_i - \Delta_i^-, E_i + \Delta_i^+]$, then the interval $[\underline{M}_i, \overline{M}_i]$ of possible values of $M_i = E[v_i^2]$ is equal to $[E_i^2, E_i^2 + \Delta_i^- \cdot \Delta_i^+]$.*

We want to apply these results to the variables $v_i = \Delta x_i'$ for which $E_i = E[\Delta x_i'] = 0$. As a result, the enclosure $\mathbf{E}_{\text{orig}}$ for the range $[\underline{E}, \overline{E}]$ of $E = E[\Delta y]$ takes the following form:

$$\mathbf{E}_{\text{orig}} = a_0' - \sum_i a_{ii}' \cdot [0, \Delta_i^- \cdot \Delta_i^+] +$$

$$\sum_{i=1}^{n} \sum_{j \neq i} a_{ij}' \cdot [-\min(\Delta_i^- \cdot \Delta_j^+, \Delta_i^+ \cdot \Delta_j^-), \min(\Delta_i^- \cdot \Delta_j^-, \Delta_i^+ \cdot \Delta_j^+)].$$

Comment. Alternatively, we can represent the matrix a_{ij}' in terms of its eigenvalues λ_k' and the corresponding unit eigenvectors $e_k' = (e_{k1}', \ldots, e_{kn}')$, then

$$\Delta y = a_0' + \sum_{i=1}^{n} a_i' \cdot \Delta x_i' + \sum_{k=1}^{n} \lambda_k' \cdot \left(\sum_{i=1}^{n} e_{ki}' \cdot \Delta x_i' \right).$$

Each variable $\Delta x_i'$ has 0 mean and range $[-\Delta_i^-, \Delta_i^+]$. Thus, for the linear combination $\sum_{i=1}^{n} e_{ki}' \cdot \Delta x_i'$, the mean is equal to 0, and the range is

$$\sum_{i=1}^{n} e_{ki}' \cdot [-\Delta_i^-, \Delta_i^+].$$

In other words, the range is equal to $[-\delta_k^-, \delta_k^+]$, where

$$\delta_k^- = \sum_{i:e_{ki}'<0} |e_{ki}'| \cdot \Delta_i^+ + \sum_{i:e_{ki}'>0} e_{ki}' \cdot \Delta_i^-;$$

$$\delta_k^+ = \sum_{i:e_{ki}'<0} |e_{ki}'| \cdot \Delta_i^- + \sum_{i:e_{ki}'>0} e_{ki}' \cdot \Delta_i^+.$$

As a result, we get a different enclosure $\mathbf{E}_{\text{new}}$ for the range $[\underline{E}, \overline{E}]$ of $E = E[\Delta y]$:

$$\mathbf{E}_{\text{new}} = a_0' + \sum_{k=1}^{n} \lambda_k \cdot [0, \delta_k^- \cdot \delta_k^+].$$

Even Faster Computations are Possible. According to Proposition 1, in general, to compute the range for E, it is sufficient to perform 5 multiplications: one to compute $E_1 \cdot E_2$ and 4 to compute 4 products $\Delta_1^\pm \cdot \Delta_2^\pm$. In particular, for $v_i = \Delta x_i'$, since we have $E_i = 0$, we only need 4 multiplications.

It is possible to follow the ideas behind the fast algorithm for interval multiplication [2,3] and reduce the number of multiplications by one.

For this, first, we compare Δ_1^- with Δ_1^+ and Δ_2^- with Δ_2^+. As a result, we get $2 \times 2 = 4$ different comparison results. In all 4 cases, we can avoid at least one multiplication in the formula (8); indeed:

- if $\Delta_1^- \le \Delta_1^+$ and $\Delta_2^- \le \Delta_2^+$, then $\min(\Delta_1^- \cdot \Delta_2^-, \Delta_1^+ \cdot \Delta_2^+) = \Delta_1^- \cdot \Delta_2^-$;
- if $\Delta_1^- \le \Delta_1^+$ and $\Delta_2^+ \le \Delta_2^-$, then $\min(\Delta_1^- \cdot \Delta_2^+, \Delta_1^+ \cdot \Delta_2^-) = \Delta_1^- \cdot \Delta_2^+$;
- if $\Delta_1^+ \le \Delta_1^-$ and $\Delta_2^- \le \Delta_2^+$, then $\min(\Delta_1^- \cdot \Delta_2^+, \Delta_1^+ \cdot \Delta_2^-) = \Delta_1^+ \cdot \Delta_2^-$;
- if $\Delta_1^+ \le \Delta_1^-$ and $\Delta_2^+ \le \Delta_2^-$, then $\min(\Delta_1^- \cdot \Delta_2^-, \Delta_1^+ \cdot \Delta_2^+) = \Delta_1^+ \cdot \Delta_2^+$.

Proof of Proposition 1. Let us start with the expression for $\overline{E}$. When $p_1 \ge p_2$, the expression (7) takes the following simplified form:

$$\overline{E} = p_2 \cdot \overline{v}_1 \cdot \overline{v}_2 + (p_1 - p_2) \cdot \overline{v}_1 \cdot \underline{v}_2 + (1 - p_1) \cdot \underline{v}_1 \cdot \underline{v}_2.$$

Grouping together terms proportional to p_1 and p_2, we conclude that

$$\overline{E} = p_2 \cdot \overline{v}_1 \cdot (\overline{v}_2 - \underline{v}_2) + p_1 \cdot (\overline{v}_1 - \underline{v}_1) \cdot \underline{v}_2 + \underline{v}_1 \cdot \underline{v}_2.$$

Substituting the expression defining p_i into this formula, we conclude that

$$\overline{E} = \overline{v}_1 \cdot (E_2 - \underline{v}_2) + (E_1 - \underline{v}_1) \cdot \underline{v}_2 + \underline{v}_1 \cdot \underline{v}_2.$$

Grouping the last two terms, we get

$$\overline{E} = \overline{v}_1 \cdot (E_2 - \underline{v}_2) + E_1 \cdot \underline{v}_2.$$

Finally, substituting $\underline{v}_i = E_i - \Delta_i^-$ and $\overline{v}_i = E_i + \Delta_i^+$, we conclude that

$$\overline{E} = (E_1 + \Delta_1^+) \cdot \Delta_2^- + E_1 \cdot (E_2 - \Delta_2^-) = E_1 \cdot \Delta_2^- + \Delta_1^+ \cdot \Delta_2^- + E_1 \cdot E_2 - E_1 \cdot \Delta_2^-,$$

$$\overline{E} = E_1 \cdot E_2 + \Delta_1^+ \cdot \Delta_2^-. \tag{9}$$

Similarly, if $p_1 \le p_2$, we get

$$\overline{E} = E_1 \cdot E_2 + \Delta_1^- \cdot \Delta_2^+. \tag{10}$$

The condition $p_1 \ge p_2$, i.e., $(E_1 - \underline{v}_1)/(\overline{v}_1 - \underline{v}_1) \ge (E_2 - \underline{v}_2)/(\overline{v}_2 - \underline{v}_2)$, can be equivalently described as

$$(E_1 - \underline{v}_1) \cdot (\overline{v}_2 - \underline{v}_2) \ge (E_2 - \underline{v}_2) \cdot (\overline{v}_1 - \underline{v}_1),$$

i.e., in terms of E_i and $\Delta_i^\pm$, as

$$\Delta_1^- \cdot (\Delta_2^- + \Delta_2^+) \ge \Delta_2^- \cdot (\Delta_1^- + \Delta_1^+),$$

or, equivalently, as

$$\Delta_1^- \cdot \Delta_2^+ \ge \Delta_1^+ \cdot \Delta_2^-.$$

Since this condition determines whether we have an expression (9) or (10), we thus get the desired formula for $\overline{E}$.

For $\underline{E}$, we similarly consider two cases: $p_1 + p_2 \ge 1$ and $p_1 + p_2 < 1$. In the first case, we have

$$\underline{E} = (p_1 + p_2 - 1) \cdot \overline{v}_1 \cdot \overline{v}_2 + (1 - p_2) \cdot \overline{v}_1 \cdot \underline{v}_2 + (1 - p_1) \cdot \underline{v}_1 \cdot \overline{v}_2,$$

i.e.,

$$\underline{E} = p_1 \cdot (\overline{v}_1 - \underline{v}_1) \cdot \overline{v}_2 + p_2 \cdot \overline{v}_1 \cdot (\overline{v}_2 - \underline{v}_2) + \underline{v}_1 \cdot \overline{v}_2 + \overline{v}_1 \cdot \underline{v}_2 - \overline{v}_1 \cdot \overline{v}_2.$$

Substituting the expressions for p_i, we conclude that

$$\underline{E} = (E_1 - \underline{v}_1) \cdot \overline{v}_2 + \overline{v}_1 \cdot (E_2 - \underline{v}_2) + \underline{v}_1 \cdot \overline{v}_2 + \overline{t}_1 \cdot \underline{v}_2 - \overline{v}_1 \cdot \overline{v}_2,$$

i.e.,

$$\underline{E} = E_1 \cdot \overline{v}_2 + \overline{v}_1 \cdot E_2 - \overline{v}_1 \cdot \overline{v}_2.$$

Finally, substituting the expressions $\underline{v}_i = E_i - \Delta_i^-$ and $\overline{v}_i = E_i + \Delta_i^+$, we conclude that

$$\underline{E} = E_1 \cdot E_2 - \Delta_1^+ \cdot \Delta_2^+. \tag{11}$$

Similarly, if $p_1 + p_2 < 1$, then

$$\underline{E} = p_1 \cdot \overline{v}_1 \cdot \underline{v}_2 + p_2 \cdot \underline{v}_1 \cdot \overline{v}_2 + (1 - p_1 - p_2) \cdot \underline{v}_1 \cdot \underline{v}_2,$$

i.e.,

$$\underline{E} = p_1 \cdot (\overline{v}_1 - \underline{v}_1) \cdot \underline{v}_2 + p_2 \cdot \underline{v}_1 \cdot (\overline{v}_2 - \underline{c}_2) + \underline{v}_1 \cdot \underline{v}_2.$$

Substituting the expressions for p_i, we conclude that

$$\underline{E} = (E_1 - \underline{v}_1) \cdot \underline{v}_2 + \underline{v}_1 \cdot (E_2 - \underline{v}_2) + \underline{v}_1 \cdot \underline{v}_2,$$

i.e., that

$$\underline{E} = E_1 \cdot \underline{v}_2 + \underline{v}_1 \cdot (E_2 - \underline{v}_2).$$

Finally, substituting the expressions $\underline{v}_i = E_i - \Delta_i^-$ and $\overline{v}_i = E_i + \Delta_i^+$, we conclude that

$$\underline{E} = E_1 \cdot E_2 - \Delta_1^- \cdot \Delta_2^-. \tag{12}$$

The inequality $p_1 + p_2 \geq 1$ can be reformulated as

$$(E_1 - \underline{v}_1) \cdot (\overline{v}_2 - \underline{v}_2) + (E_2 - \underline{v}_2) \cdot (\overline{v}_1 - \underline{v}_1) \geq (\overline{v}_1 - \underline{v}_1) \cdot (\overline{v}_2 - \underline{v}_2),$$

i.e., subtracting the first product from both sides, as

$$(E_2 - \underline{v}_2) \cdot (\overline{v}_1 - \underline{v}_1) \geq (\overline{v}_1 - E_1) \cdot (\overline{v}_2 - \underline{v}_2),$$

or, in terms of $\Delta_i^{\pm}$, as

$$\Delta_2^- \cdot (\Delta_1^- + \Delta_1^+) \geq \Delta_1^+ \cdot (\Delta_2^- + \Delta_2^+),$$

i.e., equivalently,

$$\Delta_1^- \cdot \Delta_2^- \geq \Delta_1^+ \cdot \Delta_2^+.$$

Since this condition determines whether we have an expression (11) or (12), we thus get the desired formula for $\overline{E}$.

The proposition is proven.

Proof of Proposition 2. It is known that the second moment M_i is equal to $V_i + E_i^2$, where V_i is the variance, i.e., the second moment of the auxiliary variable $v \stackrel{\text{def}}{=} v_i - E_i$ for which $E[v] = 0$ and for which the range of possible values is equal to $[-\Delta_i^-, \Delta_i^+]$. Thus, to prove the theorem, it is sufficient to prove that if we have a random variable v with a known mean $E[v] = 0$ and a known range $[-\Delta^-, \Delta^+]$, then the interval $[\underline{M}, \overline{M}]$ of possible values of $M = E[v^2]$ is equal to $[0, \Delta^- \cdot \Delta^+]$.

The second moment is always non-negative, so $\underline{M} \geq 0$. It is possible that v is identically 0, in which case $M = 0$; thus, $\underline{M} = 0$.

Among all distributions with 0 mean located on the interval $[-\Delta^-, \Delta^+]$, we want to find a distribution for which the second moment is the largest. For discrete distributions, that attains values $x_1, \ldots, x_n$ with probabilities $p_1, \ldots, p_n$, this means that we must maximize the expression $\sum p_i \cdot x_i^2$ under the constraints $p_i \geq 0$, $\sum p_i = 1$, and $\sum p_i \cdot x_i = 1$.

Once the values x_i are fixed, this constraint optimization problem becomes a linear programming problem with 2 equality constraints; according to the general properties of linear programming problems, the maximum is attained when at most 2 of the variables p_i are non-zero (see [6] for detailed description). Thus, to find the value $\overline{M}$, it is sufficient to consider distributions located at only two points x_1 and x_2. Since the average is 0, one of these values should be negative, and another positive. Let us denote the negative value by $-x^-$ and the positive one by x^+, and the corresponding probabilities by p^- and p^+.

Since $p^- + p^+ = 1$, we get $p^+ = 1 - p^-$. Thus, from the condition

$$p^- \cdot (-x^-) + (1 - p^-) \cdot x^+ = 0,$$

i.e., equivalently, $p^-(x^- + x^+) = x^+$, we conclude that $p^- = x^+/(x^- + x^+)$ – and hence, that $p^+ = 1 - p^- = x^-/(x^- + x^+)$. Therefore,

$$M = p^- \cdot (x^-)^2 + p^+ \cdot (x^+)^2 =$$

$$\frac{x^+}{x^- + x^+} \cdot (x^-)^2 + \frac{x^-}{x^- + x^+} \cdot (x^+)^2 = \frac{x^- \cdot x^+ \cdot (x^- + x^+)}{x^- + x^+} = x^- \cdot x^+.$$

This expression is a strictly increasing function of both its variables $x^-, x^+ \geq 0$. Thus, its maximum under the constraints $x^- \leq \Delta^-$ and $x^+ \leq \Delta^+$ is attained when $x^- = \Delta^-$ and $x^+ = \Delta^+$, and the corresponding value is exactly the one we described. The proposition is proven.

3 Discussion

What if Measurements are Actually Independent? In the above text, we assumed that the measurement errors Δx_i can be correlated, and we have no information about possible correlations. In some cases, however, we do know that the variables Δx_i are independent. Under this additional assumption of independence, what is the possible range of $E[\Delta y]$?

This question is the easiest to answer under the formula (5). Indeed, in this case, $E[\Delta x'_i] = 0$, and for $i \neq j$, we have $E[\Delta x'_i \cdot \Delta x'_j] = E[\Delta x'_i] \cdot E[\Delta x'_j] = 0 \cdot 0 = 0$; so, we conclude that

$$E = E[\Delta y] = a'_0 + \sum_{i=1}^{n} a'_{ii} \cdot E[(\Delta x'_i)^2].$$

From Proposition 2, we already know how to compute the range of possible values of $E[(\Delta x'_i)^2]$ for $i = 1, \ldots, n$. So, in the case of independence, the interval of possible values of E is equal to

$$[\underline{E}, \overline{E}] = a'_0 + \sum_{i=1}^{n} a'_{ii} \cdot [0, \Delta_i^- \cdot \Delta_i^+].$$

Remaining Open Problem. In practice, in addition to the first moments $E[\Delta x_i]$, we often know the second moments $E[\Delta x_i \cdot \Delta x_j]$ of the corresponding distributions. If we know the second moments, then, of course, computing the first moment of the quadratic expression y is easy, since y is a linear combination of terms Δx_i and $\Delta x_i \cdot \Delta x_j$. However, in this case, a natural next question remains open: what can we say about the second moment of y?

Acknowledgment. This work was supported in part by the National Science Foundation grants 1623190 (A Model of Change for Preparing a New Generation for Professional Practice in Computer Science), HRD-1834620 and HRD-2034030 (CAHSI Includes), and by the AT&T Fellowship in Information Technology.

It was also supported by the program of the development of the Scientific-Educational Mathematical Center of Volga Federal District No. 075-02-2020-1478, and by a grant from the Hungarian National Research, Development and Innovation Office (NRDI).

The authors are thankful to the anonymous referees for valuable suggestions.

References

1. Belohlavek, R., Dauben, J.W., Klir, G.J.: Fuzzy Logic and Mathematics: A Historical Perspective. Oxford University Press, New York (2017)
2. Hamzo, C., Kreinovich, V.: On average bit complexity of interval arithmetic. Bull. Eur. Assoc. Theor. Comput. Sci. (EATCS) **68**, 153–156 (1999)
3. Heindl, G.: An improved algorithm for computing the product of two machine intervals. Interner Bericht IAGMPI- 9304, Fachbereich Mathematik, Gesamthochschule Wuppertal (1993)
4. Jaulin, L., Kiefer, M., Didrit, O., Walter, E.: Applied Interval Analysis, with Examples in Parameter and State Estimation, Robust Control, and Robotics. Springer, London (2001)
5. Klir, G., Yuan, B.: Fuzzy Sets and Fuzzy Logic. Prentice Hall, Upper Saddle River, New Jersey (1995)
6. Kreinovich, V.: Probabilities, intervals, what next? Optimization problems related to extension of interval computations to situations with partial information about probabilities. J. Global Optim. **29**(3), 265–280 (2004)
7. Kreinovich, V., Ferson, S., Ginzburg, L.: Exact upper bound on the mean of the product of many random variables with known expectations. Reliable Comput. **9**(6), 441–463 (2003)

8. Kubica, B.J.: Interval Methods for Solving Nonlinear Constraint Satisfaction, Optimization, and Similar Problems: From Inequalities Systems to Game Solutions. Springer, Cham, Switzerland (2019)
9. Kuznetsov, V.P.: Interval Statistical Models. Radio i Svyaz Publ, Moscow (1991). (in Russian)
10. Mayer, G.: Interval Analysis and Automatic Result Verification. de Gruyter, Berlin (2017)
11. Mendel, J.M.: Uncertain Rule-Based Fuzzy Systems: Introduction and New Directions. Springer, Cham, Switzerland (2017)
12. Moore, R.E., Kearfott, R.B., Cloud, M.J.: Introduction to Interval Analysis. SIAM, Philadelphia (2009)
13. Nguyen, H.T., Walker, C.L., Walker, E.A.: A First Course in Fuzzy Logic. Chapman and Hall/CRC, Boca Raton, Florida (2019)
14. Novák, V., Perfilieva, I., Močkoř, J.: Mathematical Principles of Fuzzy Logic. Kluwer, Boston, Dordrecht (1999)
15. Rabinovich, S.G.: Measurement Errors and Uncertainty: Theory and Practice. Springer Verlag, New York (2005)
16. Walley, P.: Statistical Reasoning with Imprecise Probabilities. Chapman & Hall, New York (1991)
17. Zadeh, L.A.: Fuzzy sets. Inf. Control **8**, 338–353 (1965)

Formal Descriptive Modeling for Self-verification of Fuzzy Network Systems

Owen Macmann[1(✉)], Rick Graves[2], and Kelly Cohen[1]

[1] University of Cincinnati, Cincinnati, US
owen.macmann@gmail.com, cohenky@mail.uc.edu
[2] Air Force Research Laboratory, Dayton, US
rick.graves.1@us.af.mil

Abstract. The ongoing development of deep learning systems has generated novel use cases with abundant potential. However, identifying the criteria under which these use cases are developed remains an intractable problem from the point of view of verification and validation of these systems. This problem is notorious for traditional neural networks. Fuzzy-neural networks, genetic-fuzzy trees, and neuro-fuzzy inference systems are types of fuzzy networks that represent a bridge in the gap between fuzzy inference and traditional inference models popularly used in deep learning systems. In recent years, research into the development of fuzzy network systems has revealed that these possess an innate capability for verification and validation (V&V), even for non-deterministically derived systems e.g. genetic-fuzzy trees, in a way that traditional neural networks do not. This paper presents a framework referred to as formal descriptive modeling (FDM) for extracting these criteria from a fuzzy network of any shape or size and trained under any conditions. A model use case is presented in the form of V&V of a sample fuzzy network designed to administer controls to a flight sub-system for a material transfer problem with certain model requirements. The extraction and identification of internal system criteria, application of those to external human-derived design criteria, and the methodology for deriving the logical principles defining those criteria are demonstrated in reference to the sample problem.

1 Introduction

Neural networks have garnered significant attention in recent years due to their potential to process complex data and perform a wide range of tasks. Despite their impressive capabilities, however, traditional neural networks have limitations related to verification and validation, which can be a significant concern for applications where the consequences of errors can be severe. For example, recent studies have shown that traditional neural networks can be vulnerable to adversarial attacks, which can result in significant errors and incorrect decisions [1,2]. Additionally, a lack of transparency and interpretability in neural network design can make it difficult to understand how a given network arrived at a particular decision, making it challenging to verify and validate its performance.

These challenges have been particularly notable in aeronautical engineering, where the consequences of errors in control systems can be catastrophic. For example, in 2019,

K. Cohen et al. (Eds.): NAFIPS 2023, LNNS 751, pp. 301–311, 2023.
https://doi.org/10.1007/978-3-031-46778-3_28

the Boeing 737 MAX crashes were attributed, in part, to issues with the plane's control systems, which were based on neural networks [3,4]. This has raised concerns about the use of neural networks in safety-critical applications and highlighted the need for alternative approaches.

In response, there has been a growing interest in fuzzy networks, which represent a bridge between traditional inference models and fuzzy inference models used in deep learning systems. Fuzzy-neural networks, genetic-fuzzy trees, and neuro-fuzzy inference systems are types of fuzzy networks that offer greater transparency and interpretability in design, making them more amenable to verification and validation. Recent studies have shown that fuzzy networks can offer innate capabilities for verification and validation, even in non-deterministically derived systems [5–7]. Talpur et al. [5] and de Campos Souza [6] provide comprehensive reviews of deep neuro-fuzzy systems and their potential applications, highlighting their efficacy in constructing intelligent systems with high accuracy and interpretability across various domains. Fuzzy networks offer a significant advantage over traditional neural networks, as they are more conducive to V&V, a crucial factor in safety-critical applications. Additionally, fuzzy networks bridge traditional and fuzzy inference models, potentially improving the accuracy and reliability of intelligent systems. Arnett's dissertation [7] is particularly relevant and presents a novel approach for iteratively increasing the complexity of fuzzy systems during optimization to improve accuracy and verifiability. This applies even to an extraordinarily deep and complex hierarchical fuzzy system, and is an analysis that traditional neural networks are not exposed to. Implementations like Eclipse Capella [9] already exist to utilize these properties as part of a model-based systems engineering (MBSE) solution.

The challenges associated with the limitations of traditional neural networks related to verification and validation have been well-documented, particularly in aeronautical engineering. Fuzzy networks represent an alternative approach that offers greater transparency and interpretability in design, altogether more conducive to verification and validation. The Formal Descriptive Modeling (FDM) framework presented in this paper provides a transparent and reliable means of designing fuzzy networks for safety-critical applications. It extracts the criteria for verification and validation and derives the logical implementation necessary for on-board assessment of those criteria.

2 Background

Fuzzy inference systems use linguistic terms to define fuzzy rules, which are logical rules governing information flow. Defuzzification is needed to obtain useful output for engineering systems. The fuzziness arises from expressing mathematically precise concepts in imprecise language.

For physical engineering systems like helicopters, dynamic models can describe their behavior. Hypothetical systems require exchanging precision for prescription, with technical requirements serving as imprecise characterizations. Verification and validation, historically used to assess systems, involve simulation, formal methods, and testing. Formal methods rely on accurate requirements, translating logic into verifiable terms, but necessitate skilled requirements engineers. Finite state machines, produced from this effort, help engineers with complex systems.

Natural language processing (NLP) techniques, such as ChatGPT, can analyze requirements and identify complications. However, their black-box nature makes them unreliable for safety-critical applications.

Formal methods offer a framework for verifying systems using mathematically rigorous languages like that of fuzzy inference systems (FIS). Combining seemingly contradictory rules is managed through defuzzification. By short-circuiting defuzzification and introducing new fuzzy linguistic terms, a fuzzy network system can be modeled linguistically.

The methodology of formal descriptive modeling (FDM) described by this paper extends the principles of formal methods to the realm of fuzzy network systems, taking advantage of exactly this property, and illustrating how real system requirements and formal model descriptions are elicited from fuzzy networks of any complexity.

The UNLESS clause presented in the approach bears some resemblance to fuzzy expert system shells, as both aim to capture the inherent coupling between rules in a rulebase [8]. Fuzzy expert system shells typically use a set of heuristics or domain-specific knowledge to guide the inference process, while FDM uses the UNLESS clause to explicitly define the conditions under which certain rules are overridden or modified. This allows for a more structured and formalized representation of the dependencies between rules in a fuzzy system.

3 Methodology

In Fig. 1, three types of fuzzy networks are demonstrated. Each network consists of a group of FIS (Fuzzy Inference Systems), typically with two inputs and a single output. The correlation of the input to the output is based on a set of fuzzy rules, which have antecedents (input variables) and consequents (output variable). The antecedents are connected using fuzzy logic operators (like AND, OR, and NOT). The inputs and

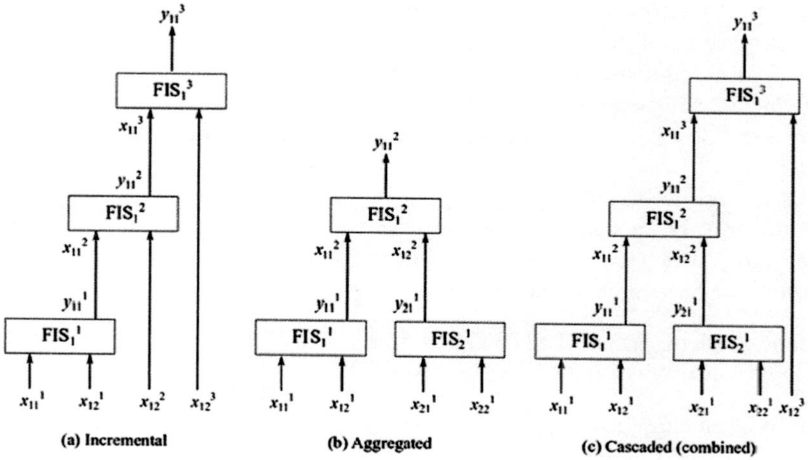

Fig. 1. Various architectures of fuzzy networks/trees

outputs also have membership functions, which separate their domains into smaller parts. The membership functions are used to show the degree of membership of each input variable in the input space. Every fuzzy rule explains how the output changes due to the degree of membership of the MF for the current values of the input variables.

A helicopter's cyclic and collective controls govern its direction and altitude, respectively. These inputs can be modeled using fuzzy logic, with the pilot's control inputs fuzzified into membership functions which activate the FIS. The outputs of the FIS are defuzzified to generate a crisp output, which may refresh the pilot's desired attitude or altitude, or define the desired control deflection directly.

In the case of the genetic-fuzzy tree (GFT), a fuzzy network which combines genetic algorithms and FIS trees to generate an optimized fuzzy decision tree, the input variables are supplied in a cascading fashion to a decision tree of FIS systems, and a genetic algorithm is used to determine the optimal parameters and configuration of the tree. The optimized tree will then map the inputs to the outputs of the entire system. In the case of aerodynamic controls, optimized systems such as these have been shown to provide very good tracking, responsiveness, and robustness against unusual or mitigating conditions.

As has already been shown by Arnett [7], each FIS or group of FIS can be shown to satisfy the requirements of formal verification. It only becomes a matter of transcribing the requirements, and performing the analysis. Even so, we are still in need of a formal verification expert to perform this analysis. But suppose we could use the GFT's ability to operate on formal logic directly so that it can self-verify. How might we utilize our formal requirements to directly enforce those requirements within the GFT, and so confirm through simple functionality that the system is behaving as expected?

Two things are necessary to achieve this. The first is a full codification of the requirements as logical input variables to the system. This is bound to increase the complexity of any GFT, but that is possibly an inescapable consequence of designing the GFT to self-verify (as the GFT must "make room" for the verification bits).

The second is that the entire GFT must be consistently and reliably explainable as a list of logical requirements, without any defuzzification necessary at all. The consistency of this requirement ensures that a formal description will always apply, and the elimination of the defuzzy step allows the logical requirements to be treated as Boolean logical requirements rather than fuzzy logical ones.

Dealing with this second point, to eliminate the defuzzification step, we propose it is only necessary to amend every fuzzy rule statement with a corollary statement dubbed "UNLESS."

Let's use the well-known fuzzy tipper example to illustrate use of the UNLESS clause numerically. In this example, we have two input variables: food quality and service quality, which will be used to determine the tip amount.

Assume we have the following two rules:

1. IF food quality is EXCELLENT THEN tip is GENEROUS
2. IF service quality is POOR THEN tip is AVERAGE

Now, we can introduce the UNLESS clause to create a coupling between these rules:

1. IF food quality is EXCELLENT THEN tip is GENEROUS UNLESS service quality is POOR

Suppose we have a specific case where food quality is $8.5/10$ and service quality is $2.5/10$. Using a standard triangular membership function, the degree of membership for the food quality being EXCELLENT (e.g., μ(EXCELLENT)) might be 0.7, and for the service quality being POOR (e.g., μ(POOR)) might be 0.8.

Applying the rule with the UNLESS clause, we would first evaluate the degree of membership for the UNLESS condition, which is $\mu(\text{POOR}) = 0.8$. Since the UNLESS condition is met (i.e., the service quality is POOR), we will not use the GENEROUS tip rule. Instead, we will use the second rule: IF service quality is POOR THEN tip is AVERAGE.

Let's say the AVERAGE tip is 15% and the degree of membership for this rule (μ(AVERAGE)) is 0.8. We would then apply defuzzification, such as the centroid method, to calculate the final tip percentage:

$$\text{Tip percentage} = \mu(\text{AVERAGE}) * \text{AVERAGE tip} = 0.8 * 15\% = 12\%$$

In this example, the tip would be 12% of the total bill due to the coupling between the rules created by the UNLESS clause. This demonstrates how the UNLESS clause can be used with actual numbers in a fuzzy tipper example to account for the interaction between different aspects of a decision-making process.

Altogether, the logical requirements list looks like Table 1, referred to as the fuzzy network behavior matrix. The essential point to remember is that the unless statement qualifies every fuzzy rule according to the other rules and their consequents. This does not extend merely to the end of a single FIS, but as in the fuzzy tipper example, may extend to the rest of the system. We can also suppose that some statements are actually redundant, and condense them into fewer statements with the same meaning. The approach for doing so is generalized from formal logical treatments of such statement condensation: say you had fuzzy statements each with the antecedents "If service is bad AND ambiance is bad" and "If service is bad AND mood is bad," and both have the same consequent, "Tip is bad," and the same qualifier, "(Unless) Food is good," the antecedent statements can be combined to form: "If service is bad AND (ambiance is bad OR mood is bad)."

In the case of cascaded FIS with sequential outputs or output-input pairs, there are now abstract variables which are "midputs" between systems and have their own fuzzy rules as well. These can still be described and codified in exactly the same manner as other inputs and outputs. It then becomes possible to define this abstract midput in the context of the wider system logic.

The potential of formal descriptive modeling for this specific use case is exactly what is proposed by this paper as the solution to the problem of understanding the abstract reasoning of a neural network's hidden layers, its internal criteria that otherwise cannot be readily understood. In principle, it is possible to also trace every output to an input in a deep neural network as well. This is how backpropagation essentially works. But lacking the inherent capability for fuzzy statement generation, it is only possible to describe each neuron according to its sigmoidal or similar function. The reader is left to ponder the relative difficulty in describing those sigmoids, their inferences, and their necessary "unless" statements, and compare that to a fuzzy network.

Through this method, it is possible to characterize one or multiple FIS in a single network. It is possible to characterize all the FIS, and indeed the entire system function,

Table 1. If-Then-Unless Fuzzy Network Behavior Matrix

IF (Inputs)	THEN (Outputs)	UNLESS
If food is good AND service is good	Tip is good	Ambiance OR Mood is poor
If food is okay OR service is okay	Tip is average	Ambiance OR Mood is poor
If ambiance is good AND mood is good	Tip is good	Food OR Service is poor
If ambiance is okay OR mood is okay	Tip is average	Food OR Service is poor
If service is bad AND (ambiance is bad OR mood is bad)	Tip is bad	Food is good

even for a large fuzzy network. And regardless of the method undertaken to define the rules and tune the membership functions, regardless of whether the system is Mamdani, Sugeno, Type-1, or Type-2, the approach prescribed by FDM will extract every rule and pare them down to a minimum necessary description that describes the actual internal logic and criteria of the fuzzy network system.

Now that we can fully characterize the fuzzy network system in descriptive, accurate language, it becomes necessary to codify every requirement as, itself, a set of logical inputs to the system. The approach is to add FIS to the system that apply the validation rules to specific inputs, outputs, or midputs of the system. The trouble here is to determine how to characterize a system with validation FIS. Doesn't the characterization change when the system itself changes? The answer is: Not necessarily. We do not have to change anything about how the system handles the fuzzy logic it was handling already. What we have done is introduced new reasoning on its own, separate portion of the output layer. The fuzzy network could be extended at this point using the validation bits as inputs to a new layer, or as fuzzy network outputs that characterize certain metered elements of the fuzzy network system under examination. In the case of the fuzzy tipper, let's say it was desirable to guarantee that good service produced a good tip, regardless of other consequences. We can already see from the description we laid out that this will not be the case, but this is just to illustrate. We can then examine the fuzzy behavior matrix of this modified network, including our validation statement, and confirm that essential behaviors have not changed; or even if they have changed, say because we wish to use one of our validation FIS in a new cascade, we can confirm the exact nature and specification of the change.

To acquire this logical implementation, it is first necessary to assess the real requirements of such a system, including the requirements of integration and performance validation, and translate these into a logical implementation for self-verification. We might consider how these requirements would be generated and presented in an MBSE software suite such as Capella, using a method known as the Arcadia method. [9] The Arcadia method is a systems engineering approach developed by Thales for the design, development, and validation of complex systems. It aims to provide a comprehensive framework for the definition, specification, and management of requirements throughout a system's lifecycle. With this method and using Capella, requirements are generated in a "Requirements Analysis" phase, which establishes testable criteria for the entire system and all its requirements. The logical implementation could now be derived; let's consider our previous example, that we should wish that "Good service *guarantees* a good tip." A statement of this nature is expected to be produced by the specification phase. Then, via the Arcadia method's allocation phase, we might find that this requirement applies to the Service logic subsystem – that is, all FIS which directly operate on Service as an input, and all FIS which cascade from that. These are the FIS fundamentally involved in the treatment of Service. Hence, the Service logic subsystem scopes to all FIS thus involved. It is apparent, due to the convergent and cascading nature of fuzzy network inputs, that FIS close to the terminal FIS are more likely to belong to multiple subsystem scopes.

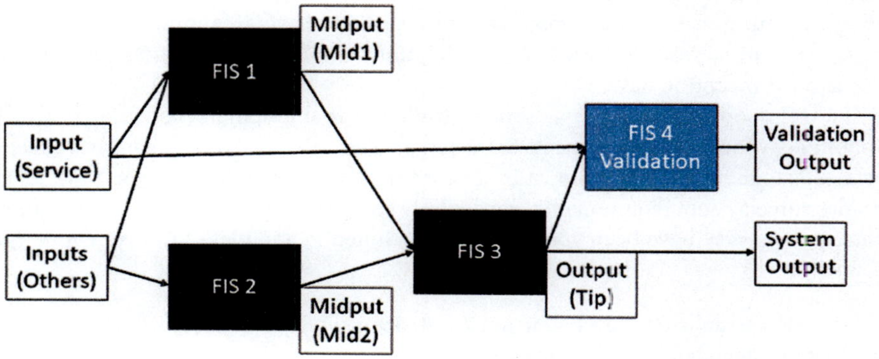

Fig. 2. Representation of FIS network with validation FIS

We now possess a scope for our requirement, and this scope is where we will introduce the validation bits required to validate our requirement. In this context, we will refer to the fuzzy descriptive statements belonging to these validation bits as "guarantees." The validation bit for the requirement is one new output produced by one new FIS, referred to by its descriptive statement, i.e., "Good service guarantees a good tip." This output is generated from two inputs: one, the raw Service input, and the other, the Tip output. An example of how this inclusion looks from a network perspective is represented in Fig. 2. This validation bit output will be made only applicable under the conditions specified; so, the rulebase must be fixed and apply the validation criteria to those inputs. The MF shapes and dimensions and the output and input domains are variable, and can be optimized by any method so long as the rule is held inviolate. The rulebase is simple: "If service is good and tip is good, the bit is validated." The validation output is binary: validated or not. This final output belongs to the output layer of the fuzzy network, and can be integrated in further data cascades or fuzzy networks as desired; but ultimately, are fully separate outputs produced by the fuzzy network and which serve to measure the constant validation of their validation statements.

In the next section, we will use the case study of the material transfer problem to demonstrate how to produce a logical implementation for self-verification of a real safety-critical aerospace system.

4 Safety-Critical System Implementation

This paper explores the mid-flight two-craft material transfer problem in various scenarios. We propose a models-based systems engineering solution, where technical requirements are verified using FDM within a fuzzy network system controlling the subsystem. This ensures integration with overarching aircraft systems without negative interactions or mitigates them through design or training doctrine.

Our solution features a telescoping delivery chute system from the delivering aircraft's center of gravity. With a payload delivery bay, ball-and-socket joint, and payload-release hatch, the chute extends up to 10 m with interior rail tracks for controlled payload sliding. An infrared laser-and-sensor targeting system assists in aiming at the receiver aircraft's reception bay. A flexible magnetic joint secures the connection between the chute and the reception bay.

The telescoping joint is a self-controlled variable-length pendulum, which autonomously locks on and connects to the target. This self-controlling feature is exactly what needs to be verified for compatibility with the larger integration needs. Now, let us consider some system requirements, particularly some specific operation safety requirements, such as may have been generated and instituted in a model of the system design in Capella:

- The payload delivery bay hatch must not open while a delivery is occurring, to prevent unintended release of the payload.
- The telescoping chute system must not extend or retract while the delivery bay hatch is open, to prevent entanglement or interference with the payload or personnel.
- The system must have fail-safes in place to prevent accidental release of the payload or disengagement of the connection between the two aircraft.
- The system must be designed to minimize the risk of collision or damage to either aircraft during the material transfer process.
- The system must be capable of detecting and responding to any abnormal or emergency situations, such as sudden changes in weather conditions or equipment malfunctions.

With these safety requirements, we can identify the core of what will eventually serve as validation bits, for a self-verifying safety-critical fuzzy network system. We refer to Table 3, which lists control switches and Boolean bits that would be used by the flight control and autopilot systems to implement the controls for the material transfer system. In this table, every bit is listed and described according to its designation, its intended function, the trigger for that bit's activation, and the corresponding "Disabled if True" statement that should be observable in the final implementation as an "Unless" statement. This Unless statement, unlike the fuzzy behavior matrix Unless, does not describe what is already true in the system but what is required to be true; hence we would generate the fuzzy guarantee validation matrix for these requirements and implement it in these statements per Table 2.

The statements in Table 2 are validation statements, and each is associated with a validation FIS. With this approach, we generated a GFT to dynamically stabilize the telescope through the use of two aerodynamic "control vanes", offering full lateral and

Table 2. Fuzzy Guarantee Validation Matrix

IF (Input)	AND (Unless/Output)	Output	Else
If payload delivery bay hatch is open	Delivery is occurring	Error	No Error
If payload delivery bay hatch is open	Telescoping chute is extending	Error	No Error
If chute extending	Unsafe flight conditions detected	Error	No Error
If payload released into chute	Unsafe flight conditions detected	Error	No Error

longitudinal control in addition to elongation and contraction control. We sought to verify that the GFT's control characteristics could be actively verified against some additional criteria. The resulting system was a hybrid of autonomously-generated controls and human-generated, integrated validation oversight. We noted that the validation outputs could be used as inputs to the wider integration logic, so that if the autonomous portion of the network was found to de-validate the validation bit, it was possible to a.) verify with certainty that the condition was being invalidated within the network, and to b.) use the validation bit itself as an input to another element of the integration system. Hence, the augmented system not only achieves validation of specific requirements within the network, but also provides a logical gate and line of defense against validation errors for the integration of the network system.

Let's now consider a specific example to derive a proof of correctness or validation. In the material transfer problem, we have a specific requirement:"The payload delivery bay hatch must not open while a delivery is occurring, to prevent unintended release of the payload". To formally verify this requirement, we can use the FIS and FDM approach. First, we represent the requirement as a rule in the FIS:

$$R_1 : \text{IF delivery is occurring THEN hatch must not open} \tag{1}$$

Now, we need to ensure that no other rule in the FIS contradicts this requirement. Suppose we have the following additional rules in the FIS:

$$R_2 : \text{IF delivery is not occurring AND hatch is closed THEN hatch can open} \tag{2}$$

$$R_3 : \text{IF delivery is not occurring AND hatch is open THEN hatch can close} \tag{3}$$

Notice that these rules do not contradict the requirement, as they only dictate the behavior of the hatch when a delivery is not occurring. To provide a proof of correctness, we then use the following approach:

1. Formalize the rule conditions and actions using fuzzy sets and membership functions.
2. Check that the fuzzy sets corresponding to "delivery is occurring" and "hatch must not open" in rule R_1 have non-zero membership values for all relevant inputs during the operation of the fuzzy network system.

3. Ensure that the aggregation of rules R_1, R_2, and R_3 do not lead to contradictory conclusions under any circumstances. For example, if the membership value of "delivery is occurring" is high, the membership value of "hatch must not open" should also be high.
4. Verify that the defuzzification process results in crisp output values that conform to the requirement.

Table 3. Material transfer system control switches, crisp Boolean variables

Designation	Function	Trigger[a]	Disabled if True[b]
SYS_ARM	System Armed	SYS_ARM_ON	SYS_ETS
SYS_ETS	Emergency Termination System	SYS_ETS_ON	None[c]
CTRL_TX	Control transmission line	SYS_ARM	None
CTRL_RX	Control feedback reception line	SYS_ARM	None
BAY_OPEN	Opens delivery bay to load or check payload	BAY_SW_OPEN, SYS_ARM	PLD_SND, BAY_LOCK
BAY_CLOSED	Closes delivery bay	BAY_SW_CLOSED, SYS_ARM	None
BAY_LOAD	Loads delivery bay with new payload	BAY_OPEN, BAY_SW_LOAD, SYS_ARM	BAY_CLOSED
BAY_LOCK	Locks bay hatch during delivery operation or chute extension	PLD_SND, CHUTE_EXTEND	PLD_SND_OFF
TARGET_LOCK	Infrared sensor confirms target coordinates and locks-on	ACQ_TARGET	TURB_CANCEL
CHUTE_EXTEND	Extends delivery chute towards target	EXTEND_SW, TARGET_LOCK, SYS_ARM	BAY_OPEN, ACQ_TARGET, TURB_CANCEL
PLD_SND	Payload release into chute	PLD_SND_ON, BAY_CLOSED, PLD_RDY, MADE_CONTACT, SYS_ARM	PLD_SND_OFF, TURB_CANCEL
PLD_SENT	Confirms payload has been successfully delivered	PLD_SND, RECEIVER_CATCH	None
TURB_CANCEL	If unsafe flight conditions detected, freeze extension of chutes	Automatic Control	None
ERROR_CHK	Sends "1" if error checker finds an error	$\vee$(BAY_OPEN_ERR, PAYLOAD_ERR)	None

[a] Under all of the stated conditions ($\wedge$), trigger the associated function and Boolean to "1"
[b] Under any of the stated conditions ($\vee$), disable the associated function and Boolean to "0"
[c] Always allowed, or always available

5 Conclusions

This paper has presented the concept of formal descriptive modeling (FDM) and proposed it as a powerful tool for verifying complex systems, especially deep learning networks and fuzzy networks. We have demonstrated the suitability of FDM for aerospace applications by applying it to a sample material transfer subsystem problem, with a set of specific requirements. By using FDM to verify the subsystem's technical requirements within a fuzzy network system, we have demonstrated the potential for FDM to provide a comprehensive and effective means of ensuring safe and reliable operation of complex fuzzy network systems in various industries, including aerospace. We also noted in our analysis that it would be possible to assess and define the actual criteria of dynamic stability using FDM as well.

We propose that the methodology of FDM is suitable as part of an MBSE approach to verification and validation of "intelligent" and deep learning systems. This paper demonstrates how FDM can be integrated with MBSE tools and approaches to verify the internal criteria of a fuzzy network, and implement continuous self-verification within a fuzzy network.

References

1. Szegedy, C., et al.: Intriguing properties of neural networks. arXiv preprint arXiv:1312.6199 (2013)
2. Goodfellow, I.J., Shlens, J., Szegedy, C.: Explaining and harnessing adversarial examples. arXiv preprint arXiv:1412.6572 (2014)
3. Boeing. 737 Max Software System Safety Analysis Report (2019). https://www.boeing.com/
4. Makó, S., Pilat, M., Šváb, P., Kozuba, J., Cicváková, M.: Evaluation of MCAS System. Acta Avionica J. 21–28 (2020). https://doi.org/10.35116/aa.2020.0003
5. Talpur, N., Abdulkadir, S.J., Alhussian, H., et al.: Deep neuro-fuzzy system application trends, challenges, and future perspectives: a systematic survey. Artif. Intell. Rev. **56**, 865–913 (2023). https://doi.org/10.1007/s10462-022-10188-3
6. Souza, P.V.C.: Fuzzy neural networks and neuro-fuzzy networks: a review of the main techniques and applications used in the literature. Appl. Soft Comput. **92**, 106275 (2020). https://doi.org/10.1016/j.asoc.2020.106275
7. Arnett, T.J.: Iteratively Increasing Complexity During Optimization for Formally Verifiable Fuzzy Systems [Doctoral dissertation, University of Cincinnati]. OhioLINK Electronic Theses and Dissertations Center (2019). http://rave.ohiolink.edu/
8. Zimmermann, H.-J.: Fuzzy Set Theory and Its Applications. Springer, Berlin (1996)
9. Roques, P.: Systems Architecture Modeling with the Arcadia Method: A Practical Guide to Capella (1st ed.). Elsevier (2017). ISBN: 9781785481680 (hardcover), 9780081017920 (ebook)

How People Make Decisions Based on Prior Experience: Formulas of Instance-Based Learning Theory (IBLT) Follow from Scale Invariance

Palvi Aggarwal, Martine Ceberio, Olga Kosheleva, and Vladik Kreinovich(✉)

University of Texas at El Paso, 500 W. University, El Paso, TX 79968, USA
{paggarwal,mceberio,olgak,vladik}@utep.edu

Abstract. To better understand human behavior, we need to understand how people make decisions, how people select one of possible actions. This selection is usually based on predicting consequences of different actions, and these predictions are, in their turn, based on the past experience. For example, consequences that occur more frequently in the past are viewed as more probable. However, this is not just about frequency: recent observations are usually given more weight that past ones. Researchers have discovered semi-empirical formulas that describe our predictions reasonably well; these formulas form the basis of the Instance-Based Learning Theory (IBLT). In this paper, we show that these semi-empirical formulas can be derived from the natural idea of scale invariance.

1 Formulation of the Problem

How Do People Make Decisions? To properly make a decision, i.e., to select one of the possible actions, we need to predict the consequences of each of these actions. To predict the consequences of each action, we take into account past experience, in which we know the consequences of similar actions. Often, at different occasions, the same action led to different consequences. So, we cannot predict what exactly will be the consequence of each action. At best, for each action, we can try to predict the probability of different consequences.

In this prediction, we take into account the frequency with which each consequence occurred in the past. We also take into account that situations change, so more recent observations should be given more weight than the ones that happened long ago. To better understand human behavior, we need to know how people take all this into account.

Semi-empirical Formulas. By performing experiments and by analyzing the resulting data, researchers found some semi-empirical formulas that provide a very good description of the actual human behavior [4,5] (see also [2]). These formulas form the basis of the Instance-Based Learning Theory (IBLT).

In the first approximation, when we only consider completely different consequences, these formulas have the following form. For each possible action, to estimate the probability p_i of each consequence i, we first estimate the activation A_i of this consequence as

K. Cohen et al. (Eds.): NAFIPS 2023, LNNS 751, pp. 312–319, 2023.
https://doi.org/10.1007/978-3-031-46778-3_29

$$A_i = \ln\left(\sum_j (t - t_{i,j})^{-d}\right), \tag{1}$$

where:

- t is the current moment of time (i.e., the moment of time at which we make a decision),
- the values $t_{i,1}$, $t_{i,2}$, etc. are past moments of time at which the same action led to consequence i, and
- $d > 0$ is a constant – depending on the decision maker.

Based on these activation values, we estimate the probability p_i as

$$p_i = \frac{\exp(c \cdot A_i)}{\sum\limits_k \exp(c \cdot A_k)}, \tag{2}$$

where c is another constant depending on the decision maker, and the summation in the denominator is over all possible consequences k.

Challenge. How can we explain why these complex formulas properly describe human behavior?

What We Do in this Paper. In this paper, we show that these formulas can be actually derived from the natural idea of scale invariance.

2 Our Explanation

Analysis of the Problem. If time was not the issue, then the natural way to compare different consequences i would be by comparing the number of times n_i that the i-th consequence occurred in the past:

$$n_i = \sum_j 1. \tag{3}$$

In this formula, for each past observation of the i-th consequence, we simply add 1. In other words, all observations are assigned the same weight.

As we have mentioned, it makes sense to provide larger weight to more recent observations and smaller weight to less recent ones. In other words, instead of adding 1s, we should add some weights depending on the time $\Delta t = t - t_{i,j}$ that elapsed since the observation. Let us denote the dependence on the weight on time by $f(\Delta t)$. This function should be decreasing with Δt: the more time elapsed, the smaller the weight. In these terms, the simplified formula (3) should be replaced by the following more adequate formula

$$n_i = \sum_j f(t - t_{i,j}). \tag{4}$$

Based on the corresponding values n_i – describing the time-adjusted number of observations – we need to predict the corresponding probabilities p_i. The more frequent the consequence, the higher should be its probability. At first glance, it may seem that we can simply take $p_i = g(n_i)$ for some increasing function $g(z)$. However, this will not work, since the sum of the probabilities of different consequences should be equal to 1. To make sure that this sum is indeed 1, we need to "normalize" the values $g(n_i)$, i.e., to divide each of them by their sum:

$$p_i = \frac{g(n_i)}{\sum\limits_k g(n_k)}. \tag{5}$$

Remaining Question. The above formulas (4) and (5) leave us with a natural question: which functions $f(t)$ and $g(z)$ better describe human behavior?

Our Main Idea: Scale-Invariance. To answer this question, let us take into account that the numerical value of time duration – as well as the numerical values of many other physical quantities – depends on the choice of the measuring unit. If we replace the original unit for measuring time by a new unit which is $\lambda > 0$ times smaller, then all numerical values of time intervals get multiplied by λ: $t \mapsto \lambda \cdot t$. For example, if we replace minutes by seconds, then all numerical values are multiplied by 60, so that, e.g., 2 min becomes 120 s.

In many physical (and other) situations, there is no physically preferred unit for measuring time intervals x. This means that the formulas should remain the same if we "re-scale" t by choosing a different measuring unit, i.e., by replacing all numerical values t with $t' = \lambda \cdot t$. This "remains the same" is called scale invariance [3,6].

Now, we are ready to formulate our main result.

Proposition. *Let $f(x)$ and $g(x)$ by continuous monotonic functions. Then, the following conditions are equivalent to each other:*

- *for each $\lambda > 0$, the values of p_i as described by the formulas (4) and (5) will remain the same if we replace t and $t_{i,j}$ with $t' = \lambda \cdot t$ and $t'_{i,j} = \lambda \cdot t_{i,j}$;*
- *the dependence (4)–(5) is described by the formulas (1)–(2).*

Proof.

1°. Let us first show that if we re-scale all the time values in the formulas (1)–(2), the probabilities remain the same.

Indeed, in this case,

$$\sum_j (t' - t'_{i,j})^{-d} = \lambda^{-d} \cdot \sum_j (t - t_{i,j})^{-d}$$

and thus, the new value A'_i of the activity is:

$$A'_i = \ln\left(\sum_j (t' - t'_{i,j})^{-d}\right) = \ln\left(\sum_j (t - t_{i,j})^{-d}\right) + \ln(\lambda^{-d}) = A_i + \ln(\lambda^{-d}).$$

Thus, we have

$$c \cdot A_i' = c \cdot A_i + c \cdot \ln(\lambda^{-d}),$$

and $\exp(c \cdot A_i') = C \cdot \exp(c \cdot A_i)$, where we denoted $C \stackrel{\text{def}}{=} \exp(c \cdot \ln(\lambda^{-d}))$. Hence, the new expression for probability takes the form

$$p_i' = \frac{\exp(c \cdot A_i')}{\sum\limits_k \exp(c \cdot A_k')} = \frac{C \cdot \exp(c \cdot A_i)}{C \cdot \sum\limits_k \exp(c \cdot A_k)}.$$

If we divide both the numerator and the denominator of the right-hand side by the same constant C, we conclude that $p_i' = p_i$, i.e., that the probabilities indeed do not change.

2°. So, to complete our proof, it is sufficient to prove that if the expressions (4)–(5) leads to scale-invariant probabilities, then the dependence (4)–(5) is described by the formulas (1)–(2).

2.1°. Let us first find what we can deduce from scale-invariance about the function $g(x)$.

To do that, let us consider the case when we have two consequences, one of which was observed only once, and the other one was observed m times for some $m > 1$. Let us also assume that all the observations occurred at the same time Δt moments in the past, so that $t - t_{i,j} = \Delta t$. In this case, $n_1 = f(\Delta t)$, $n_2 = m \cdot n_1$, and the formula for the probability p_1 takes the form

$$p_1 = \frac{g(f(\Delta t))}{g(m \cdot f(\Delta t)) + g(f(\Delta t))}. \tag{6}$$

By re-scaling time, we can replace Δt with any other value, and this should not change the probabilities. Thus, the formula (6) should retain the same value for all possible values of $z = f(\Delta t)$. In other words, the ratio

$$\frac{g(z)}{g(m \cdot z) + g(z)}$$

should not depend on z, it should only depend on m. Hence, its inverse

$$\frac{g(m \cdot z) + g(z)}{g(z)} = \frac{g(m \cdot z)}{g(z)} + 1$$

should also depend only on m, and therefore, that we should have

$$\frac{g(m \cdot z)}{g(z)} = a(m)$$

for some function $a(m)$. So, we should have $g(m \cdot z) = a(m) \cdot g(z)$ for all z and m.

In particular, for $z' = z/m'$ for which $m' \cdot z' = z$, we should have $g(m' \cdot z') = a(m') \cdot g(z')$, i.e., $g(z) = a(m') \cdot g(z/n')$, hence $g(z/n') = (1/a(m')) \cdot g(z)$. So, for each rational number $r = m/m'$, we should have

$$g(r \cdot z) = g(m \cdot (z/m')) = a(m) \cdot g(z/m') = a(m) \cdot (1/a(m')) \cdot g(z).$$

In other words, for every z and for every rational number r, we should have

$$g(r \cdot z) = a(r) \cdot g(z), \tag{7}$$

where we denoted $a(r) \stackrel{\text{def}}{=} a(m) \cdot (1/a(m'))$.

By continuity, we can conclude that the formula (7) should hold for all real values r, not necessarily for rational values. It is known that every continuous solution to the functional Eq. (7) is the power law, i.e., it has the form $y = A \cdot x^a$ for some constants A and a; see, e.g., [1]. Thus, we conclude that

$$g(z) = A \cdot z^a. \tag{8}$$

2.2°. We can simplify the resulting expression (8) even more.

Indeed, substituting the expression (8) into the formula (5), we conclude that

$$p_i = \frac{A \cdot n_i^a}{A \cdot \sum_k n_k^a}.$$

We can simplify this expression if we divide both the numerator and the denominator by the same constant A. Then, we get the following simplified formula

$$p_i = \frac{n_i^a}{\sum_k n_k^a} \tag{9}$$

that corresponds to the function $g(z) = z^a$. So, without loss of generality, we can conclude that $g(z) = z^a$.

2.3°. Let us now find out what we can deduce from scale-invariance about the function $f(t)$.

For this purpose, let us consider two consequences each of which was observed exactly once, one of which was observed 1 time unit ago and the other one was observed t_0 time units ago. Then, according the formulas (4) and (9), the predicted probability p_1 should be equal to

$$p_1 = \frac{(f(1))^a}{(f(1))^a + (f(t_0))^a}.$$

By scale-invariance, this probability should not change if we multiply both time intervals by λ, so that $1 \mapsto \lambda$ and $t_0 \mapsto \lambda \cdot t_0$:

$$\frac{(f(1))^a}{(f(1))^a + (f(t_0))^a} = \frac{(f(\lambda))^a}{(f(\lambda))^a + (f(\lambda \cdot t_0))^a}.$$

The equality remains valid if we take the inverses of both sides:

$$\frac{(f(1))^a + (f(t_0))^a}{(f(1))^a} = \frac{(f(\lambda))^a + (f(\lambda \cdot t_0))^a}{(f(\lambda))^a},$$

subtract 1 from both sides, resulting in:

$$\frac{(f(t_0))^a}{(f(1))^a} = \frac{(f(\lambda \cdot t_0))^a}{(f(\lambda))^a},$$

and raise both sides to the power $1/a$:

$$\frac{f(t_0)}{f(1)} = \frac{f(\lambda \cdot t_0)}{f(\lambda)}.$$

The left-hand side of this equality does not depend on λ, it depends only on t_0. Thus, the right-hand side should also depend only on t_0, i.e., we should have

$$\frac{f(\lambda \cdot t_0)}{f(\lambda)} = F(t_0),$$

for some function $F(t_0)$. Multiplying both sides by $f(\lambda)$, we conclude that

$$f(\lambda \cdot t_0) = F(t_0) \cdot f(\lambda)$$

for all $t_0 > 0$ and $\lambda > 0$.

We have already mentioned that every continuous solution to this functional equation has the form

$$f(t) = B \cdot t^b \tag{10}$$

for some constants B and b. Since the function $f(t)$ is decreasing, we have $b < 0$.

2.4°. We can simplify the expression (10) even more.

Indeed, substituting the expression (10) into the formula (4), we get

$$n_i = B \cdot \sum_j (t - t_{i,j})^b,$$

i.e., $n_i = B \cdot a_i$, where we denoted

$$a_i \stackrel{\text{def}}{=} \sum_k (t - t_{i,j})^b. \tag{11}$$

Substituting the formula $n_i = B \cdot a_i$ into the formula (9), we get

$$p_i = \frac{B^a \cdot a_i^a}{B^a \cdot \sum\limits_k a_k^a}.$$

We can simplify this expression if we divide both the numerator and the denominator by the same constant B^a. Then, we get the following simplified formula

$$p_i = \frac{a_i^a}{\sum\limits_k a_k^a}$$

that corresponds to using a_i instead of n_i, i.e., in effect, to using the function $f(t) = t^b$. So, without loss of generality, we can conclude that $f(t) = t^b$. For this function $f(t)$, we have

$$p_i = \frac{n_i^a}{\sum\limits_k n_k^a}. \tag{12}$$

2.5°. Let us show that for $f(t) = t^b$ for $b < 0$ and $g(z) = z^a$, we indeed get the expression (1)–(2).

Indeed, for $f(t) = t^b$, the formula (4) takes the form

$$n_i = \sum_j (t - t_{i,j})^b,$$

i.e., the form

$$n_i = \sum_j (t - t_{i,j})^{-d},$$

where we denoted $d \stackrel{\text{def}}{=} -b$. Thus, the expression (1) takes the form $A_i = \ln(n_i)$. So, the expression $\exp(c \cdot A_i)$ in the empirical formula (2) takes the form

$$\exp(c \cdot A_i) = \exp(c \cdot \ln(n_i)) = (\exp(\ln(n_i))^c = n_i^c.$$

Thus, the formula (2) takes the form

$$p_i = \frac{n_i^c}{\sum\limits_k n_k^c}.$$

One can see that this is exactly our formula (11), the only difference is that the parameters that is denoted by c in the formula (2) is denoted a in the formula (12).

Thus, we have indeed explained the empirical formulas (1) and (2). The proposition is proven.

Acknowledgments. This work was supported in part by the National Science Foundation grants 1623190 (A Model of Change for Preparing a New Generation for Professional Practice in Computer Science), HRD-1834620 and HRD-2034030 (CAHSI Includes), EAR-2225395, and by the AT&T Fellowship in Information Technology.

It was also supported by the program of the development of the Scientific-Educational Mathematical Center of Volga Federal District No. 075-02-2020-1478, and by a grant from the Hungarian National Research, Development and Innovation Office (NRDI).

References

1. Aczél, J., Dhombres, J.: Functional Equations in Several Variables. Cambridge University Press, Cambridge (2008)
2. Cranford, E.A., Gonzalez, C., Aggarwal, P., Tambe, M., Cooney, S., Lebiere, C.: Towards a cognitive theory of cyber deceoption. Cogn. Sci. **45**, e13013 (2021)

3. Feynman, R., Leighton, R., Sands, M.: The Feynman Lectures on Physics. Addison Wesley, Boston (2005)
4. Gonzalez, C., Dutt, V.: Instance-based learning: integrating sampling and repeated decisions from experience. Psychol. Rev. **118**(4), 523–551 (2011)
5. Gonzalez, C., Lerch, J.F., Lebiere, C.: Instance-based learning in dynamic decision making. Cogn. Sci. **27**, 591–635 (2003)
6. Thorne, K.S., Blandford, R.D.: Modern Classical Physics: Optics, Fluids, Plasmas, Elasticity, Relativity, and Statistical Physics. Princeton University Press, Princeton (2021)

Integrity First, Service Before Self, and Excellence: Core Values of US Air Force Naturally Follow from Decision Theory

Martine Ceberio[1], Olga Kosheleva[2], and Vladik Kreinovich[1](✉)

[1] Department of Computer Science, University of Texas at El Paso 500 W. University, El Paso, TX 79968, USA
{mceberio,vladik}@utep.edu

[2] Department of Teacher Education, University of Texas at El Paso, 500 W. University, El Paso, TX 79968, USA
olgak@utep.edu

Abstract. By analyzing data both from peace time and from war time, the US Air Force came with three principles that determine success: integrity, service before self, and excellence. We show that these three principles naturally follow from decision theory, a theory that describes how a rational person should make decisions.

1 Formulation of the Problem

Empirical fact. Based on its several decades of both peace-time and war-time experience, the US Air Force has come up with the three major principles that determine success (see, e.g., [9]):

- Integrity,
- Service before Self, and
- Excellence.

Empirically, of these three criteria, integrity is the most important one.

Natural question. How can we explain this empirical observation?

What we do in this paper. In this paper, we show that these three principles naturally follow from decision theory, a theory that describes how a rational person should make decisions.

Structure of the paper. We start, in Sect. 2, with a brief reminder of decision theory. In Sect. 3, we show how this leads to the general principles of decision making. Finally, in Sect. 4, we show that these general principles are, in effect, exactly the above three principles of the US Air Force.

K. Cohen et al. (Eds.): NAFIPS 2023, LNNS 751, pp. 320–324, 2023.
https://doi.org/10.1007/978-3-031-46778-3_30

2 Decision Making: A Brief Reminder

What we want: a description in commonsense terms. When we make a decision, we want to select the best of all possible decisions.

Let us describe this in precise terms. To describe the above commonsense description in precise terms, we need:

- to describe, in precise terms, which decisions are possible and which are not, and
- to describe, in precise terms, what we mean by "the best".

What we mean by a possible decision: the notion of constraints. To describe which decisions are possible and which are not means to describe the set S of possible decisions. This set is usually described by *constraints* – properties that all possible decisions must satisfy.

For example, if we want to select the best design for a building, the constraints are:

- limits on the cost,
- the requirements that this building should withstand winds (and, if relevant, earthquakes) typical for this area, etc.

What we mean by "the best": optimization. Sometimes, we have a numerical characteristic $f(x)$ that describes the relative quality of different possible decisions x. For example, for a company, this characteristic is the expected profit.

Between any two possible decisions x and x', we should select the one for which the value of the objective function is larger. Corresponding, we say that a possible decision x is *the best* (*optimal*) if the value $f(x)$ is larger than or equal to the value $f(x')$ for any other possible decision x'. The problem of finding such optimal x is known as *optimization*.

What if we do not know the objective function? In some situations, we have an intuitive idea of which decisions are better, but we do not know a function that describes our preferences. Decision theory (see, e.g., [1,2,4–8]) shows that in such situations, we can still describe preferences by an appropriate numerical function. To do that, we need to select two alternatives:

- an alternative A_+ which is better than the consequence of any of the possible decisions, and
- an alternative A_- which is worse than the consequence of any of the possible decisions.

Then, for each value p from the interval $[0,1]$, we can think of a "lottery" $L(p)$ in which:

- we get A_+ with probability p and
- we get A_- with the remaining probability $1-p$.

For each possible decision x, we can ask the user to compare the consequences of this decision with lotteries $L(p)$ corresponding to different values p.

- For small $p \approx 0$, the lottery $L(p)$ is close to A_- and is, thus, worse than the consequences of the decision x: let us denote it by $L(p) < x$.
- For p close to 1, the lottery $L(p)$ is close to A_+ and is, thus, better than the consequences of the decision$x$: $L(p) < x$.

As we continuously change p from 0 to 1, at some point, there should be a switch from $L(p) < x$ to $x < L(p)$. The corresponding threshold point

$$u(x) = \sup\{p : L(p) < x\} = \inf\{p : x < L(p)\}$$

is know as the *utility* of x. In this sense, the consequences of the decision x are equivalent to the lottery $L(u(x))$ in which we get the very good alternative A_+ with the probability $u(x)$ and we get A_- with the remaining probability.

If we compare two lotteries $L(p)$ and $L(p')$, then, of course, the lottery in which the very good alternative A_+ appears with the larger probability is better. Since each alternative x is equivalent to a lottery $L(u(x))$ in which the very good alternative A_+ appears with the probability $u(x)$, we can thus conclude:

- that between any two possible decisions x and x', the decision maker will select the one with the larger value of the utility, and
- that the best decision is the one that has the largest value of the utility.

In other words, decisions are equivalent to optimizing the utility function $u(x)$.

Comments.

- We can get the value $u(x)$ by bisection: first we compare x with the lottery $L(0.5)$ and thus find out whether $u(x) \in [0, 0.5]$ or $u(x) \in [0.5, 1]$; then, we compare x with the lottery corresponding to the midpoint of the resulting interval, etc. At any given moment, we only have an interval containing $u(x)$ – i.e., we only know $u(x)$ with some uncertainty. This way, after k steps, we determine $u(x)$ with accuracy $2^{-(k+1)}$. Thus, for each desired accuracy $\varepsilon > 0$, after a few iterations $k = \lceil |\log_2(\varepsilon)| \rceil - 1$, we will find the value $u(x)$ with the desired accuracy.
- Of course, decision theory described the ideal solution, when the decision maker is perfectly rational: e.g., if the decision maker prefers A to B and B to C, he/she should also prefer A to C. It should be mentioned that decisions of actual decision makers are not always rational in this sense; see, e.g., [3].

Summarizing: resulting description of the decision making problem.

- *What we have:* we have a set S, and we have a function $f(x)$ that maps elements of this set to real numbers.
- *What we want:* we want to find the element $x \in S$ for which the value of the function $f(x)$ is the largest possible, i.e., for which $f(x) \geq f(x')$ for all $x' \in S$.

3 Resulting General Principles of Decision Making

Ideal problem and realistic solutions. In some cases, we have:

- the exact description of the set S,
- the exact description of the objective function $f(x)$, and
- the exact description of the possible decision x which is optimal with respect to the given objective function.

However, such situations are rare.

In practice:

- we may only have an approximate description of the set S of possible solutions,
- we may only have an approximate description of the objective function $f(x)$ – since this function, as we have mentioned, often needs to be elicited from the decision maker, and at each stage of this elicitation, we only get an approximate value of utility, and
- optimization algorithms may only provide an approximate solution to the optimization problem.

How the difference between ideal and realistic solution affects the quality of the decision. Let us analyze how the above difference between the ideal and the realistic solution to the corresponding optimization problem affects the quality of the resulting decision.

- If the generated solution is not actually possible, this "solution" is useless. From this viewpoint, satisfying constraints is the most important thing.
- Once we make sure that we limit ourselves to possible solutions, we need to make sure that the optimized function should be correct. This is more important than having an effective optimization technique – since even if we perform perfect optimization with respect to this wrong objective function, the resulting decision will not be optimal with respect to the *desired* objective function.
- Finally, the optimization technique should be effective – otherwise, the selected decision will not be as good as it could be.

4 The General Principles of Decision Making Are, in Effect, Exactly the Three Principles of the US Air Force

Let us show that the above general principles of decision making indeed correspond to the above three principles of the US Force: integrity, service (before self), and excellence.

Constraint satisfaction means integrity. As we have mentioned, the most important principle of decision making is that all constraints should be satisfied. This is exactly what is usually mean by integrity: according to Wikipedia, it means "a consistent and uncompromising adherence to strong moral and ethical principles and values".

This principle is, as we have mentioned, the most important in decision making – and it is indeed listed first in the usual description of the three principles of the UA Air Force.

Correctness of objective function means service (before self). For decisions involving a group of people, correctness of the objective function means that this objective function should perfectly reflect the needs of this group – and it should reflect the needs of the decision maker only to the extent that these needs are consistent with the group needs. This is what is meant by service: when the interests of other are valued before one's own interests.

This principle is second in importance in decision making – and it is indeed listed second in the usual description of the three principles of the UA Air Force.

Effectiveness of solving the corresponding optimization problem means excellence. Excellence (but not perfection) means that we need to try to our best to find solutions that are as good as possible, and that we must be good at this task.

Acknowledgment. This work was supported in part by the National Science Foundation grants 1623190 (A Model of Change for Preparing a New Generation for Professional Practice in Computer Science), HRD-1834620 and HRD-2034030 (CAHSI Includes), EAR-2225395, and by the AT&T Fellowship in Information Technology.

It was also supported by the program of the development of the Scientific-Educational Mathematical Center of Volga Federal District No. 075-02-2020-1478, and by a grant from the Hungarian National Research, Development and Innovation Office (NRDI).

The authors are greatly thankful to Dr. Heather Wilson for valuable discussions.

References

1. Fishburn, P.C.: Utility Theory for Decision Making. Wiley, New York (1969)
2. Fishburn, P.C.: Nonlinear Preference and Utility Theory. The John Hopkins Press, Baltimore (1988)
3. Kahneman, D.: Thinking, Fast and Slow. Farrar, Straus, and Giroux, New York (2011)
4. Kreinovich, V.: Decision making under interval uncertainty (and beyond). In: Guo, P., Pedrycz, W. (eds.) Human-Centric Decision-Making Models for Social Sciences. SCI, vol. 502, pp. 163–193. Springer, Heidelberg (2014). https://doi.org/10.1007/978-3-642-39307-5_8
5. Luce, R.D., Raiffa, R.: Games and Decisions: Introduction and Critical Survey. Dover, New York (1989)
6. Nguyen, H.T., Kosheleva, O., Kreinovich, V.: "Decision making beyond Arrow's 'impossibility theorem', with the analysis of effects of collusion and mutual attraction". Int. J. Intell. Syst. **24**(1), 27–47 (2009)
7. Nguyen, H.T., Kreinovich, V., Wu, B., Xiang, G.: Computing Statistics under Interval and Fuzzy Uncertainty. Springer, Heidelberg (2012). https://doi.org/10.1007/978-3-642-24905-1
8. Raiffa, H.: Decision Analysis. McGraw-Hill, Columbus (1997)
9. US Air Force, A Profession of Arms: Our Core Values (2022). https://www.doctrine.af.mil/Portals/61/documents/Airman_Development/BlueBook.pdf

Conflict Situations are Inevitable When There are Many Participants: A Proof Based on the Analysis of Aumann-Shapley Value

Sofia Holguin and Vladik Kreinovich(✉)

Department of Computer Science, University of Texas at El Paso, 500 W. University, El Paso, TX 79968, USA
seholguin2@miners.utep.edu, vladik@utep.edu

Abstract. When collaboration of several people results in a business success, an important issue is how to fairly divide the gain between the participants. In principle, the solution to this problem is known since the 1950s: natural fairness requirements lead to the so-called Shapley value. However, the computation of Shapley value requires that we can estimate, for each subset of the set of all participants, how much gain they would have gained if they worked together without others. It is possible to perform such estimates when we have a small group of participants, but for a big company with thousands of employers this is not realistic. To deal with such situations, Nobelists Aumann and Shapley came up with a natural continuous approximation to Shapley value – just like a continuous model of a solid body helps, since we cannot take into account all individual atoms. Specifically, they defined the Aumann-Shapley value as a limit of the Shapley value of discrete approximations: in some cases this limit exists, in some it does not. In this paper, we show that, in some reasonable sense, for almost all continuous situations the limit does not exist: we get different values depending on how we refine the discrete approximations. Our conclusion is that in such situations, since computing of fair division is not feasible, conflicts are inevitable.

1 Formulation of the Problem

Collaboration is often beneficial. In many practical tasks, be it menial or intellectual tasks, it is beneficial for several people to collaborate. This way, every participant is focusing on the task in which he/she is most skilled while tasks at which this participant is not very skilled are performed by those who are better in these tasks. In such situations, in general, the more people participate, the better the result.

Question: how to divide the resulting gain. Often, the resulting gain is financial: the company gets a profit, a research group gets a bonus or an award, etc. A natural question is: what is the fair way to divide this gain between the participants.

Shapley value: a description of a fair division. There is a known answer to this question, the answer originally produced by the Nobelist Lloyd Shapley [3–5]; let us describe this answer.

K. Cohen et al. (Eds.): NAFIPS 2023, LNNS 751, pp. 325–330, 2023.
https://doi.org/10.1007/978-3-031-46778-3_31

Let n denote the number of collaborators, and let us number them by numbers from 1 to n. To describe the contribution of each participant, for each subset $S \subseteq N \stackrel{\text{def}}{=} \{1,\ldots,n\}$, we can estimate the gain $v(S)$ that participants from the set S would have gained if they worked on this project without any help from the others. Based on the function $v(S)$ – that assigns, to each subset S, the gain value $v(S)$ – we need to determine how to divide the overall gain $v(N)$ between the participants, i.e., how to come up with the values $x_i(v)$ for which

$$x_1(v)+\cdots+x_n(v)=v(N).$$

Shapley introduced natural requirements on the function $x_i(v)$. First, this function should not depend on the numbers that we assign to the participants: if we start with a different participant etc., each participant should receive the same portion as before. Second, people may participate in two different collaborative projects, corresponding to functions $u(S)$ and $v(S)$. As a result of the first project, each participant i gets $x_i(u)$; as the result of the second project, this participant gets $x_i(v)$. Thus, the overall amount gained by the i-th participant is $x_i(u)+x_i(v)$. Alternatively, we can view the two projects as two parts of one big project. In this case, for each set S, the gain $w(S)$ is equal to the sum of their gains in the two parts: $w(S)=u(S)+v(S)$. Based on this overall project, the i-th participant should get the value $x_i(w)=x_i(u+v)$. It is reasonable to require that the portion assign to the i-th participant should not depend on whether we treat two projects separately or as two parts of a big project. In other words, we should have $x_i(u+v)=x_i(u)+x_i(v)$. This property is known as *additivity*.

Shapley has shown that these two natural requirements uniquely determine the function $x_i(v)$: namely, we should have

$$x_i(v)=\sum_{S:\, i\notin S}\frac{|S|!\cdot(n-|S|-1)!}{n!}\cdot(v(S\cup\{i\})-v(S)),$$

where $|S|$ denotes the number of elements in the set S. This formula is known as *Shapley value*.

Continuous approximation: idea. As we have mentioned, for large number of participants, using the above formula to compute the Shapley value is not feasible. Such situations – when a large number of objects makes computations difficult – are common in science. For example, we know that a solid body consists of molecules, and we know the equations that describe how the molecules interact. However, for a large number of molecules forming a body it is not realistic to take all the molecules into account. Instead, we use a continuous approximation: we assume that the matter is uniformly distributed, and perform computations based on this assumption.

Continuous approximation: towards a precise formulation of the problem. In the continuous approximation, the set of participants forms an area A in 1-D or 2-D or multi-dimensional space. For example, if we want to consider the best way to divide the budget surplus in the state of Texas between projects benefiting local communities, we can view the set of participation as the 2-D area of the state. If in the same task, we want to treat people with different income level differently, it make sense to consider the set of participants as a 3-D region, in which the first two parameters are geographic coordinates, and the third parameter is the income. If we take more parameters into account, we get an area of larger dimension.

As we mentioned earlier, to describe the situation, we assign, to each subset S of the original set of participants, the value $v(S)$ describing how much participants from this set can gain if they act together without using any collaboration from others. In the continuous approximation, we consider reasonable (e.g., measurable) subsets S of the original area A. We also consider reasonable functions $v(S)$, e.g., functions of the type $F(v_1(S),\ldots,v_k(S))$, where $F(z_1,\ldots,z_k)$ is a continuous function, and each v_i has the form $v_i(S) = \int_S f_i(x)\,dx$ for some bounded measurable function $f_i(x)$. Let us denote the set of all such functions by V.

This set V is closed under addition and under a multiplication by a positive number, i.e., if $u,v \in V$ and $\alpha > 0$, then $u+v \in V$ and $\alpha \cdot v \in V$.

Continuous approximation: towards a natural solution. In the original formulation, based on the known values $v(S)$ corresponding to different sets $S \subseteq N$, we decide what portion $x_i(v)$ of the gain $v(N)$ to allocated to each participant i. Once we decide on this, to each group $S \subseteq N$, we thus allocate the sum $x_S(v)$ of values x_i allocated to all the members of this group:

$$x_S(v) = \sum_{i \in S} x_i(v).$$

How can we extend this formula to the continuous case? To do this, we can use the experience of physicists who use the continuous approximation to predict the properties of a solid body. Their technique – see, e.g., [2,6] – is to discretize the space, i.e., to divide the area occupied by the solid body into small cells, to approximate the behavior of each physical quantity inside a cell by a few parameters (e.g., we assume that this value is constant throughout this small cell), and to solve the corresponding finite-parametric problem. In other words, the usual idea is to provide a discrete approximation to the continuous approximation.

The Nobelists Aumann and Shapley proposed, in effect, to apply the same idea to the continuous approximation to the gain-dividing problem [1]. To approximate the continuous game with a sequence of discrete games, they proposed to consider sequences of partitions $P^{(1)}, P^{(2)}, \ldots, P^{(k)}, \ldots$ each of which divides the area A into a finite number of disjoint measurable sub-areas $A_1^{(k)}, \ldots, A_{n_k}^{(k)}$, and that satisfy the following two properties:

- the next division $P^{(k+1)}$ is obtained from the previous division $P^{(k)}$ by sub-dividing each of the sets $A_i^{(k)}$, and
- for every two elements $a \neq b$ from the area A, there exists a number k for which the partition $P^{(k)}$ allocates these two elements to different sub-areas.

Such sequences of partitions are called *admissible*.

For each partition $P^{(k)}$ for an admissible sequence, we consider sets S consisting of the corresponding sub-areas, i.e., sets S of the type

$$S(s) = \bigcup_{i \in s} A_i^{(k)}$$

for some set $s \subseteq \{1,\ldots,n_k\}$. In this discrete-approximation-to-continuous-approximation scheme, we get, for each k, a situation with n_k participants for which the gain of each subset $s \subseteq \{1,\ldots,n_k\}$ is described by the value $v(S(s))$. Based on these values, we can compute, for each of these participants, the Shapley value $x_i^{(k)}(v)$.

For each measurable subset $S \subseteq A$ and for each partition $P^{(k)}$, we can thus find the approximate lower bound for the amount allocated to S as the sum

$$x_S^{(k)}(v) = \sum_{i:\, A_i^{(k)} \subseteq S} x_i^{(k)}(v).$$

Some functions $v(S)$ have the following nice property: for each set S, no matter what admissible sequence of partitions $P^{(k)}$ we take, the values $x_S^{(k)}(v)$ tend the same limit. This limit $x_S(v)$ is called the *Aumann-Shapley value* corresponding to the function $v(S)$.

Sometimes, the Aumann-Shapley value exists, sometimes, it does not exist. For some functions $v(S)$, the sequences $x_S^{(k)}(v)$ always converge. So, if the original situation with a large number of participants can be described by such function $v(S)$, then, to fairly allocate gains to all the participants, we can use an appropriate simplified approximation for which computing such a fair allocation is feasible.

On the other hand, it is known that for some functions $v(S)$, the sequences $x_S^{(k)}(v)$ do not converge. For example, Example 19.2 from [1] shows that the limits do not exist already for the following simple function on the domain $A = [0,1]$:

$$v(S) = \min\left(\int_S \chi_{[0,\frac{1}{3})}\,dx, \int_S \chi_{[\frac{1}{3},\frac{2}{3})}\,dx, \int_S \chi_{[\frac{2}{3},1]}\,dx\right),$$

where χ_I is the characteristic function of the set I, i.e., $\chi_I(x) = 1$ if $x \in I$ and $\chi_I(x) = 0$ is $x \notin I$. For such functions v, we cannot use a simplified approximation.

And since for large number of participants, direct computation of the Shapley value – i.e., of the fair division – is not feasible, this means that for such functions $v(S)$, we cannot feasibly compute a division that everyone would recognize as fair. Thus, in such situations, conflicts are inevitable.

Natural question. A natural question is: which of these two situations occurs in real life?

What we do in this paper. In this paper, we come up with an answer to this question.

2 Discussion and the Resulting Conclusion

Main idea behind our analysis. To get an answer to the above question, we will use the following principle that physicists use (see, e.g., [2, 6]): that if some phenomenon occurs in almost all situations ("almost all" in some reasonable sense), then they conclude that this phenomenon occurs in real life as well.

For example, if you flip a coin many times and select 1 when it falls head and 0 when it falls tail, in principle, you can get any sequence of 0s and 1s. However, we know that for almost all sequences of 0s and 1s, the frequency of 1s tends to 1/2. So, physicists conclude (and experiments confirm this) that when we flip a coin many times, the frequency of 1s tends to 1/2. This is not just what physicists do, this is common sense: if you go to a casino and the roulette ends up on red (as opposed to black) 30 times in a row, you will naturally conclude that the roulette is biased.

Similarly: it is, in principle, possible that due to random thermal interactions between all the molecules in a cat's body, all the molecules will start moving up, and the poor cat will start rising in the air. However, the probability of this event is practically 0. In almost all cases, this is not possible, so physicists conclude that this is not possible in real life.

An even simpler example: if we run some stochastic process twice and we get the exact same result in both cases, this would mean that something is wrong: indeed, once we have the first result r_1, the second result r_2 can, in principle, take any real value. Only for one of these values r_1, out of continuum many, we can have $r_2 = r_1$. Thus, for almost all possible values r_2 (with one exception), we have $r_2 \neq r_1$; so, we conclude that in real life, we will have $r_2 \neq r_1$.

In our analysis, we will take into account additivity and homogeneity of the Aumann-Shapley value. It is known that several properties of the Shapley value can be extended to the Aumann-Shapley value: we just need to take into account that, in contrast to the Shapley value – which is always defined – the Aumann-Shapley value is only defined for some functions $v(S)$. As we have mentioned, the Shapley value has the additivity property: for every two functions $u(S)$ and $v(S)$, we have $x_i(u+v) = x_i(u) + x_i(v)$. Also, it has the following homogeneity property: for every $\alpha > 0$, we have $x_i(\alpha \cdot v) = \alpha \cdot x_i(v)$. In the limit, these two properties leads to the following properties of the Aumann-Shapley (AS) value:

- if the AS value exists for functions u and v, then AS value exists for $u+v$, and

$$x_S(u+v) = x_S(u) + x_S(v);$$

- if the AS value exists for functions u and $u+v$, then AS value exists for v, and

$$x_S(u+v) = x_S(u) + x_S(v);$$

- if the AS value exists for v, then for each $\alpha > 0$, the AS value exists for $\alpha \cdot v$, and

$$x_S(\alpha \cdot v) = \alpha \cdot x_S(v).$$

What we can conclude from these properties. Let v_0 be a function for which the AS value does not exist. For each function $v \in V$, we can consider a 1-D set $L(v) \stackrel{\text{def}}{=} \{v + t \cdot v_0 \,|\, v + t \cdot v_0 \in V\}$, where t denotes an arbitrary real number. For all $t \geq 0$, we have $v + t \cdot v_0 \in V$, so this set contains a whole half-line, i.e., it contains continuum many points. Let us prove that in this set, we can have at most one real value t for which the AS value exist.

The proof is by contradiction. Indeed, if we have two values $t_1 < t_2$ for which the functions $v + t_i \cdot v_0$ have AS value, then, by additivity, the AS value exists for the difference $(v + t_2 \cdot v_0) - (v + t_1 \cdot v_0) = (t_2 - t_1) \cdot v_0$ and thus, by homogeneity, for the function

$(t_2 - t_1)^{-1} \cdot ((t_2 - t_1) \cdot v_0) = v_0$. However, we have selected v_0 for which the AS value does not exist. The resulting contradiction proves our statement.

This means that for almost all functions v, the AS value does not exist. One can easily check that for two functions v and v', if the corresponding sets $L(v)$ and $L(v')$ have a common element, then these sets coincide. Thus, we can divide the whole set V of all possible functions $v(S)$ into non-intersecting 1-D subsets $L(v)$ corresponding to different v. In each such subset, out of continuum many points, there is at most one point for which the AS value exists.

When some property holds only for one point on the whole half-line, this means, intuitively, that in almost all cases, this property is not satisfied. This is exactly what we wanted to prove.

Corollary: reminder. We have shown that for almost all functions $v(S)$, the AS value does not exist – i.e., that the Shapley values corresponding to discrete few-player approximations do not converge. Following the above physicists' principle, we conclude that in real life, these values do not converge. Thus, in situations with many participants, we cannot use this approximation idea.

And since for large number of participants, direct computation of the Shapley value (i.e., of the fair division) is not feasible, this means that for such functions $v(S)$, we cannot feasibly compute a division that everyone would recognize as fair. Thus, in such situations, conflicts are inevitable.

Acknowledgment. This work was supported in part by the National Science Foundation grants 1623190 (A Model of Change for Preparing a New Generation for Professional Practice in Computer Science), HRD-1834620 and HRD-2034030 (CAHSI Includes), EAR-2225395, and by the AT&T Fellowship in Information Technology.

It was also supported by the program of the development of the Scientific-Educational Mathematical Center of Volga Federal District No. 075-02-2020-1478, and by a grant from the Hungarian National Research, Development and Innovation Office (NRDI).

References

1. Aumann, R.J., Shapley, L.S.: Values of Non-Atomic Games. Princeton University Press, Princeton (1974)
2. Feynman, R., Leighton, R., Sands, M.: The Feynman Lectures on Physics. Addison Wesley, Boston (2005)
3. Maschler, M., Solan, E., Zamir, S.: Game Theory. Cambridge University Press, Cambridge (2020)
4. Shapley, L.S.: Notes on the n-Person Game - II: The Value of an n-Person Game, RAND Corporation Research Memorandum RM-670, Santa Monica, California (1951). https://www.rand.org/content/dam/rand/pubs/research_memoranda/2008/RM670.pdf
5. Shapley, L.S.: A value of n-person games. In: Kuhn, H.W., Tucker, A.W. (eds.) Contributions to the Theory of Games, pp. 307–317. Princeton University Press, Princeton (1953)
6. Thorne, K.S., Blandford, R.D.: Modern Classical Physics: Optics, Fluids, Plasmas, Elasticity, Relativity, and Statistical Physics. Princeton University Press, Princeton (2021)

Computing at Least One of Two Roots of a Polynomial is, in General, Not Algorithmic

Vladik Kreinovich(✉)

Department of Computer Science, University of Texas at El Paso, 500 W. University, El Paso, TX 79968, USA
vladik@utep.edu

Abstract. In our previous work, we provided a theoretical explanation for an empirical fact that it is easier to find a unique root than the multiple roots. In this short note, we strengthen that explanation by showing that finding one of many roots is also difficult.

1 Introduction

Background. Before we start designing an algorithm for some class of problems, it is desirable to know whether a general algorithm is indeed possible for this class.

Among these problems are solving systems of equations (in particular, equations), finding optima of a given function on a given box, etc.

It is known that there exists an algorithm which is applicable to every system $f_1(x_1,\ldots,x_n)=0$, ..., $f_m(x_1,\ldots,x_n)=0$ with computable functions f_i which has exactly one solution on a given box $\mathbf{x}_1 \times \ldots \times \mathbf{x}_n$, and which computes this solution – i.e., produces, for every $\varepsilon > 0$, an ε-approximation to this solution [4–6].

It is also known that no algorithm is possible which is applicable to every computable system which has exactly two solutions and which would return *both* solutions [6]. The proof shows that such an algorithm is not possible even for computable polynomial equations.

Mathematical Comment. This algorithmic impossibility is due to the fact that we allow computable polynomials; for polynomials with rational (or even algebraic) coefficients, solution problems are algorithmically decidable; see, e.g., [1,6,8,9].

Practical Comment. This result is in good accordance with the empirical fact that in general, it is easier to find a point $(x_1,\ldots,x_n)$, in which a given system of equations has a unique solution than when this system has several solutions; see, e.g., [3].

Formulation of the Problem. A natural question is: since in the general two-roots case, we cannot return *both* roots, maybe we can return at least one of them?

K. Cohen et al. (Eds.): NAFIPS 2023, LNNS 751, pp. 331–334, 2023.
https://doi.org/10.1007/978-3-031-46778-3_32

Comment. This is actually possible in the example from [6].

What was Known. It is known that no such algorithm is possible for *general* computable functions [4,5]. This construction requires a computable function which is more complex than a polynomial.

What We Plan to Do. In this paper, we show that already for computable *polynomial* equations, it is impossible to compute even one of the roots.

In this proof, we will use the polynomials with the smallest possible number of variables and of the smallest possible degree. We also prove a similar result for optimization problems.

2 Results

Definitions. In order to precisely formulate this result, we need to recall the definition of a computable number. Crudely speaking, a real number is computable if it can be computed with an arbitrary accuracy.

Definition 1. (see, e.g., [2,6]) *A real number x is called* computable *if there exists an algorithm that, given an integer k, returns a rational number r_k for which*

$$|x - r_k| \le 2^{-k}.$$

By a *computable polynomial*, we mean a polynomial with computable coefficients.

According to our promise, we will prove this result for the case of the smallest possible number of variables: one.

Proposition 1. *No algorithm is possible that is applicable to any computable polynomial function $f(x)$ with exactly two roots, and returns one of these roots.*

Proof. Our proof will use the known fact that no algorithm is possible for detecting whether a given constructive real number α is non-negative or non-positive [2,7].

For every computable real number α, we can form a polynomial

$$f_\alpha(x) = [(x-1)^2 + \alpha] \cdot [(x+1)^2 - \alpha].$$

This polynomial is equal to 0 if one of the two factors is equal to 0.

- When $\alpha = 0$, this polynomial $f_\alpha(x)$ has exactly two roots: 1 and -1.
- When $\alpha > 0$, the first factor is positive, so $f_\alpha(x) = 0$ if and only if $(x+1)^2 = \alpha$, hence $x + 1 = \pm\sqrt{\alpha}$. So, for such α, the polynomial $f_\alpha(x)$ has exactly two roots:

$$x = -1 \pm \sqrt{\alpha}.$$

- Similarly, when $\alpha < 0$, the polynomial $f_\alpha(x)$ has exactly two roots:

$$x = 1 \pm \sqrt{|\alpha|}.$$

If we could compute one of roots, then by computing this root with enough accuracy and comparing it with 1, we could tell whether this root is close to 1 or to -1. According to our description of the roots, if this root is close to 1, then $\alpha \le 0$; if this root is close to -1, then $\alpha \le 0$. Since, as we have mentioned, we cannot check whether $\alpha \ge 0$ or $\alpha \le 0$, we thus cannot return one of the roots. The proposition is proven.

Comment. In the proof, we used a 4th degree polynomial. Let us give reasons why we cannot use a polynomial of a lower degree. Since we need polynomials with two roots, we must use polynomials of degree at least 2. If a quadratic polynomial has exactly 2 roots, then we can find these rules by using a standard formula, so these roots are easy to compute.

For a cubic polynomial $f(x)$, the only way to have exactly two real roots is to have one double root. At this root, the derivative $f'(x)$ is equal to 0 – and the solution to the quadratic equation $f'(x) = 0$ can be easily found.

Proposition 2. *No algorithm is possible which is applicable to any computable polynomial $f(x)$ that attains its minimum at exactly two points, and returns one of these points.*

Proof. It is sufficient to consider $f_\alpha^2(x)$, where $f_\alpha(x)$ is the polynomial from the previous proof; this new polynomial is always non-negative, and it attains its minimum 0 if and only $f_\alpha(x) = 0$.

Acknowledgments. This work was supported in part by the National Science Foundation grants 1623190 (A Model of Change for Preparing a New Generation for Professional Practice in Computer Science), and HRD-1834620 and HRD-2034030 (CAHSI Includes), and by the AT&T Fellowship in Information Technology.

It was also supported by the program of the development of the Scientific-Educational Mathematical Center of Volga Federal District No. 075-02-2020-1478, and by a grant from the Hungarian National Research, Development and Innovation Office (NRDI).

The author is thankful to Michael Beeson for the formulation of the problem and stimulating advice.

References

1. Basu, S., Pollack, R., Roy, M.-F.: Algorithms in Real Algebraic Geometry. Springer-Verlag, Berlin (2006). https://doi.org/10.1007/3-540-33099-2
2. Beeson, M.: Foundations of Constructive Mathematics: Metamathematical Studies. Springer, Berlin (1985). https://doi.org/10.1007/978-3-642-68952-9
3. Kearfott, R.B.: Rigorous Global Search: Continuous Problems. Kluwer, Dordrecht (1996)
4. Kreinovich, V.: Uniqueness implies algorithmic computability. In: Proceedings of the 4th Student Mathematical Conference, Leningrad University, Leningrad, pp. 19–21 (1975) (in Russian)
5. Kreinovich, V.: Categories of space-time models, Ph.D. dissertation, Novosibirsk, Soviet Academy of Sciences, Siberian Branch, Institute of Mathematics (1979) (in Russian)

6. Kreinovich, V., Lakeyev, A., Rohn, J., Kahl, P.: Computational Complexity and Feasibility of Data Processing and Interval Computations. Kluwer, Dordrecht (1998)
7. Kushner, B.A.: Lectures on Constructive Mathematical Analysis. American Mathematical Soc., Providence, RI (1985)
8. Mishra, B.: Computational real algebraic geometry. In: Handbook on Discreet and Computational Geometry, CRC Press (1997)
9. Tarski, A.: A decision method for elementary algebra and geometry. In: Caviness, B.F., Johnson, J.R. (eds.) Quantifier Elimination and Cylindrical Algebraic Decomposition. Texts and Monographs in Symbolic Computation, 2nd edn., pp. 24–84. Springer, Vienna (1951). https://doi.org/10.1007/978-3-7091-9459-1_3

Towards Decision Making Under Interval Uncertainty

Juan A. Lopez and Vladik Kreinovich(✉)

Department of Computer Science, University of Texas at El Paso, 500 W. University, El Paso, TX 79968, USA
jalopez44@miners.utep.edu, vladik@utep.edu

Abstract. In many real-life situations, we need to make a decision. In many cases, we know the optimal decision in situations when we know the exact value of the corresponding quantity x. However, often, we do not know the exact value of this quantity, we only know the bounds on the value x – i.e., we know the interval containing x. In this case, we need to select a decision corresponding to some value from this interval. The selected value will, in general, be different from the actual (unknown) value of this quantity. As a result, the quality of our decision will be lower than in the perfect case when we know the value x. Which value should we select in this case? In this paper, we provide a decision-theory-based recommendation for this selection.

1 Introduction

Situation. In many real-life situations, we need to make a decision.

The quality of the decision usually depends on the value of some quantity x. For example, in construction:

- the speed with which the cement hardens depends on the humidity, and
- thus, the proportions of the best cement mix depend on the humidity.

In practice, we often do not know the exact value of the corresponding quantity. For example, in the case of the pavement:

- while we can accurate measure the current humidity,
- what is really important is the humidity in the next few hours.

For this future value, at best, we only know the bounds, i.e., we only know the interval $[\underline{x}, \overline{x}]$ that contains the actual (unknown) value x. In other words, we have a situation of interval uncertainty; see, e.g., [2,4,7,8],

Problem. To select a decision, we need to select some value x_0 from this interval and make the decision corresponding to this selection. Which value x_0 should we select?

What We Do in This Paper. In this paper, we describe a solution to this problem.

Comment. Results from this paper first appeared in [5].

K. Cohen et al. (Eds.): NAFIPS 2023, LNNS 751, pp. 335–337, 2023.
https://doi.org/10.1007/978-3-031-46778-3_33

2 Our Solution

General Idea. In such situations of interval uncertainty, the ideal case is when the selected value x_0 is exactly equal to the actual value x. When these two values differ, i.e., when $x < x_0$ or $x > x_0$, the situation becomes worse.

In both cases when $x < x_0$ and when $x > x_0$, we have losses, but we often have two different reasons for a loss.

- If the humidity will be larger than expected, the hardening of the cement will take longer and we will lose time (and thus, money).
- In contrast, if the humidity is lower than expected, the cement will harden too fast, and the pavement will not be as stiff as it could be. So we will not get a premium for a good quality road (and we may even be required to repave some road segments).

In both cases, the larger the difference $|x - x_0|$, the larger the loss.

Possibility of Linearization. The interval $[\underline{x}, \overline{x}]$ is usually reasonable narrow, so the difference is small. In this case, the dependence of the loss on the difference can be well approximated by a linear expression; so:

- when $x < x_0$, the loss is $\alpha_- \cdot (x_0 - x)$ for some α_-, and
- when $x > x_0$, the loss is $\alpha_+ \cdot (x - x_0)$ for some α_+.

Resulting Formula for the Worst-Case Loss.

- When $x < x_0$, the worst-case loss is when x is the smallest: $\alpha_- \cdot (x_0 - \underline{x})$.
- When $x > x_0$, the worst-case loss is when x is the largest: $\alpha_+ \cdot (\overline{x} - x_0)$.

In general, the worst-case loss is the largest of these two:

$$w(x_0) = \max(\alpha_- \cdot (x_0 - \underline{x}), \alpha_+ \cdot (\overline{x} - x_0)).$$

Range (Interval) of Possible Values of the Loss. The best-case loss is 0 – when we guessed the value x correctly. In this case, all we know is that the loss is somewhere between 0 and $w(x_0)$. So, the gain is somewhere between $\underline{g} = -w(x_0)$ and $\overline{g} = 0$.

Let us Use Hurwicz Criterion. In situations where we only know the interval of possible values of the gain, decision theory recommends to use Hurwicz optimism-pessimism criterion to make a decision [1,3,6], i.e.:

- to select some value $\alpha > 0$ and
- then to select an alternative for which the value $g \stackrel{\text{def}}{=} \alpha \cdot \overline{g} + (1 - \alpha) \cdot \underline{g}$ is the largest possible.

In our case, $g = -(1 - \alpha) \cdot w(x_0)$, so maximizing g simply means selecting the value x_0 for which $w(x_0)$ is the smallest.

Let us Solve the Resulting Optimization Problem. Here, the value $\alpha_- \cdot (x_0 - \underline{x})$ increases with x_0, while the value $\alpha_+ \cdot (\overline{x} - x_0)$ decreases with x_0. Thus, the function $w(x_0)$ – which is the minimum of these two expressions:

- decreases until the point $\widetilde{x}$ at which these two expressions coincide, and
- then increases.

So, the minimum of the worst-case loss $w(x_0)$ is attained at the point $\widetilde{x}$ for which $\alpha_- \cdot (\widetilde{x} - \underline{x}) = \alpha_+ \cdot (\overline{x} - \widetilde{x})$, i.e., for $\widetilde{x} = \widetilde{\alpha} \cdot \overline{x} + (1 - \widetilde{\alpha}) \cdot \underline{x}$. Here, we denoted

$$\widetilde{\alpha} \stackrel{\text{def}}{=} \frac{\alpha_+}{\alpha_+ + \alpha_-}.$$

Comment. Interestingly, we get the same expression as with the Hurwicz criterion!

Acknowledgments. This work was supported in part by the National Science Foundation grants 1623190 (A Model of Change for Preparing a New Generation for Professional Practice in Computer Science), HRD-1834620 and HRD-2034030 (CAHSI Includes), EAR-2225395, and by the AT&T Fellowship in Information Technology.

It was also supported by the program of the development of the Scientific-Educational Mathematical Center of Volga Federal District No. 075-02-2020-1478, and by a grant from the Hungarian National Research, Development and Innovation Office (NRDI).

The authors are thankful to all the participants of the 29th Joint UTEP/NMSU Workshop on Mathematics, Computer Science, and Computational Science (Las Cruces, New Mexico, USA, April 1, 2023) for valuable suggestions.

References

1. Hurwicz, L.: Optimality criteria for decision making under ignorance, Cowles Commission Discussion Paper, Statistics, No. 370 (1951)
2. Jaulin, L., Kiefer, M., Didrit, O., Walter, E.: Applied Interval Analysis, with Examples in Parameter and State Estimation, Robust Control, and Robotics. Springer, London (2001). https://doi.org/10.1007/978-1-4471-0249-6
3. Kreinovich, V.: Decision making under interval uncertainty (and beyond). In: Guo, P., Pedrycz, W. (eds.) Human-Centric Decision-Making Models for Social Sciences. SCI, vol. 502, pp. 163–193. Springer, Heidelberg (2014). https://doi.org/10.1007/978-3-642-39307-5_8
4. Kubica, B.J.: Interval Methods for Solving Nonlinear Constraint Satisfaction, Optimization, and Similar Problems: from Inequalities Systems to Game Solutions. Springer, Cham, Switzerland (2019). https://doi.org/10.1007/978-3-030-13795-3
5. Lopez, J.A., Kreinovich, V.: Towards decision making under interval uncertainty. Abstracts of the NMSU/UTEP Workshop on Mathematics, Computer Science, and Computational Science, Las Cruces, New Mexico (2023)
6. Luce, R.D., Raiffa, R.: Games and Decisions: Introduction and Critical Survey. Dover, New York (1989)
7. Mayer, G.: Interval Analysis and Automatic Result Verification. de Gruyter, Berlin (2017)
8. Moore, R.E., Kearfott, R.B., Cloud, M.J.: Introduction to Interval Analysis. SIAM, Philadelphia (2009)

What Do Goedel's Theorem and Arrow's Theorem have in Common: A Possible Answer to Arrow's Question

Miroslav Svítek[1], Olga Kosheleva[2], and Vladik Kreinovich[2](✉)

[1] Faculty of Transportation Sciences, Czech Technical University in Prague, Konviktska 20, CZ-110 00 Prague 1, Prague, Czech Republic
svitek@fd.cvut.cz

[2] University of Texas at El Paso, 500 W. University, El Paso, TX 79968, USA
{olgak,vladik}@utep.edu

Abstract. Kenneth Arrow, the renowned author of the Impossibility Theorem that explains the difficulty of group decision making, noticed that there is some commonsense similarity between his result and Goedel's theorem about incompleteness of axiomatic systems. Arrow asked if it is possible to describe this similarity in more precise terms. In this paper, we make the first step towards this description. We show that in both cases, the impossibility result disappears if we take into account probabilities. Namely, we take into account that we can consider probabilistic situations, that we can make probabilistic conclusions, and that we can make probabilistic decisions (when we select different alternatives with different probabilities).

1 Formulation of the Problem

Need to Consider the Axiomatic Approach. Both Goedel's and Arrow's results are based on axioms. So, to analyze possible similarities between these two results, let us recall where the axiomatic approach came from.

Already ancient people had to deal the measurements of lengths, angles, areas, and volumes: this was important in constructing building, it was important for deciding how to re-mark borders between farms after a flood, etc.

The experience of such geometric measurements led to the discovery of many interesting relations between all these quantities. For example, people empirically discovered what we now call Pythagoras theorem – that in a right triangle, the square of the hypothenuse is equal to the sum of the square of the sides.

At first, this was just a collection of interesting and useful relations. After a while, people noticed that some of these relations can be logically deduced from others. Eventually, it turned out that it is enough to state a small number of these relations – these selected relations became known as *axioms* – so that every other geometric relation, every other geometric fact, every other statement about geometric objects can be deduced from these axioms.

K. Cohen et al. (Eds.): NAFIPS 2023, LNNS 751, pp. 338–343, 2023.
https://doi.org/10.1007/978-3-031-46778-3_34

In general, it was believed that for each geometric statement, we can either deduce this statement or its negation from the axioms of geometry.

Can Axiomatic Method be Applied to Arithmetic? The success of axiomatic method in geometry led to a natural idea of using this method in other disciplines as well. Since geometry is part of mathematics, a natural idea is to apply axiomatic method to other parts of mathematics, starting with statements about natural numbers.

The corresponding axioms have indeed been formulated in the 19th century. Most mathematicians believed that – similarly to geometry – for each statement about natural numbers, we can either deduce this statement or its negation from these axioms. OK, some mathematicians were not 100% sure that the available axioms would be sufficient for this purpose, but they were sure that in this case, we can achieve a similar result if we add a few additional axioms.

Goedel's Impossibility Result. Surprisingly, it turned out that the desired complete description of natural numbers is not possible: no matter what axioms we select:

- either there is a statement for which neither this statement not its negation can be derived from the axioms,
- or the set of axioms is inconsistent – i.e., for each statement, we can deduce *both* this statement and its negation from these axioms.

In other words, it is not possible to formulate the set of axioms about natural numbers that would satisfy the natural condition of completeness. This result was proven by Kurt Goedel [5].

In particular, Goedel proved that the incompleteness can already be shown for statements of the type $\forall n\, P(n)$ for algorithmically decidable formulas $P(n)$.

Arrow's Impossibility Result. In addition to mathematics, there is another area of research where we have reasonable requirements: namely, human behavior. Already Baruch Spinoza tried to describe human behavior in axiomatic terms; see, e.g., [14].

In the 20th century, researchers continued such attempts. In particular, attempts were made to come up with a scheme of decision making that would satisfy natural fairness restrictions. Similar to natural numbers, at first, researchers hoped that such a scheme would be possible. However, in 1951, the future Nobelist Kenneth Arrow proved his famous Impossibility Theory: that it is not possible to have an algorithm that, given preferences of several participants would come up with a group decision that would satisfy natural fairness requirements; see, e.g., [1]. For this result, Arrow was awarded the Nobel prize.

What Do Goedel's and Arrow's Impossibility Theorems have in Common: Arrow's Question. From the commonsense viewpoint, these two results are somewhat similar: they are both about impossibility of satisfying seemingly reasonable conditions. However, from the mathematical viewpoint, these two results are very different: the formulations are different, the proofs are different, etc.

The fact that from the commonsense viewpoint, these results are similar made Arrow conjecture that there must be some mathematical similarity between these two results as well; see, e.g., [2].

What We Do in This Paper. In this paper, we show that, yes, there is some mathematical similarity between these two results.

Our result just scratches the surface, but we hope that it will lead to discovering deeper and more meaningful similarities.

2 Our Main Idea and the Resulting Similarity

Implicit Assumption Underlying Arrow's Result. Arrow's result implicitly assumed that all our decisions are deterministic. But is this assumption always true?

But is This Assumption True? Not Really. The fact that this assumption is somewhat naive can be best illustrated by the known argument about a donkey – first described by a philosopher Buridan. According to this argument, a donkey placed between two identical heaps of hay will not be able to select one of them and will, thus, die of hunger.

Of course, the real donkey will not die, it will select one of the heaps at random and start eating the juicy hay. Similarly, a human when facing a fork in a road – without any information about possible paths, will randomly select one of the two directions. When two friends meet for dinner and each prefers his own favorite restaurant, they will probably flip a coin to decide where to eat.

In other words, people not only make deterministic decisions, they sometimes make probabilistic decisions, i.e., they select different actions with different probabilities.

Let Us Take Probabilities into Account When Describing Preferences and Decisions. Since probabilistic decisions are possible, we need to take them into account – both when we describe preferences and when we select a decision. We will show that taking into account – that leads to so-called *decision theory* [3,4,6,7,10–12] – helps to resolve the paradox of Arrow's theorem, namely, leads to a reasonable way to select a joint decision. From this viewpoint:

- Arrow's theorem means that if we only know preferences between *deterministic* alternatives, then we cannot consistently select a *deterministic* fair joint action;
- however, we will show that if we take into account preferences between *probabilistic* options and allow *probabilistic* joint decisions, then a fair solution is possible.

Let Us Take Probabilities into Account When Describing Preferences. First, we need to show how we can describe the corresponding preferences. Let us select two alternatives:

- a very good alternative A_+ which is better than anything that we will actually encounter and
- a very bad alternative A_- which is worse than anything that we will actually encounter.

Then, for each real number p from the interval $[0, 1]$, we can consider a probabilistic alternative ("lottery") $L(p)$ in which we get A_+ with probability p and A_- with the remaining probability $1 - p$. For each actual alternative A, we can ask the user to compare this alternative A with lotteries $L(p)$ corresponding to different probabilities p.

Here:

- For small p, the lottery is close to A_- and is, thus, worse than A; we will denote it by $L(p) < A$.
- For probabilities close to 1, the lottery $L(p)$ is close to A_+ and is, thus, better than A: $A < L(p)$.

Also:

- If $L(p) < A$ and $p' < p$, then clearly $L(p') < A$.
- Similarly, if $A < L(p)$ and $p < p'$, then $A < L(p')$.

One can show that in this case, there exists a threshold value u such that:

- for all $p < u$, we have $L(p) < A$, while
- for all $p > u$, we have $A < L(p)$:

$$\inf\{p : L(p) < A\} = \sup\{p : A < L(p)\}.$$

In this case, for arbitrarily small $\varepsilon > 0$, we have $L(u - \varepsilon) < A < L(u - \varepsilon)$. Since in practice, we can only set up probability with some accuracy, this means that, in effect, the alternative A is equivalent to $L(u)$. This threshold value u is called the *utility* of the alternative A. Utility of A is denoted by $u(A)$.

Utility is Not Uniquely Determined. The numerical value of the utility depends on the selection of the two alternatives A_- and A_+. One can show that if we select two different alternatives A'_- and A'_+, then the new utility value is related to the original utility value by a linear transformation $u'(A) = a \cdot u(A) + b$ for some $a > 0$ and b.

Utility of a Probabilistic Alternative. One can also show that the utility of a situation in which we get alternatives A_i with probabilities p_i is equal to

$$p_1 \cdot u(A_1) + p_2 \cdot u(A_2) + \ldots$$

How to Make a Group Decision. Suppose now that n participants need to make a joint decision. There is a status quo state A_0 – the state that will occur if we do not make any decision. Since utility of each participant is defined modulo a linear transformation, we can always apply a shift $u(A) \mapsto u(A) - u(A_0)$ and thus, get the utility of the status quo state to be equal to 0. Therefore, preferences of each participant i can be described by the utility $u_i(A)$ for which $u_i(A_0) = 0$.

These utilities are defined modulo linear transformations $u_i(A) \mapsto a_i \cdot u_i(A)$ for some a_i. It therefore makes sense to require that the group decision making should not change if we thus re-scale each utility value. It turns out that the only decision making that does not change under this re-scaling means selecting an alternative A with the largest product of the utilities $u_1(A) \cdot \ldots \cdot u_n(A)$. This result was first shown by a Nobelist John Nash and is thus known as Nash's bargaining solution; see, e.g., [7–9]. If we allow probabilistic combinations of original decisions, then such a solution is, in some reasonable sense, unique.

Adding Probabilities to Arrow's Setting: Conclusion. One can show that Nash's bargaining solution satisfies all fairness requirements. So, in this case, we indeed have a solution to the group decision problem – exactly what, according to Arrow's result, is not possible if we do not take probabilities into account.

What About Goedel's Theorem? Goedel's theorem states that, no matter what finite list of axioms we choose, for some true statements of the type $\forall n\, P(n)$, we will never deduce the truth of this statement from these axioms. Let us look at this situation from the commonsense viewpoint.

The fact that the statement $\forall n\, P(n)$ is true means that for all natural numbers n, we can check that the property $P(n)$ is true. We can check it for $n = 0$, we can check it for $n = 1$, etc., and this will be always true. And this is exactly how we reason about the properties of the real world:

- someone proposes a new physical law,
- we test it many times,
- every time, this law is confirmed,
- so we start believing that this law is true.

The more experiments we perform, the higher our subjective probability that this law is true. If a law is confirmed in n experiments, statistics estimates it as $1 - 1/n$; see, e.g., [13].

So, yes, we can never become 100% sure (based on the axioms) that the statement $\forall n\, P(n)$ is true, but are we ever 100% sure? Even for long proofs, there is always a probability that we missed a mistake – and sometimes such mistakes are found. We usually trust results of computer computations, but sometimes, some of the computer cells goes wrong, and this makes the results wrong – such things also happened.

So, in this case too, allowing probabilistic outcomes also allows us to make practically definite conclusions – contrary to what happens when we are not taking probabilities into account, in which case Goedel's theorem shows that conclusions are, in general, not possible.

Acknowledgments. This work was supported in part by the National Science Foundation grants 1623190 (A Model of Change for Preparing a New Generation for Professional Practice in Computer Science), HRD-1834620 and HRD-2034030 (CAHSI Includes), EAR-2225395, and by the AT&T Fellowship in Information Technology.

It was also supported by the program of the development of the Scientific-Educational Mathematical Center of Volga Federal District No. 075-02-2020-1478, and by a grant from the Hungarian National Research, Development and Innovation Office (NRDI).

One the authors (VK) is greatly thankful to Graciela Chichilnisky for attracting his attention to Arrow's question, and to all the participants of the Workshop on the Applications of Topology to Quantum Theory and Behavioral Economics (Fields Institute for Research in Mathematical Sciences Toronto, Canada, March 23–24, 2023) for valuable discussions.

References

1. Arrow, K.J.: Social Choice and Individual Values. Wiley, New York (1951)
2. Chichilnisky, G.: The topology of quantum theory and social choice. Presentation at the Workshop on the Applications of Topology to Quantum Theory and Behavioral Economics, Fields Institute for Research in Mathematical Sciences Toronto, Canada, 23–24 March (2023)
3. Fishburn, P.C.: Utility Theory for Decision Making. John Wiley & Sons Inc., New York (1969)
4. Fishburn, P.C.: Nonlinear Preference and Utility Theory. The John Hopkins Press, Baltimore, Maryland (1988)
5. Gödel, K.: Über formal unentscheidbare Sätze der principia mathematica und verwandter systeme I. Monatshefte für Mathematik und Physik **38**(1), 173–198 (1931). English translation in: Feferman, S. (ed.), Kurt Gödel Collected Works, vol. 1, Oxford University Press, Oxford, UK, pp. 144–195 (1986)
6. Kreinovich, V.: Decision making under interval uncertainty (and beyond). In: Guo, P., Pedrycz, W. (eds.) Human-Centric Decision-Making Models for Social Sciences. SCI, vol. 502, pp. 163–193. Springer, Heidelberg (2014). https://doi.org/10.1007/978-3-642-39307-5_8
7. Luce, R.D., Raiffa, R.: Games and Decisions: Introduction and Critical Survey. Dover, New York (1989)
8. Nash, J.: The bargaining problem. Econometrica **18**(2), 155–162 (1950)
9. Nguyen, H.P., Bokati, L., Kreinovich, V.: New (simplified) derivation of Nash's bargaining solution. J. Adv. Comput. Intell. Intell. Inf. (JACIII) **24**(5), 589–592 (2020)
10. Nguyen, H.T., Kosheleva, O., Kreinovich, V.: Decision making beyond Arrow's 'impossibility theorem', with the analysis of effects of collusion and mutual attraction. Int. J. Intell. Syst. **24**(1), 27–47 (2009)
11. Nguyen, H.T., Kreinovich, V., Wu, B., Xiang, G.: Computing Statistics under Interval and Fuzzy Uncertainty. Springer Verlag, Berlin, Heidelberg (2012). https://doi.org/10.1007/978-3-642-24905-1
12. Raiffa, H.: Decision Analysis. McGraw-Hill, Columbus, Ohio (1997)
13. Sheskin, D.J.: Handbook of Parametric and Nonparametric Statistical Procedures. Chapman and Hall/CRC, Boca Raton, Florida (2011)
14. Spinoza, B.: Ethics. Cambridge University Press, Cambridge, UK (2018)

High-Impact Low-Probability Events are Even More Important Than it is Usually Assumed

Aaron Velasco[1], Olga Kosheleva[2], and Vladik Kreinovich[3](✉)

[1] Department of Earth, Environmental, and Resource Sciences, University of Texas at El Paso, 500 W. University, El Paso, TX 79968, USA
aavelasco@utep.edu

[2] Department of Teacher Education, University of Texas at El Paso, 500 W. University, El Paso, TX 79968, USA
olgak@utep.edu

[3] Department of Computer Science, University of Texas at El Paso, 500 W. University, El Paso, TX 79968, USA
vladik@utep.edu

Abstract. A large proportion of undesirable events like earthquakes, floods, tornados occur in zones where these events are frequent. However, a significant number of such events occur in other zones, where such events are rare. For example, while most major earthquakes occur in a vicinity of major faults, i.e., on the border between two tectonic plates, some strong earthquakes also occur inside plates. We want to mitigate all undesirable events, but our resources are limited. So, to allocate these resources, we need to decide which ones are more important. For this decision, a natural idea is to use the product of the probability of the undesirable event and possible damage caused by this event. A natural way to estimate probability is to use the frequency of such events in the past. This works well for high-probability events like earthquakes in a seismic zone near a fault. However, for high-impact low-probability events the frequency is small and, as a result, the actual probability may be very different from the observed frequency. In this paper, we show how to take this difference between frequency and probability into account. We also show that if we do take this difference into account, then high-impact low-probability events turn out to be even more important than it is usually assumed.

1 Introduction

High-Probability vs. Low-Probability Events. Many undesirable events happen all the time. It is important to prepare for these events, to mitigate the future damage as much as possible.

Some of these events have a reasonable high probability. For example, earthquakes regularly happen in California, hurricanes regularly happen in Florida and other states, tornadoes regularly happen in North Texas, floods regularly happen around big rivers, etc. Such events happen all the time with varying strength, and everyone understands that we need to be prepared for situations when the strength will becomes high, causing potentially disastrous consequences.

K. Cohen et al. (Eds.): NAFIPS 2023, LNNS 751, pp. 344–349, 2023.
https://doi.org/10.1007/978-3-031-46778-3_35

Undesirable events are not confined to zones where such events have a reasonably high frequency. For example, while the majority of major earthquakes occur in seismic zones, where earthquakes are a common occurrence, some major earthquakes occur in areas where strong earthquakes are very rare, where the last serious earthquake may have occurred thousands of years ago – and the only reason we know about it is indirectly, by the effect of this strong past earthquake on the geology of the region.

Low-Probability Events are Important. The problem with high-impact low-probability events is that, in contrast to events in high-probability zones where most people are prepared, people in low-probability events are mostly unprepared. For example, in California, since medium-size earthquakes happen there all the time, building codes require that buildings be resistant to (at least medium-strength) earthquakes, and within each building, most shelves are attached to the walls, so that they do not cause extra damage when the earthquake hits. In contrast, in low-probability zones, none of these measures are implemented. As a result, when the undesirable event happens – with even medium strength – it causes much more damage than a similar-strength event in high-probability zones.

How Should We Allocate Resources: Current Approach. The fact that we need to take into account high-impact low-probability events is well understood. Of course, it is not realistically possible to take into account all possible low-probability events, so we need to allocate resources to the most important events. Traditional way to decide on the importance of an event is to multiply its probability by the damage it may cause – this idea is in perfect accordance with the decision theory.

Remaining Problem. The problem of the current approach to resource allocation is how to estimate the corresponding probability. For high-probability events, events that occur reasonably frequently, we can estimate it as the frequency of observed events: e.g., if a major flood happens, on average, every 10 years, we estimate the yearly probability of this flood as 1/10. For high-probability events, this estimate is based on a large number of observations and is, therefore, reasonably accurate.

In principle, we can apply the same approach to low probability events – and this is exactly how such events are analyzed now. However, when events are rare, the sample of such events is very small, and it is known that for small samples, the difference between observed frequency and actual probability can be large. So, the natural questions are:

- How can we take this difference into account? and
- If we do, what will be the consequences?

What We Do in This Paper. In this paper:

- We show how the above difference can be taken into account.
- We also show that if we take this difference into account, then high-impact low-frequency events become even more important than it is usually assumed.

2 Current Way of Allocating Resources: Justification, Description, and Limitations

Justification: Let Us Use Decision Theory. There is a whole science of rational decision making, known as *decision theory*; see, e.g., [1–7]. According to decision theory, preferences of a rational person can be described by assigning, to each possible alternative x, a real number $u(x)$ called *utility*, so that the decision maker prefers alternative x to alternative y if and only if the alternative x has higher utility: $u(x) > u(y)$.

In situations when we have different outcomes with different utilities u_i and different probability p_i, the equivalent utility u of the corresponding situation is equal to the expected value of utility:

$$u = p_1 \cdot u_1 + p_2 \cdot u_2 + \ldots$$

In particular, if we consider a disaster with potential damage d (and thus, utility $-d$) and probability p, then the utility of not taking this potential disaster into account is equal to $p \cdot (-d) = -p \cdot d$. According to the above-mentioned notion of utility, this means that we need to primarily allocate resources to situations in which this negative utility is the worst, i.e., in which the product $p \cdot d$ is the largest.

Description. The above analysis leads to the way resources are allocated now:

- For each possible zone, we compute the product $p \cdot d$ of the probability p of the undesirable event and the damage d that would be caused by this event.
- The larger this product, the higher the priority of this zone.

Comment. This way, many low-probability zones get funding: in these zones, the probability p is lower than in the high-probability zones, but, as we explained in the previous section, the potential damage can be much higher, since such zones are usually unprepared for the undesirable event (or at least much less prepared).

How the Corresponding Probabilities are Estimated. To use the usual techniques, we need to estimate, for each zone, the probability p of the undesirable event. In statistics, the usual way to estimate probability is take the frequency with which this even happened in the past.

Example 1. If in 200 years of record, the major Spring flood occurred 20 times, we estimate the probability of the flood as

$$20/200 = 0.1 = 10\%.$$

Example 2. If in some other area, a similar flood happened only twice during the 200 years, we estimate the probability of flooding in this areas as

$$2/200 = 0.01 = 1\%.$$

Limitations. As we have mentioned, the main limitation of this approach is that it does not take into account that the frequency is only as approximate value of the probability. It is therefore desirable to come up with a more adequate technique, a technique that would take this difference into account.

3 Coming Up with a More Adequate Technique for Allocating Resources

What do We Know About the Difference Between Probability and Frequency. According to statistics (see, e.g., [8]), if we estimate the probability based on n observations, then, for large n, the difference between the frequency f and probability p is normally distributed with mean $\mu = 0$ and standard deviation

$$\sigma = \sqrt{\frac{p \cdot (1-p)}{n}}.$$

We do not know the exact probability p, we only know its approximate value f. By using this approximate value instead of p, we can estimate the above standard deviation as

$$\sigma \approx \sqrt{\frac{f \cdot (1-f)}{n}}.$$

In general, for a normal distribution, with confidence 95% all random values are located within the 2-sigma interval $[\mu - 2\sigma, \mu + 2\sigma]$. In our case, this means that the actual probability can be somewhere in the interval

$$\left[f - 2 \cdot \sqrt{\frac{f \cdot (1-f)}{n}}, f + 2 \cdot \sqrt{\frac{f \cdot (1-f)}{n}}\right]. \tag{1}$$

How to Take This Difference into Account When Making a Decision. As we have mentioned, all we know about the probability p is that it is located somewhere on the interval (1). This probability may be smaller than f, it may be larger that the frequency f. Disaster preparedness means preparing for the worst possible scenario. So, it makes sense to consider the worst-case probability:

$$\overline{p} = f + 2 \cdot \sqrt{\frac{f \cdot (1-f)}{n}}. \tag{2}$$

So, the idea is to use this higher probability instead of the frequency when comparing the importance of different zones. In other words, instead of using the products $p \cdot d$ for $p = f$ (as in the traditional approach), we need to use the products $\overline{p} \cdot d$, where $\overline{p}$ is determined by the formula (2).

4 How This Will Affect Our Ranking of High-Impact Low-Probability Events

Examples. Before providing a general analysis, let us first illustrate, on the above two examples, how our estimate for probability will change if we use the new technique.

Example 1. In this case,

$$\sqrt{\frac{f\cdot(1-f)}{n}}=\sqrt{\frac{0.1\cdot 0.9}{200}}=\sqrt{0.00045}\approx 0.02.$$

Thus, $\overline{p}\approx 0.1+2\cdot 0.02=0.14$. This probability is somewhat larger than the frequency (actually, 40% larger) but it is in the same range as the frequency.

Example 2. In this case,

$$\sqrt{\frac{f\cdot(1-f)}{n}}=\sqrt{\frac{0.01\cdot 0.99}{200}}=\sqrt{0.00005}\approx 0.007.$$

Thus, $\overline{p}\approx 0.01+2\cdot 0.007=0.024$. This probability is more than twice larger than the frequency—actually, here, the 2-sigma term is larger than the original frequency.

Analysis and the Resulting Quantitative Recommendations. An important difference between these two examples is that:

- in the first example, the frequency is larger than the 2-sigma term, so the estimate $\overline{p}$ is of the same order as frequency, while
- in the second example, the 2-sigma term is larger than the original frequency, so the estimate $\overline{p}$ is of the same order as the 2-sigma term.

The borderline between these two cases if when the two terms are equal, i.e., when

$$f=2\cdot\sqrt{\frac{f\cdot(1-f)}{n}}. \tag{3}$$

Here, the frequency f is much smaller than 1, so $1-f\approx 1$ and thus, the formula (3) takes the following simplified form:

$$f=2\cdot\sqrt{\frac{f}{n}}. \tag{4}$$

By squaring both sides, we get

$$f^2=4\cdot\frac{f}{n},$$

i.e., equivalently,

$$f=\frac{4}{n}.$$

Thus:

- If we have more than 4 past events of this type, then the estimate $\overline{p}$ is of the same order as frequency. In this case, the usual estimate for the product $p\cdot d$ works OK.
- However, if we have fewer than 4 past events of this type, then the estimate $\overline{p}$ is much larger than the frequency and thus, the usual estimate for the product leads to a drastic underestimation of this event's importance.

Quantitative Consequences. The new estimate $\overline{p}$ for the probability is, in general, larger than the usual estimate f. Thus, the new value $\overline{p} \cdot d$ is larger than the value $f \cdot d$ estimated by the traditional method.

How larger? The ratio between these two products is equal to

$$\frac{\overline{p} \cdot d}{f \cdot d} = \frac{\overline{p}}{f} = \frac{f + 2 \cdot \sqrt{\frac{f \cdot (1-f)}{n}}}{f} = 1 + 2 \cdot \sqrt{\frac{1-f}{f \cdot n}} = 1 + \frac{2}{\sqrt{n}} \cdot \sqrt{\frac{1}{f} - 1}. \quad (5)$$

We can see that as the frequency f decreases, this ratio grows – and this ratio tends to infinity as f tends to 0. So indeed, the use of this new techniques increase the product corresponding to low-probability events much higher than for high-probability events – and thus, makes high-impact low-probability events even more important.

Acknowledgments. This work was supported in part by the National Science Foundation grants 1623190 (A Model of Change for Preparing a New Generation for Professional Practice in Computer Science), HRD-1834620 and HRD-2034030 (CAHSI Includes), EAR-2225395, and by the AT&T Fellowship in Information Technology.

It was also supported by the program of the development of the Scientific-Educational Mathematical Center of Volga Federal District No. 075-02-2020-1478, and by a grant from the Hungarian National Research, Development and Innovation Office (NRDI).

References

1. Fishburn, P.C.: Utility Theory for Decision Making. John Wiley & Sons Inc., New York (1969)
2. Fishburn, P.C.: Nonlinear Preference and Utility Theory. The John Hopkins Press, Baltimore, Maryland (1988)
3. Kreinovich, V.: Decision making under interval uncertainty (and beyond). In: Guo, P., Pedrycz, W. (eds.) Human-Centric Decision-Making Models for Social Sciences. SCI, vol. 502, pp. 163–193. Springer, Heidelberg (2014). https://doi.org/10.1007/978-3-642-39307-5_8
4. Luce, R.D., Raiffa, R.: Games and Decisions: Introduction and Critical Survey. Dover, New York (1989)
5. Nguyen, H.T., Kosheleva, O., Kreinovich, V.: "Decision making beyond Arrow's 'impossibility theorem', with the analysis of effects of collusion and mutual attraction". Int. J. Intell. Syst. **24**(1), 27–47 (2009)
6. Nguyen, H.T., Kreinovich, V., Wu, B., Xiang, G.: Computing Statistics Under Interval and Fuzzy Uncertainty: Applications to Computer Science and Engineering. Springer, Heidelberg (2012). https://doi.org/10.1007/978-3-642-24905-1
7. Raiffa, H.: Decision Analysis. McGraw-Hill, Columbus, Ohio (1997)
8. Sheskin, D.J.: Handbook of Parametric and Nonparametric Statistical Procedures. Chapman and Hall/CRC, Boca Raton, Florida (2011)

People Prefer More Information About Uncertainty, but Perform Worse When Given This Information: An Explanation of the Paradoxical Phenomenon

Jieqiong Zhao[1], Olga Kosheleva[2], and Vladik Kreinovich[3](✉)

[1] School of Computing and Augmented Intelligence, Arizona State University, 699 S Mill Ave, Tempe, AZ 85281, USA
jieqiong.zhao@asu.edu

[2] Department of Teacher Education, University of Texas at El Paso, 500 W. University, El Paso, TX 79968, USA
olgak@utep.edu

[3] Department of Computer Science, University of Texas at El Paso, 500 W. University, El Paso, TX 79968, USA
vladik@utep.edu

Abstract. In a recent experiment, decision makers were asked whether they would prefer having more information about the corresponding situation. They confirmed this preference, and such information was provided to them. However, strangely, the decisions of those who received this information were worse than the decisions of the control group – that did not get this information. In this paper, we provide an explanation for this paradoxical situation.

1 Formulation of the Problem

When Making a Decision, it is Desirable to have as Much Information as Possible. To make a decision, it is desirable to have as much information about the decision making situation as possible: additional information can help make a better decision.

This desirability has been confirmed by many polls; see, e.g., [17].

This Includes the Need for More Information About Uncertainty. In complex situations, to make decisions, we usually use computers. In general, computers process numbers. Thus, the information about a situation usually consists of several numbers, e.g., the values of the corresponding physical quantities. These values usually come from measurements – it could be direct measurement or so-called indirect measurement, i.e., processing of measurement results.

Measurements are never absolutely accurate, there is always a non-zero difference $\Delta x \stackrel{\text{def}}{=} \widetilde{x} - x$ between the measurement result $\widetilde{x}$ and the actual (unknown) value x of the corresponding quantity; see, e.g., [14]. This difference is known as the *measurement error*.

In view of the desirability to get as much information as possible, it is desirable to provide the users not only the measurement results, but also with the information about

K. Cohen et al. (Eds.): NAFIPS 2023, LNNS 751, pp. 350–356, 2023.
https://doi.org/10.1007/978-3-031-46778-3_36

the measurement error. The desirability of this information was indeed confirmed by the polls; see, e.g., [17].

What do We Know About the Measurement Errors: Probabilistic and Interval Uncertainty. Ideally, we should know what are the possible values of the measurement error, and what is the frequency (probability) with which each of these values will appear in the actual measurements. In other words, ideally, we should know the probability distribution of the measurement errors.

To determine this probability distribution, we need to *calibrate* the measuring instrument, i.e., to compare the values $\widetilde{x}_k$ measured by this instrument and the values x'_k measured (for the same actual quantity) by a much more accurate measuring instrument (known as *standard*). The standard measuring instrument (SMI) is so much more accurate that, in comparison with the measurement errors of the tested measuring instrument (MI), we can safely ignore the measurement errors of the SMI and assume that SMI's measurement results x'_k are equal to the corresponding actual values x_k. Thus, the differences $\widetilde{x}_k - x'_k$ can be safely assumed to equal to the values of the measurement errors $\widetilde{x}_k - x_k$. So, from the resulting sample of these differences, we can determine the desired probability distribution.

This procedure is rather time-consuming and expensive – since the use of the complex standard measuring instruments is not cheap. As a result, in practice, such a detailed calibration is often not performed. Instead, the manufacturer of the measuring instruments provides a guaranteed upper bound Δ on the absolute value of the measurement error: $|\Delta x| \le \Delta$. Under this information, after we perform the measurement and get the measurement result $\widetilde{x}$, the only information that we have about the actual (unknown) value of the measured quantity is that this value lies somewhere in the interval $[\underline{x}, \overline{x}] \stackrel{\text{def}}{=} [\widetilde{x} - \Delta, \widetilde{x} + \Delta]$. This uncertainty is known as *interval uncertainty*; see, e.g., [4, 6, 8, 9].

Probability Distribution of the Measurement Error is Often Gaussian. In many practical cases, the measurement error is caused by the large number of small independent factors. It is known that the probability distribution of the joint effect of many independent small factors is close to Gaussian; the corresponding mathematical result is known as the Central Limit Theorem; see, e.g., [16]. Thus, we can safely assume that the measurement error is distributed according to the normal (Gaussian) distribution.

This assumption is in good accordance with the empirical data, according to which in the majority of the cases, the distribution is indeed Gaussian [12, 13].

The Gaussian distribution is uniquely determined by two parameters: mean μ and standard deviation σ. By testing the measuring instrument – i.e., by comparing its results with the results of a much more accurate ("standard") measuring instrument, we can determine the bias – as the arithmetic average of the observed differences between the measurements by two instruments:

$$\mu \approx \frac{1}{n} \cdot \sum_{k=1}^{n} (\widetilde{x}_k - x'_k).$$

Once we know μ, we can recalibrate the scale – i.e., subtract this mean value μ from each measurement result, i.e., replace each value $\widetilde{x}$ with $\widetilde{x}' \stackrel{\text{def}}{=} \widetilde{x} - \mu$. After that, the mean will be 0.

Also, we can determine the standard deviation as the mean squared value of the (re-calibrated) difference

$$\sigma \approx \sqrt{\frac{1}{n} \cdot \sum_{k=1}^{n} (\widetilde{x}'_k - x'_k)^2}.$$

So, it makes sense to assume that the measurement error is normally distributed with mean 0 and the known standard deviation σ.

Confidence Intervals. It is known that with high probability, all the values of the normally distributed random variable are located within the interval

$$[\mu - k_o \cdot \sigma, \mu + k_0 \cdot \sigma]$$

for an appropriate k_0: for $k_0 = 2$, this is true with probability 95%, for $k_0 = 3$, this is true with probability 99.9%, and for $k_0 = 6$, this is true with probability $1 - 10^{-8}$. These intervals are particular cases of *confidence intervals*. These intervals are what is often supplied to the users as a partial information about the probability distributions.

What Happened When Decision Makers Received Confidence Intervals: Paradoxical Situation. Since the decision makers expressed the desire to receive information about uncertainty, they were supplied, in addition to measurement results, with the corresponding confidence intervals. And here, a strange thing happened. One would expect that this additional information would help the decision makers make better decisions – or at least did not degrade the quality of their decisions. However, in reality, the decisions of those users who got this additional information were worse than the decisions make by users from the control group – that did not receive this information. How can we explain this paradoxical phenomenon?

2 Our Explanation

Preliminaries. To understand the above paradoxical phenomenon, let us recall how rational people make decisions. Let us start with the case of full information, when we know all the probabilities.

How Rational People Make Decisions Under Full Information: Reminder. According to decision theory (see, e.g., [1,2,5,7,10,11,15]), preferences of each rational person can be described by assigning, to each alternative x, a numerical value $u(x)$ called its *utility*. To assign these numerical values, one need to select two alternatives: a very good alternative A_+ which is better than anything that the decision maker will actually encounter, and a very bad alternative A_- which is worse than anything that the decision maker will actually encounter. Once these two alternatives are selected, we can form, for each value p from the interval $[0,1]$, a lottery $L(p)$ in which we get A_+ with probability p and A_- with the remaining probability $1 - p$.

For any actual alternative x, when p is close to 0, the lottery $L(p)$ is close to A_- and is, thus, worse than x: $A_- < x$. When p is close to 1, the lottery $L(p)$ is close to A_+ and is, thus, better than x: $x < A_+$. When we move from 0 to 1, there is a threshold value at which the relation $L(p) < x$ is replaced with $x < L(p)$. This threshold value $u(x)$ is what is called the utility of x. By definition of utility, each alternative x is equivalent, to the decision maker, to a lottery $L(u(x))$ in which we get A_+ with probability $u(x)$ and we get A_- with the remaining probability $1 - u(x)$.

In general, the larger the probability p of getting the very good alternative A_+, the better the lottery. Thus, if we need to select between several lotteries of this type, we should select the lottery with the largest values of the probability p. Since each alternative x is equal to the lottery $L(p)$ with probability $p = u(x)$, this means that we should always select the alternative with the largest value of utility.

What is the utility of an action in which we get n possible outcomes, for each of which we know its probability p_i and its utility u_i? By definition of utility, each outcome is equivalent to a lottery $L(u_i)$ in which we get A_+ with probability u_i and A_- with probability $1 - u_i$. Thus, to the user, the action is equivalent to a 2-stage lottery in which we first select each i with probability p_i and then, depending on what i we selected on the first stage, select A_+ with probability u_i and A_- with probability $1 - u_i$. As a result of this two-stage lottery, we get either A_+ or A_-, and the probability u of getting A_+ is determined by the formula of full probability:

$$u = p_1 \cdot u_1 + \ldots + p_n \cdot u_n.$$

Thus, by definition of utility, this value u is the utility of the action under consideration. The right-hand side of the formula for u is the expected value of the utility. So, we can conclude that the utility of an action with random consequences is equal to the expected value of the utility of different consequences.

Comment. The numerical value of the utility depends on the selection of the alternatives A_+ and A_-. If we select a different pair A'_+ and A'_-, then we will have different numerical values $u'(x)$. It turns out that for every two pairs of alternatives, there exist real numbers $a > 0$ and b for which, for each x, we have $u'(x) = a \cdot u(x) + b$. In other words, utility is defined modulo a linear transformation.

This is similar to the fact that, e.g., the numerical value of the moment of time also depends on what starting point we use to measure time and on what measuring unit we use, and all the scales are related to each other by an appropriate linear transformation $t' = a \cdot t + b$ for some $a > 0$ and b.

How Rational People Make Decisions Under Interval Uncertainty. As we have mentioned, in many practical situations, we only know the values of the quantities (that describe the state of the world) with interval uncertainty. In this case, we can describe the consequences – and their utility – also under interval uncertainty. In other words, for each possible decision x, instead of the exact value $u(x)$ of the corresponding utility, we only know the interval $[\underline{u}(x), \overline{u}(x)]$ of possible utility values.

How can we make a decision in this case? To make a decision, we need to assign, to each interval $[\underline{u}, \overline{u}]$, an equivalent numerical value $u(\underline{u}, \overline{u})$. As we have mentioned, utility is defined modular a linear transformation. There is no fixed selection of the alternatives

A_+ and A_-, so it makes sense to require that the function $u(\underline{u},\overline{u})$ remains the same for all the scales, i.e., that if $u = u(\underline{u},\overline{u})$, then for all $a > 0$ and b, we should have $u' = u(\underline{u}',\overline{u}')$, where $u' = a \cdot u + b$, $\underline{u}' = a \cdot \underline{u} + b$, and $\overline{u}' = a \cdot \overline{u} + b$.

Let us denote $\alpha \stackrel{\text{def}}{=} u(0,1)$. If we know that the utility is between 0 and 1, then the situation is clearly better (or at least as good) than when utility is 0, and worse (or at least as good) then when utility is 1. Thus, we must have $0 \le \alpha \le 1$. Every interval $[\underline{u},\overline{u}]$ can be obtained from the interval $[0,1]$ by a linear transformation $u \mapsto a \cdot u + b$ for $a = \overline{u} - \underline{u}$ and $b = \underline{u}$. Thus, due to invariance, from $\alpha = u(0,1)$, we can conclude that

$$u(\underline{u},\overline{u}) = a \cdot \alpha + b = \alpha \cdot (\overline{u} - \underline{u}) + \underline{u} = \alpha \cdot \overline{u} + (1-\alpha) \cdot \underline{u}.$$

This formula was first proposed by the Nobelist Leo Hurwicz and is thus known as *Hurwicz optimism-pessimism criterion*; see, e.g., [3,5,7]. The name comes from the fact that for $\alpha = 1$, this means only taking into account the best-case value $\overline{u}$ – the case of extreme optimism, while for $\alpha = 0$, this means only taking into account the worst-case value $\underline{u}$ – the case of extreme pessimism. The value α is different for different decision makers, depending on their level of optimism and pessimism.

Finally, An Explanation. Now we are ready to produce the desired explanation. Let us consider the simplest possible setting, when the decision maker is directly provided with the information about his/her utility $u(x)$ of each possible decision x:

- the estimate $\widetilde{u}(x)$ and
- the information that the difference $\Delta u(x) \stackrel{\text{def}}{=} \widetilde{u}(x) - u(x)$ between the estimated utility $\widetilde{u}(x)$ and the actual utility value $u(x)$ (corresponding to the actual – unknown – values of the corresponding quantities) is distributed according to normal distribution with 0 mean and standard deviation σ.

In this *ideal* case, when the decision maker knows this distribution, his/her equivalent utility of each possible decision x is equal to the expected value of the random utility value $u(x) = \widetilde{u}(x) - \Delta u(x)$, i.e., to the value $\widetilde{u}(x)$.

This was the ideal case. In the above-mentioned experiment, we never report the whole distribution to the decision maker. Instead, we report either a single value $\widetilde{u}(x)$ or the confidence interval $[\widetilde{u}(x) - k_0 \cdot \sigma, \widetilde{u}(x) + k_0 \cdot \sigma]$.

In the first case, when we supply no information about uncertainty, the decision maker uses the provided value $\widetilde{u}(x)$ in his/her decisions. It so happens that this value is exactly what we would get if we knew the exact distributions. In other words, in this case, the decision maker makes an optimal decision.

On the other hand, if we provide the decision maker with the confidence interval, the decision maker – using Hurwicz criterion – will assign, to each possible decision x, an equivalent value

$$\alpha \cdot (\widetilde{u}(x) + k_0 \cdot \sigma) + (1-\alpha) \cdot (\widetilde{u}(x) - k_0 \cdot \sigma) = \widetilde{u}(x) + (2\alpha - 1) \cdot k_0 \cdot \sigma.$$

Thus, for almost all possible values α from the interval $[0,1]$ – with the only exception of the value $\alpha = 0.5$ – this value will be different from the optimal value (corresponding to the full information). So, the decision based on such values will be not as good as the optimal decision – and this is exactly what we observe in the above-described seemingly paradoxical experiment.

Comment. Note that the worsening of the decision happens when we provide the decision make with *partial* information about uncertainty. If we provide the decision maker with full information, the decision will, of course, be optimal.

Acknowledgments. This work was supported in part by the National Science Foundation grants 1623190 (A Model of Change for Preparing a New Generation for Professional Practice in Computer Science), HRD-1834620 and HRD-2034030 (CAHSI Includes), EAR-2225395, and by the AT&T Fellowship in Information Technology.

It was also supported by the program of the development of the Scientific-Educational Mathematical Center of Volga Federal District No. 075-02-2020-1478, and by a grant from the Hungarian National Research, Development and Innovation Office (NRDI).

References

1. Fishburn, P.C.: Utility Theory for Decision Making. John Wiley & Sons Inc., New York (1969)
2. Fishburn, P.C.: Nonlinear Preference and Utility Theory. The John Hopkins Press, Baltimore, Maryland (1988)
3. Hurwicz, L.: Optimality Criteria for Decision Making Under Ignorance, Cowles Commission Discussion Paper, Statistics, No. 370 (1951)
4. Jaulin, L., Kieffer, M., Didrit, O., Walter, E.: Interval analysis. In: Applied Interval Analysis. Springer, London (2001). https://doi.org/10.1007/978-1-4471-0249-6_2
5. Kreinovich, V.: Decision making under interval uncertainty (and beyond). In: Guo, P., Pedrycz, W. (eds.) Human-Centric Decision-Making Models for Social Sciences. SCI, vol. 502, pp. 163–193. Springer, Heidelberg (2014). https://doi.org/10.1007/978-3-642-39307-5_8
6. Kubica, B.J.: Interval Methods for Solving Nonlinear Constraint Satisfaction, Optimization, and Similar Problems: From Inequalities Systems to Game Solutions. Springer, Cham, Switzerland (2019). https://doi.org/10.1007/978-3-030-13795-3
7. Luce, R.D., Raiffa, R.: Games and Decisions: Introduction and Critical Survey. Dover, New York (1989)
8. Mayer, G.: Interval Analysis and Automatic Result Verification. de Gruyter, Berlin (2017)
9. Moore, R.E., Kearfott, R.B., Cloud, M.J.: Introduction to Interval Analysis. SIAM, Philadelphia (2009)
10. Nguyen, H.T., Kosheleva, O., Kreinovich, V.: "Decision making beyond Arrow's 'impossibility theorem', with the analysis of effects of collusion and mutual attraction". Int. J. Intell. Syst. **24**(1), 27–47 (2009)
11. Nguyen, H.T., Kreinovich, V., Wu, B., Xiang, G.: Computing Statistics Under Interval and Fuzzy Uncertainty. Springer, Heidelberg (2012). https://doi.org/10.1007/978-3-642-24905-1
12. Novitskii, P.V., Zograph, I.A.: Estimating the Measurement Errors. Energoatomizdat, Leningrad (1991). (in Russian)
13. Orlov, A.I.: How often are the observations normal? Ind. Lab. **57**(7), 770–772 (1991)
14. Rabinovich, S.G.: Measurement Errors and Uncertainty: Theory and Practice. Springer, New York (2005). https://doi.org/10.1007/0-387-29143-1
15. Raiffa, H.: Decision Analysis. McGraw-Hill, Columbus, Ohio (1997)

16. Sheskin, D.J.: Handbook of Parametric and Nonparametric Statistical Procedures. Chapman and Hall/CRC, Boca Raton, Florida (2011)
17. Zhao, J., Wang, Y., Mancenido, M., Chiou, E.K., Maciejewski, R.: Evaluating the impact of uncertainty visualization on model reliance. IEEE Trans. Vis. Comput. Graph. (to appear)

Author Index

K. Cohen et al. (Eds.): NAFIPS 2023, LNNS 751, pp. 357–358, 2023.
https://doi.org/10.1007/978-3-031-46778-3